MULTIVARIABLE FEEDBACK DESIGN

ELECTRONIC SYSTEMS ENGINEERING SERIES

Consulting Editor   **E L Dagless**
University of Bristol

OTHER TITLES IN THE SERIES

Advanced Microprocessor Architectures   *L Ciminiera and A Valenzano*

Optical Pattern Recognition Using Holographic Techniques   *N Collings*

Modern Logic Design   *D Green*

Data Communications, Computer Networks and OSI (2nd Edn)   *F Halsall*

Microwave Components and Systems   *K F Sander*

Tolerance Design of Electronic Circuits   *R Spence and R Soin*

Computer Architecture and Design   *AJ van de Goor*

MULTIVARIABLE FEEDBACK DESIGN

J.M. Maciejowski
Cambridge University
and
Pembroke College, Cambridge

Addison-Wesley Publishing Company
Wokingham, England • Reading, Massachusetts • Menlo Park, California
New York • Don Mills, Ontario • Amsterdam • Bonn
Sydney • Singapore • Tokyo • Madrid • San Juan

Many of the designations used by manufacturers and sellers to distinguish
their products are claimed as trademarks. Addison-Wesley has made every
attempt to supply trademark information about manufacturers and their
products mentioned in this book. A list of the trademark designations and
their owners appears on p. xviii.

Cover designed by Crayon Design of Henley-on-Thames
and printed by The Riverside Printing Co. (Reading) Ltd.
Typeset by Macmillan India Ltd, Bangalore 25.
Printed and bound in Great Britain by T. J. Press (Padstow), Cornwall

First printed 1989.

British Library Cataloguing in Publication Data

Maciejowski, J.M. (Jan Marian)
    Multivariable feedback design.
    1. Multivariable feedback systems. Design
    I. Title   II. Series
    629.8′312

    ISBN 0–201–18243–2

Library of Congress Cataloging in Publication Data

Maciejowski, Jan Marian.
        Multivariable feedback design/Jan Maciejowski.
            p.   cm. – (Electronic systems engineering series)
        Bibliography: p.
        Includes index.
        ISBN 0–201–18243–2
        1. Feedback control systems – Design and construction.   I.   Title.
    II. Series.
    TJ216.M23   1989   89–6552
        629.8′312–dc20   CIP

Preface

Useful techniques for the design of multivariable feedback systems have been known for at least fifteen years, yet these techniques have remained known to only a relatively small part of the community of control engineers. This book has been written in order to spread familiarity with the techniques more widely.

My objective has been to enable feedback engineers to design real systems, and the choice of material has been constantly, and I hope consistently, guided by that objective. The reader will therefore find that a complete theoretical understanding of the techniques discussed will sometimes require the consultation of other text-books or research journals, but I believe that I have given enough detail of techniques and algorithms to allow complete analyses and designs to be executed. However, this is a genuine text-book and not just a cook-book, so most of the theory required to understand each technique has been included. There is also enough theoretical development to allow the reader to go on to read the research literature relatively easily.

The basic view espoused is that sensible design of feedback systems is possible only if a frequency-domain point of view is adopted. Between about 1965 and 1980, the advantages of frequency-domain approaches were championed by Professors Rosenbrock and MacFarlane in the UK, and by Professor Horowitz in Israel, at a time when most of the academic community – particularly in the USA – regarded the frequency domain as obsolete and inherently unsuitable for the solution of multivariable problems. The instinct of those who kept faith with the frequency domain has been spectacularly vindicated during the past ten years. Not only have the rather *ad hoc* techniques of the 'British school' proved to be useful in practical design

(Chapter 4), but also the central bastion of time-domain approaches, namely 'LQG' optimal control theory, has been turned into an easily usable technique which yields sensible designs, by giving it a frequency-domain interpretation (Chapter 5). And the very latest technique, H_∞ optimal control theory, has arisen entirely as a result of frequency-domain thinking (Chapter 6).

Despite this shift towards the frequency domain, most of the 'time-domain' content of linear systems theory – state feedback, observers, minimal realizations, LQG controllers and so on – remains essential for multivariable design. Although it is necessary to think and analyse in frequency-domain terms, most computational algorithms, and many proofs of theorems, are based on state-space methods (Chapters 6 and 8).

I have omitted some material, such as multivariable root-loci and pole-placement methods, because it does not fit easily into my view of multivariable feedback design, apparently being more concerned with the shaping of transient responses than with obtaining the potential benefits of feedback. On the whole I have made such omissions without misgivings, but there are two topics which do not appear in the book and which I feel a little apprehensive about leaving out. One is the 'graph topology' introduced by Vidyasagar, which may well turn out to play a fundamental role in feedback theory; the other is the idea of 'internal-model control' introduced by Morari, which has had some impact on the process-control industries, and which is closely related to the Youla parametrization introduced in Chapter 6. But anyone wishing to learn about these topics should have no difficulty in reading the relevant literature, after reading this book, and internal-model control appears in one of the exercises.

I have also omitted material on discrete-time systems, because of lack of space. Almost everything in Chapters 1–4 and 7, and much of Chapter 8, holds for discrete-time systems defined by z-transform transfer functions or by state-space models, with some obvious modifications. The material on LQG design, in Chapter 5, can be given a nearly parallel development for discrete-time systems, but the details are considerably different. In principle, all the material on H_∞ design in Chapter 6 is applicable to discrete-time systems, since the bilinear transformation $z = 1 + s/1 - s$ can be used to transform discrete-time problems into continuous-time ones. But an explicit development of H_∞ theory for discrete-time systems is not yet available.

Some feedback and control specialists hold the view that the teaching of feedback design should be radically revised in the light of recent advances in feedback theory. They would begin, for example, with the Youla parametrization of all stabilizing feedback controllers, which is indeed a logical starting point if the subject is viewed as a branch of mathematics. In contrast with this view, I have adopted a more conservative sequence of presentation, and have begun with a review of the 'classical' techniques developed for the analysis and design of single-input, single-output feedback loops. There is a pedagogical advantage in this since the field of multivariable feedback systems can be entered painlessly by extending the 'classical' ideas. But there is a deeper reason for adopting this approach: the 'classical' treatment gives appropriate emphasis to questions which are of real significance to engineering design, but which may not appear important in a purely mathematical account of the subject. For example, there are several results in linear system theory which rely on the non-existence of plant zeros in the right half-plane (of the complex plane). An acquaintance with the classical theory makes one suspect immediately that the *location* and not the *existence* of such zeros should be significant, particularly as the non-existence of such zeros cannot be confirmed by any experimental means.

In order to facilitate comparisons between alternative design methods, each major technique presented in this book is illustrated by being applied to the *same system* – a linearized model of an aircraft, defined in the Appendix – and the design specification is similar in each case.

Readership

This book is aimed at graduate students who have taken at least one elementary course on feedback and control systems, and who have some acquaintance with linear systems theory. It is possible, although difficult, to take a course in linear systems theory at the same time as a course based on this book. A more detailed discussion of prerequisites is given in Chapter 1. The book should also be accessible to practising feedback engineers who are familiar with classical servo design and have had some exposure to 'modern' (that is, post-1960) control theory.

Most of the material has been classroom-tested in a number of graduate courses given at Imperial College, Cambridge University and, especially, the University of California at Santa Barbara, where in 1986 I gave a 40-hour course covering much of Chapters 1 to 5 and 7. There are a number of difficulties associated with teaching this material. The first is that it is essential for both instructor and students to have access to suitable software which can be run interactively. All the examples and exercises in the book have been solved using PC-Matlab (or Pro-Matlab), together with its associated *Control System Toolbox* and *Multivariable Frequency Domain Toolbox* (and, if I were solving them again, I would use the *Robust Control Toolbox* for Chapters 5 and 6). The second difficulty to be aware of is that the design exercises, such as Exercises 4.1 and 4.2, take much longer to solve than traditional homework exercises, particularly when students are familiarizing themselves with new software or if there is a shortage of computer resources. A student cannot be expected to solve more than one, or perhaps at most two, such exercises per week (assuming that the course is given at a rate of 4 hours per week, say, and that the students are taking a typical mix of courses). It is also beneficial for students to solve these exercises working in pairs. The temptation to set lots of traditional paper-and-pencil problems, and avoid the design exercises, should be resisted, however. The third difficulty is that of examining the students on the material. It seems entirely unsatisfactory to emphasize design during the course, and set computer-based design exercises for homework, but then examine the students on points of theory and grossly simplified examples in order to comply with the requirements of the traditional three-hour examination paper. The solution I have successfully adopted, both at Cambridge and at Santa Barbara, is to set the class a realistic design exercise (possibly varying the details slightly for each student) and ask them to hand in solutions by some fixed time (typically 48 hours later).

Acknowledgements

I should like to acknowledge the help of various people with the writing of this book. In writing Sections 2.2 to 2.5 I was strongly influenced by some unpublished notes written at Cambridge by A.G.J. MacFarlane, Y.-S. Hung, D.J.N. Limebeer and M.C. Smith, and my overall view of the subject has been influenced by lecture notes made available to me by M.G. Safonov. Discussions with K. Glover have left their mark on Chapter 6, and W.-Y. Ng provided detailed criticisms of Chapter 7. Alistair MacFarlane has provided me with opportunities to develop this material through teaching and research over a number of years, and has constantly provided the encouragement required to complete it. Alan Laub arranged my stay at Santa Barbara, where several sets of half-completed and semi-connected lecture notes were first transmuted into book chapters. The whole manuscript has been typed and revised by Celia Sharpe. Finally, my wife Mara and daughters Kasia and Lucy have tolerated many weekends and evenings devoted to the book rather than to them.

J.M. Maciejowski
30 September 1989
Cambridge University and
Pembroke College, Cambridge

A note on language
For reasons of simplicity, the pronoun 'he' is used to relate to both male and female throughout the book.

Contents

Preface   v

List of symbols and abbreviations   xv

1   Single-loop Feedback Design   1

    1.1   Overview and prerequisites   1
    1.2   Review of elementary feedback design   3
    1.3   A standard problem   10
    1.4   Fundamental relations   13
    1.5   The 'shape' of the solution   16
    1.6   Two approaches to design   22
    1.7   Limitations on performance   24
         1.7.1   Gain–phase relationships   24
         1.7.2   Right half-plane zeros   27
         1.7.3   Right half-plane poles   30
         1.7.4   Bode's integral theorem   31
         Summary   33
         Exercises   34
         References   36

2   Poles, Zeros and Stability of Multivariable Feedback Systems   37

    2.1   Introduction   37
    2.2   The Smith–McMillan form of a transfer-function matrix   40
    2.3   Poles and zeros of a transfer-function matrix   45
    2.4   Matrix-fraction description (MFD) of a transfer function   48
    2.5   State-space realization from a transfer-function matrix   50
    2.6   How many zeros?   52
    2.7   Internal stability   55

2.8	The generalized Nyquist stability criterion	59
2.9	The generalized inverse Nyquist stability criterion	62
2.10	Nyquist arrays and Gershgorin bands	64
2.11	Generalized stability	69
	Summary	70
	Exercises	71
	References	74

3 Performance and Robustness of Multivariable Feedback Systems **75**

3.1	Introduction	75
3.2	Principal gains (singular values)	76
3.3	The use of principal gains for assessing performance	81
3.4	Relations between closed-loop and open-loop principal gains	87
3.5	Principal gains and characteristic loci	91
3.6	Limitations on performance	94
	3.6.1 Gain–phase relationships	94
	3.6.2 Right half-plane zeros and poles	96
3.7	Transmission of stochastic signals	97
3.8	The operator norms $\|G\|_2$ and $\|G\|_\infty$	99
3.9	The use of operator norms to specify performance	101
3.10	Representations of uncertainty	102
	3.10.1 Unstructured uncertainty	102
	3.10.2 Structured uncertainty	105
	3.10.3 Uncertainty templates	111
3.11	Stability robustness	111
	3.11.1 Unstructured uncertainty	111
	3.11.2 Structured uncertainty	116
	3.11.3 Loop failures and gain variations	121
3.12	Performance robustness	124
	Summary	129
	Exercises	131
	References	135

4 Multivariable Design: Nyquist-like Techniques **137**

4.1	Introduction	137
4.2	Sequential loop closing	138
4.3	The characteristic-locus method	142
	4.3.1 Approximate commutative compensators	142
	4.3.2 Design procedure	149
4.4	Design example	155
4.5	Reversed-frame normalization	164
4.6	Nyquist-array methods	168
	4.6.1 Compensator structure	168
	4.6.2 The inverse Nyquist-array (INA) method	170
	4.6.3 The direct Nyquist-array (DNA) method	176

4.7 Achieving diagonal dominance 177
   4.7.1 Cut and try 178
   4.7.2 Perron–Frobenius theory 180
   4.7.3 Pseudo-diagonalization 186
4.8 Design example 189
   4.8.1 The design 189
   4.8.2 Analysis of the design 197
   4.8.3 Comparison with the characteristic-locus design 202
4.9 Quantitative feedback theory 203
4.10 Control-structure design 210
Summary 217
Exercises 218
References 220

5 Multivariable Design: LQG Methods 222

5.1 Introduction 222
5.2 The solution of the LQG problem 225
5.3 Performance and robustness of optimal state feedback 227
5.4 Loop transfer recovery (LTR) 231
5.5 Design procedure for square plant 235
5.6 Shaping the principal gains 235
5.7 Some practical considerations 243
5.8 Design example 244
   5.8.1 Kalman-filter design 244
   5.8.2 Recovery at the plant output 252
   5.8.3 Comparison with previous designs 258
5.9 Non-minimum-phase plant 259
Summary 261
Exercises 262
References 263

6 The Youla Parametrization and H_∞ Optimal Control 265

6.1 Introduction 265
6.2 A motivating example: sensitivity minimization 267
6.3 The H_∞ problem formulation 270
   6.3.1 Examples of H_∞ problems 270
   6.3.2 Performance robustness: an unsolved problem 273
6.4 The Youla (or Q) parametrization 274
   6.4.1 Fractional representations 274
   6.4.2 Parametrization of all stabilizing controllers 276
   6.4.3 All stabilizing controllers are observer-based 285
   6.4.4 Parametrization of closed-loop transfer functions 289
6.5 Solution of the H_∞ problem 293
   6.5.1 Equivalence to the model-matching problem 293
   6.5.2 Equivalence to the Hankel approximation problem 294
   6.5.3 1-block, 2-block and 4-block problems 295

6.6	The Hankel approximation problem	296
	6.6.1 The Hankel norm	296
	6.6.2 Glover's algorithm	298
6.7	The Glover–Doyle algorithm for general H_∞ problems	301
6.8	Design example	306
	6.8.1 Design specification	306
	6.8.2 Application of the Glover–Doyle algorithm	308
	6.8.3 Adjustment of γ and weights	310
	6.8.4 Comparison with previous designs	313
6.9	Review and comments	315
	6.9.1 The Youla parametrization	315
	6.9.2 Alternative approaches to H_∞ optimal control	315
	Summary	316
	Exercises	318
	References	323

7	**Design by Parameter Optimization**	**325**
7.1	Introduction	325
7.2	Edmunds' algorithm	326
	7.2.1 The algorithm	326
	7.2.2 Comments on Edmunds' algorithm	332
	7.2.3 A refinement of the algorithm	334
	7.2.4 Design example	336
7.3	The method of inequalities	341
7.4	Multi-objective optimization	346
	7.4.1 Formulation	346
	7.4.2 Solution	346
	7.4.3 Example	350
7.5	Conclusion	351
	Summary	352
	Exercises	353
	References	354

8	**Computer-aided Design**	**355**
8.1	Introduction	355
8.2	Elements of numerical algorithms	356
	8.2.1 Conditioning and numerical stability	356
	8.2.2 Solution of $Ax = b$ when A is square	357
	8.2.3 Householder transformations and QR factorization	359
	8.2.4 The Hessenberg form of a matrix	362
	8.2.5 The Schur form of a square matrix	363
	8.2.6 Eigenvalues	366
	8.2.7 Singular-value decomposition	366
	8.2.8 Generalized eigenvalues	367
8.3	Applications to linear systems	368
	8.3.1 Frequency-response evaluation	368
	8.3.2 Characteristic loci and principal gains	370

8.3.3 Interconnections of systems 371
8.3.4 Inverse systems 376
8.3.5 Minimal realizations 378
8.3.6 Partial-fraction decomposition 382
8.3.7 Transmission zeros 386
8.3.8 Root loci 389
8.3.9 Balanced realizations and model approximation 390
8.3.10 Algebraic Riccati equations 393
8.4 Software for control engineering 394
8.4.1 Mathematical software 394
8.4.2 Software packages 395
8.4.3 Current trends 397
Summary 399
Exercises 400
References 402

Appendix: Models used in Examples and Exercises 405

Index 409

Symbols and abbreviations

$\text{abs}(X)$	Matrix with (i,j) element $|x_{ij}|$.
$\arg z$	Argument of the complex number z.
$\underset{z}{\arg\max}\,(.)$	That value of z which maximizes $(.)$.
BD_δ	The set of block-diagonal perturbations $\text{diag}\{\Delta_1, \ldots, \Delta_2, \ldots, \Delta_n\}$, with $\|\Delta_i\|_\infty \leqslant \delta$.
$\mathbb{C}^{m \times l}$	The set of complex matrices with m rows and l columns.
CLHP, CRHP	Closed left half-plane, closed right half-plane.
$\text{cond}(G)$	Condition number, $\bar{\sigma}(G)/\underline{\sigma}(G)$.
$\text{cond}^*(G)$	$\underset{S,T}{\min}\,\text{cond}(SGT)$, where S and T are diagonal matrices.
dB	Decibels: $x\,\text{dB}$ represents a gain of $10^{x/20}$.
deg	Degree (of polynomial).
det	Determinant (of matrix).
$\text{diag}\{x_i\}$	Diagonal matrix with elements $x_1, x_2, \ldots$. If the matrix is not square, then $x_1, x_2, \ldots$ are the elements on the principal diagonal, and all other elements are zero. The x_i may themselves be matrices.
$\dim A$	Dimension of square matrix A.
DNA	Direct Nyquist Array.
$E\{x\}$	Expected (mean) value of stochastic process $\{x\}$.

e_i	The ith standard basis vector $[0 \ldots 0 \ 1 \ 0 \ldots 0]^T$, with 1 occurring in the ith position.
$G(A, B, C, D)$	Denotes that (A, B, C, D) is a state-space realization of the transfer function (matrix) G.
$G(s)$	A transfer function (matrix), frequently abbreviated to G.
$G^*(s)$	Denotes $G^T(-s)$. (But $G^H(s)$ denotes $G^T(\bar{s})$.)
$\gcd\{\cdot\}$	Greatest common divisor of the set $\{\cdot\}$.
H_∞	Set of asymptotically stable transfer functions G, with $\|G\|_\infty < \infty$.
I	Unit matrix of unspecified dimension.
I_n	Unit matrix of dimension n.
$\mathrm{Im}\{x\}$	Imaginary part of x.
INA	Inverse Nyquist Array.
j	$\sqrt{-1}$; sometimes an index, as in x_{ij}.
LHS, RHS	Left-hand side, right-hand side (of equation or inequality).
ln	Natural logarithm.
log or $\log_{10}$	Logarithm to base 10.
LQG	Linear Quadratic Gaussian.
LTR	Loop Transfer Recovery
MFD	Matrix-fraction description.
MIMO	Multi-input, multi-output.
ms(G)	Measure of skewness of G.
norm(X)	$\begin{bmatrix} \bar{\sigma}(X_{11}) & \ldots & \bar{\sigma}(X_{1m}) \\ \vdots & & \vdots \\ \bar{\sigma}(X_{m1}) & \ldots & \bar{\sigma}(X_{mm}) \end{bmatrix}$ if $X = \begin{bmatrix} X_{11} & \ldots & X_{1m} \\ \vdots & & \vdots \\ X_{m1} & \ldots & X_{mm} \end{bmatrix}$
OLHP, ORHP	Open left half-plane, open right half-plane.
QFT	Quantitative Feedback Theory.
$R^{m \times l}$	The set of real matrices with m rows and l columns.
$\mathrm{Re}\{x\}$	Real part of x.
RFN	Reversed-Frame Normalization.
SISO	Single-input, single-output.
tr(X)	Trace (spur) of matrix X, $\Sigma_i x_{ii}$.
$\Gamma(s)$	Relative gain array, with elements $\gamma_{ij}(s)$.
$\lambda_i(X)$	The ith eigenvalue of X.
$\lambda_{\max}(X), \lambda_{\min}(X)$	Largest and smallest eigenvalues of X.

$\lambda_P(X)$	Perron–Frobenius eigenvalue of X.
$\mu(G)$	Structured singular value of transfer function (matrix) G.
$\rho(X)$	Spectral radius of X, $\max_i \lvert \lambda_i(X) \rvert$.
$\sigma_i(X)$	The ith singular value of X.
$\sigma(G), \sigma(\omega)$	Principal gain (singular value) of $G(j\omega)$, the notation depending on whether dependence on the system G, or on the frequency ω, is being emphasized.
$\bar{\sigma}, \underline{\sigma}$	Largest and smallest singular values.
$\phi_{xx}(\tau)$	Autocovariance function: $E\{x(t)x^T(t+\tau)\}$.
$\Phi_{xx}(\omega)$	Power spectral density of $\{x\}$, Fourier transform of $\phi_{xx}(\tau)$.
$0_{m,l}$	Zero matrix with m rows and l columns.
$\bar{x}$	Complex conjugate of x.
X^H	Transpose of complex conjugate of matrix X, $\bar{X}^T$.
X^T	Transpose of matrix X.
$X^\dagger$	Pseudo-inverse of matrix X.
$\{x(t)\}$	A stochastic process.
$[X]_{ij}$	The (i, j) element of X, also denoted by x_{ij}.
$\lVert x \rVert$	Euclidean norm of vector, $\{x^H x\}^{1/2}$.
$\lVert X \rVert_1$	1-norm, $\max_j \Sigma_i \lvert x_{ij} \rvert$.
$\lVert X \rVert_F$	Frobenius norm, $\{\Sigma_{i,j} x_{ij}^H x_{ij}\}^{1/2} = \{\mathrm{tr}(X^H X)\}^{1/2}$.
$\lVert X \rVert_S$	Spectral or Hilbert norm of matrix, $\bar{\sigma}(X)$.
$\lVert x \rVert_2$	$\{\int_{-\infty}^{\infty} x^T(t)x(t)\,dt\}^{1/2}$, if $x(t)$ is a (real, vector-valued) signal.
$\lVert G \rVert_2$	$\{(1/2\pi)\int_{-\infty}^{\infty} \mathrm{tr}[G(j\omega)G^T(-j\omega)]\,d\omega\}^{1/2}$, if G is a transfer function (matrix).
$\lVert G \rVert_H$	Hankel norm, if G is a transfer function (matrix).
$\lVert G \rVert_\infty$	$\sup_\omega \bar{\sigma}(G(j\omega))$, if G is a transfer function (matrix).
$\lVert G \rVert_\mu$	$\sup_\omega \mu(G(j\omega))$, if G is a transfer function (matrix).
$\in$	'Is an element of'.
$*$	Element-by-element multiplication (Schur or Hadamard product).
$\otimes$	Kronecker or tensor product of matrices.
$\cup$	Union (of sets).
$\cap$	Intersection (of sets).
$\subset$	'Is a subset of'.

CHAPTER 1

Single-loop Feedback Design

1.1 Overview and prerequisites
1.2 Review of elementary feed-
    back design
1.3 A standard problem
1.4 Fundamental relations

1.5 The 'shape' of the solution
1.6 Two approaches to design
1.7 Limitations on performance
    Summary
    Exercises
    References

1.1 Overview and prerequisites

In this book it is assumed that the reader has taken a typical first course in the design of feedback systems. Such a course usually covers the Nyquist stability criterion, the use of Bode and root-locus plots, and the use of simple compensators to achieve reasonable stability margins, steady-state performance and transient response; we shall review the content of such a course very briefly (Section 1.2). We shall go on to discuss single-loop feedback design in more depth than is usual in first courses, and in the process of doing so we shall establish some nomenclature and relationships which will continue to hold when we come to examine multivariable problems in later chapters.

Chapters 2 and 3 extend the concepts and results of Chapter 1 to multivariable systems. Chapter 2 is concerned with establishing stability criteria which generalize the classical Nyquist criterion. In order to do this it is necessary to define poles and zeros of multivariable systems, and to tangle with some linear systems theory; most of the results obtained also find

1

application in later chapters. Chapter 3 deals with the analysis of performance and stability margins of multivariable systems. In addition to straightforward extensions of the results of Chapter 1, Chapter 3 presents a considerable amount of recent material on analysing the robustness of feedback systems in the face of specific disturbances or parameter variations which goes beyond the classical notions of gain and phase margins.

Chapters 4 to 7 are concerned with design techniques for multivariable feedback systems. Chapter 4 deals mostly with direct extensions of classical methods to multivariable systems. These methods have become known as the 'British school' of multivariable design, and provide the simplest and most easily comprehensible design techniques. Chapter 5 describes the use of 'linear quadratic Gaussian' control theory in such a way that sensible feedback designs are obtained. This involves analysing the resulting designs in the frequency-domain terms developed in Chapters 2 and 3. Chapter 6 moves to the very new area of H_∞ optimal control. This approach is of great current interest, and has provided some new fundamental results about feedback systems, as well as a very powerful design technique. Chapter 7 describes some methods of design by parameter optimization; these are relatively 'brute-force' approaches, but sometimes they have to be resorted to when other techniques either fail or are inapplicable, and they can produce excellent designs.

Finally, Chapter 8 discusses the software which is needed to do any analysis or design in the realm of multivariable systems.

The logical interdependence of the chapters can be represented as in the diagram below. Thus the reader interested mainly in practical design techniques may concentrate initially on Chapters 1, 2, 3, 4 and 7, while a research student may prefer Chapters 1, 2, 3 and 6. All readers are urged to look at Chapter 8, even if they skip most of the details of Sections 8.2 and 8.3.

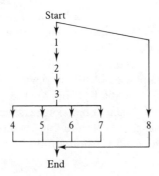

Since the basic representation of a multivariable system which we use is the transfer-function matrix – that is, a matrix whose elements are transfer functions – some familiarity with linear algebra is essential to the reader wishing to progress beyond Chapter 1. At the very least, he should be familiar with basic matrix manipulation and with eigenvalues and eigenvectors. If he is also familiar with the **singular-value decomposition**, the **rank** of a matrix, and notions such as **column span** or **invariant subspace**, he will be quite adequately prepared for the whole of the book.

A little knowledge of linear systems theory is also assumed in most chapters: the reader should know what is meant by the terms **state-space model, controllable, observable, minimal realization** and **state feedback.** Previous exposure to the theory of linear quadratic optimal control and/or Kalman filtering would also be helpful for Chapters 5 and 6, but is not essential. Readers without this background should consult Åström and Wittenmark (1984) or Franklin and Powell (1980), or (for a thorough treatment) Kailath (1980) or Chen (1984). Chapters 3 and 5 make some use of spectral densities to describe stochastic processes; spectral densities also make occasional appearances in examples and discussions in other chapters.

1.2 Review of elementary feedback design

We shall confine our attention to linear, time-invariant systems whose input/output behaviour is governed by a set of ordinary linear differential equations with constant coefficients. A simple example of such a system is

$$\ddot{y}(t) + a_1 \dot{y}(t) + a_2 y(t) = b_1 \dot{u}(t) + b_2 u(t) \tag{1.1}$$

in which $u(t)$ represents an input signal and $y(t)$ an output signal. If we apply the Laplace transform to (1.1) we obtain

$$(s^2 + a_1 s + a_2) Y(s) = (b_1 s + b_2) U(s) + (s + a_1) y(0) + a_1 \dot{y}(0) - b_1 u(0) \tag{1.2}$$

and if we set $y(0) = \dot{y}(0) = u(0) = 0$ we obtain the **transfer function**

$$\frac{Y(s)}{U(s)} = \frac{b_1 s + b_2}{s^2 + a_1 s + a_2} \tag{1.3}$$

We shall usually represent systems by transfer functions such as these. Typically, we shall use a notation such as $G(s)$ to denote a transfer function, and we shall frequently abbreviate this to G.

If the transfer function $G(s)$ is rational (that is, a ratio of polynomials), as in (1.3), then its denominator polynomial is usually the characteristic polynomial of the original differential equation, and its roots (the **characteristic roots**) therefore determine the dynamic characteristics of the system – in particular, their location in the complex plane determines the stability of the system. We say 'usually' because it is possible for cancellations to occur

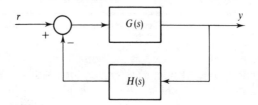

Figure 1.1   Negative-feedback connection of two linear systems.

between the numerator and denominator, so concealing some of the system's dynamics. A full elucidation of the causes and effects of such cancellations can be given by using state-space representations of systems rather than transfer functions; 'hidden dynamics' then correspond to **uncontrollable** or **unobservable modes** (Kailath, 1980). However, we do not wish to pursue this here, and shall refer rather vaguely to 'hidden modes' when necessary.

The roots of the numerator polynomial are called the **zeros** of the transfer function, and the roots of the denominator are called its **poles**. Thus the stability of a system is usually determined by its poles.

Two systems, with transfer functions $G(s)$ and $H(s)$, respectively, may be connected together in the negative-feedback configuration shown in Figure 1.1. Then, subject to certain conditions (for example, if the signals are voltages then input impedances must be much greater than output impedances), the transfer function of the closed-loop system is given by

$$\frac{Y(s)}{R(s)} = \frac{G(s)}{1 + H(s)G(s)} \tag{1.4}$$

The stability of this system is determined by the roots of the closed-loop **characteristic equation**

$$1 + H(s)G(s) = 0 \tag{1.5}$$

(or, equivalently, the closed-loop poles) provided no hidden unstable modes have been introduced by the connection of the two systems – that is, provided no cancellation of right half-plane poles and zeros has occurred when forming the product HG.

Now suppose that the designer of the feedback system has the freedom to change G or H. For any particular design he can easily check where the solutions of (1.5) lie. But if some of them are in the right half-plane, or so close to the imaginary axis as to give unacceptably resonant behaviour, then it is almost impossible to see from (1.5) how G or H should be changed for the better. At this point salvation is provided by the **Nyquist stability theorem**:

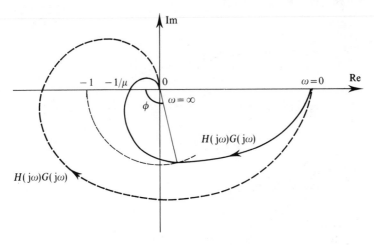

Figure 1.2 Nyquist locus of closed-loop stable system (solid curve) and unstable system (dashed curve); it is assumed that the open-loop system is stable.

> ***Theorem 1.1:*** The closed-loop system is stable if and only if the graph of $H(j\omega)G(j\omega)$, for $-\infty < \omega < \infty$, encircles the point $-1+j0$ as many times anticlockwise as $H(s)G(s)$ has right half-plane poles (provided there are no hidden unstable modes).

If the stability condition of this theorem is violated or nearly violated, it is relatively easy to see the changes which must be made to G or H.

Most commonly HG has no unstable poles, in which case the graph of $H(j\omega)G(j\omega)$, or **Nyquist locus**, of a closed-loop stable system is shown by the solid line in Figure 1.2, while that of an unstable system is shown by the dashed line; in each case only half the locus (for $0 \leqslant \omega < \infty$) is shown. These can also be displayed on a **Bode plot**, as in Figure 1.3. For the stable system the **phase margin** ϕ is shown in Figures 1.2 and 1.3, and the **gain margin** μ is shown in Figure 1.3.

If the closed-loop system is unstable, or if these stability margins are judged to be insufficient, then one or more **compensators** can be inserted into the feedback loop. One of the simplest possible compensators is a system whose transfer function is

$$K(s) = \frac{\alpha(1+sT)}{1+s\alpha T} \tag{1.6}$$

The Bode plot of this transfer function is shown in Figure 1.4, for $\alpha < 1$. If $\alpha < 1$ the compensator is said to provide **phase lead** (or **phase advance**), since $\arg K(j\omega) > 0$ for all $\omega > 0$ in this case, while if $\alpha > 1$ it is said to provide **phase**

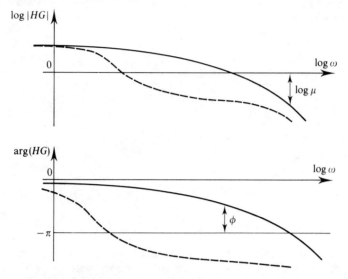

Figure 1.3   Bode plot form of Figure 1.2.

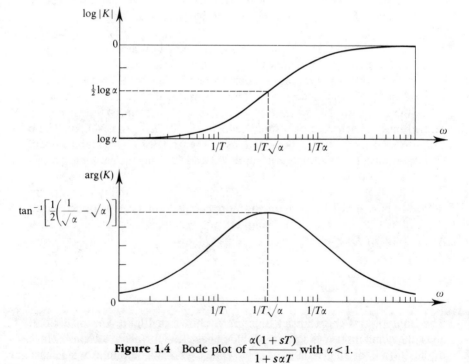

Figure 1.4   Bode plot of $\dfrac{\alpha(1+sT)}{1+s\alpha T}$ with $\alpha < 1$.

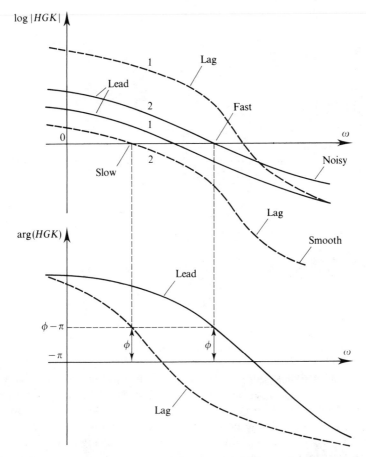

Figure 1.5 Typical Bode plots of system after compensation by phase lead (solid curves) or phase lag (broken curves); curves marked 2 are the same as those marked 1, but with adjustment of gain to achieve specified phase margin ϕ.

lag since $\arg K(j\omega) < 0$. The greatest phase change (i.e. value of $\arg K(j\omega)$) occurs at the frequency $\omega_m = 1/T\sqrt{\alpha}$, which is midway between the **corner** or **break frequencies** $\omega = 1/T$ and $\omega = 1/T\alpha$ on a Bode plot, since the frequency-axis calibration is logarithmic, and the value of this maximum phase change is

$$\arg K\left(\frac{j}{T\sqrt{\alpha}}\right) = \tan^{-1}\left[\frac{1}{2}\left(\frac{1}{\sqrt{\alpha}} - \sqrt{\alpha}\right)\right] \tag{1.7}$$

Figure 1.5 shows typical Bode plots of $H(j\omega)G(j\omega)K(j\omega)$ when K is a phase-lead (solid curves) or a phase-lag (dashed curves) compensator; it also shows them after a further gain adjustment (that is, insertion of a constant

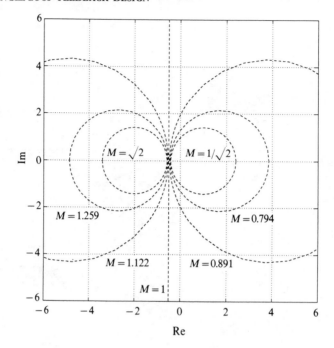

Figure 1.6   *M*-circles.

gain into the feedback loop) to achieve a specified phase margin ϕ. Although given stability margins can usually be achieved by following either a phase-lead or a phase-lag strategy, it is seen from Figure 1.5 that the phase-lead strategy increases both the **cross-over frequency**, namely the value of ω for which $|H(j\omega)G(j\omega)K(j\omega)| = 1$, and also the gain at high frequencies. This leads to faster transient responses being achieved with the phase-lead strategy, but at the expense of higher sensitivity to noise and larger signal levels appearing in the loop. In practice, compensators are frequently more complicated than the one defined by (1.6); they may exhibit both phase-lead and phase-lag characteristics (over disjoint frequency ranges), and they may have both real and complex poles and zeros.

An alternative way of specifying stability margins is to require the Nyquist locus to remain outside some neighbourhood of the point $-1 + j0$. Such neighbourhoods are usually defined by **M-circles**. These are the loci of points z in the complex plane, for which

$$\left| \frac{z}{1+z} \right| = M \tag{1.8}$$

where M is some positive real number. Some M-circles are shown in Figure 1.6; it can be seen that all those M-circles for which $M > 1$ enclose the point $-1 + j0$, and that the circles become smaller as M becomes larger. To

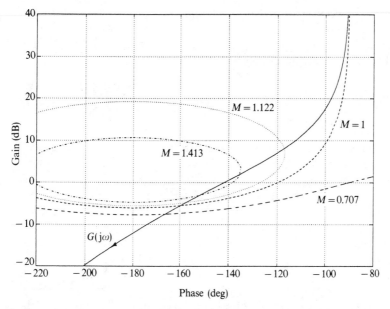

Figure 1.7 Nichols chart, showing a Nyquist locus (solid curve) and some M-circles (broken curves).

see why M-circles give useful stability margin specifications, suppose that, in Figure 1.1, we have $H(s)=1$, and that $G(s)$ already includes any compensators. Then the closed-loop gain from r to y, at frequency ω, is given by

$$\left| \frac{G(j\omega)}{1+G(j\omega)} \right| \tag{1.9}$$

So if the Nyquist locus of $G(s)$ is allowed to penetrate a high-valued M-circle over some range of frequencies, then the closed-loop frequency response will exhibit a large peak at these frequencies. This in turn indicates the presence of resonance in the closed-loop system, which is usually undesirable for practical reasons (rapid deterioration of components, heavy power consumption), and which must be caused by the presence of some lightly damped closed-loop poles close to the stability boundary.

Specifying that the Nyquist locus should remain outside the $M=\sqrt{2}$ M-circle, for example, therefore imposes some minimum degree of damping on the closed-loop poles (with the proviso, as usual, that the Nyquist locus cannot convey information about poles which are masked, or nearly masked, by zeros).

It is often useful to display a Nyquist locus on a Nichols chart, namely with $\log|G(j\omega)|$ plotted against $\arg G(j\omega)$. Figure 1.7 shows an example, with some M-circles superimposed. The M-circles are now distorted into non-circular shapes.

The approach to feedback design which has been reviewed here will serve as the basis for the multivariable techniques to be described later. Root-locus and pole-placement techniques have not been reviewed, because they will not be needed later in this book.

1.3   A standard problem

Figure 1.8 shows the form in which we shall pose the feedback design problem. The 'plant', which is assumed to be given and unalterable, is represented by the transfer function $G(s)$ and the (transform of the) disturbance signal $d(s)$, its input and output signals being $u(s)$ and $y(s)$, respectively. The function $r(s)$ represents a 'reference' or 'command' signal, which is to be followed by the plant output $y(s)$, and $m(s)$ represents measurement errors. $K(s)$ and $P(s)$ represent dynamic systems which are to be designed – we shall refer to these as the 'feedback compensator' and the 'pre-filter', respectively.

At first sight, most feedback design problems do not appear to have the structure of Figure 1.8. Disturbance signals often appear at the input of the plant, rather than (or in addition to) its output. For example, if the plant consists of an inertial load such as an aerial dish, and the input signal $u(s)$ represents the driving torque, then the major disturbance may appear as a random component of this torque. Also, transfer functions may appear in the feedback path, either because that turns out to be the only practical place to insert compensation in a particular case, or because the sensing device used to measure the plant output itself has significant dynamics. For example, if $y(s)$ represents the angular velocity about one axis of an artificial satellite, this may be measured by a rate gyro which (together with its signal-conditioning circuitry) has second-order dynamics. A further way in which the structure of a real problem may differ from that of Figure 1.8 is that there may be more than one 'loop' to design (for example, a fast velocity loop inside

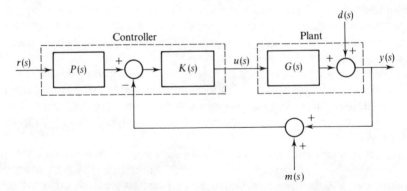

Figure 1.8   Standard feedback configuration.

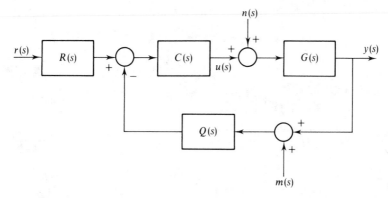

Figure 1.9 Another possible feedback configuration, often equivalent to that shown in Figure 1.8.

a slower position loop) or the compensator to be designed may be distributed around the system in a more complicated way than is shown in Figure 1.8.

However, since we have assumed the system to be linear, it is usually possible to obtain (by use of 'block-diagram algebra') a representation of the problem which does have the structure of Figure 1.8. For example, the system shown in Figure 1.9 is equivalent to that of Figure 1.8 if

$$P(s) = Q^{-1}(s)R(s)$$

$$K(s) = C(s)Q(s)$$

and

$$d(s) = G(s)n(s)$$

In this case the feedback design problem is not quite the same as it would have been had we started with Figure 1.8, because $K(s)$ and $P(s)$ are now constrained to some extent. Whether these constraints alter the problem significantly will depend on the details of $Q(s)$. So far, we have not assumed that the signals appearing in Figures 1.8 and 1.9 are single variables. They may be vectors of variables, in which case the transfer functions will be matrices. Now, if the measurement device represented by $Q(s)$ has unequal numbers of inputs and outputs, then $Q(s)$ will not be a square matrix, and $Q^{-1}(s)$ will not exist. In such a case Figures 1.8 and 1.9 are not equivalent, unless one is dealing with a 'regulator' problem, for which $r(s) \equiv 0$.

Our justification for using the structure of Figure 1.8 is that either a given problem can be recast in that structure or, if it cannot, then the general results which we shall develop will still hold, although some details are likely to change.

A multi-loop or multi-compensator problem can usually be put into the form of Figure 1.8 by allowing the transfer functions to be matrices, and

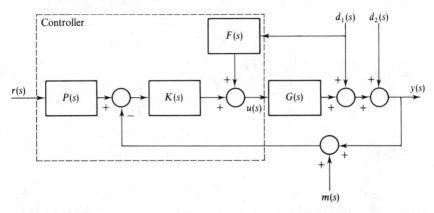

Figure 1.10   Feedback configuration with feedforward control of disturbances.

the signals to be vectors. However, it is better to decompose the problem into a sequence of single-loop problems if possible (typically, designing fast inner loops first and slow outer loops later) since a sequence of single-loop problems is usually much easier to solve than a single multivariable problem.

We assume that the measurement error m is unknown, since it could be corrected for if it were known. However, it may be permissible to assume that it can to some extent be described. Most frequently it is described as a stochastic process, with a known power spectral density (most commonly flat). Another common approach is to assume the measurement error to have a constant but unknown value – this describes systematic bias in the measurement. However, such bias can be corrected only if more than one independent measurement of the output is available, which is not the case for the configuration shown in Figure 1.8. We therefore assume that any systematic bias in the measurement has already been corrected.

The disturbance d is also assumed to be unknown. In practice, it frequently happens that one can measure or estimate part or all of a disturbance. For example, if it is caused by a change in the composition of some raw material being fed into a process, that change may be detected and measured before the material even enters the process. Suppose that

$$d(s)=d_1(s)+d_2(s)$$

and that d_1 can be measured. Then the controller should include a 'feedforward' compensator, as shown by the block $F(s)$ in Figure 1.10. If $F(s)=-G^{-1}(s)$, then the disturbance d_1 is completely cancelled by the feedforward compensator, and Figure 1.10 is again equivalent to Figure 1.8, if we take $d(s)=d_2(s)$. For several reasons, it is impossible to set $F(s)=-G^{-1}(s)$ exactly. Perhaps the most fundamental is that usually $G^{-1}(s)$ is not the transfer function of any realizable system. Also, $G(s)$ is usually known only approximately (this is often the reason for using feedback in the first place), and even if $F(s)$ only approximates $-G^{-1}(s)$, this may require impossibly large and rapid changes of the control signal $u(s)$ in response to certain disturbance signals. Finally, if the plant is either unstable, or has zeros in the right half-plane, then series compensation by $G^{-1}(s)$ would result in unstable modes which were either uncontrollable or unobservable (Kailath, 1980), and which therefore could not be stabilized. In spite of these limitations, it clearly makes sense to use any available information about disturbances in this manner, and doing so often makes the feedback design task dramatically easier. Therefore, when we start with Figure 1.8 we assume that any detailed information about disturbances has already been exploited in a feedforward scheme, and that $d(s)$ in Figure 1.8 represents the effective disturbance which remains, and which is therefore unknown.

It should be added that the 'disturbance' d can represent various sources of uncertainty in the plant. For example, suppose that we believe the true transfer function of the plant to be $G+\Delta$ rather than G, but we do not know Δ – we may have only a bound on the magnitude of Δ, for example. This can again be represented by Figure 1.8 if we take $d(s)=\Delta(s)u(s)$ (which is of course unknown, even though $u(s)$ is known).

1.4   Fundamental relations

From Figure 1.8 we see that

$$y(s)=d(s)+G(s)K(s)[P(s)r(s)-m(s)-y(s)] \tag{1.10}$$

Hence,

$$[I+G(s)K(s)]y(s)=d(s)+G(s)K(s)[P(s)r(s)-m(s)] \tag{1.11}$$

where I is the unit matrix (we have yet to restrict the discussion to single-loop systems). If we define the **return difference**

$$F_o(s)=I+G(s)K(s) \tag{1.12}$$

(the subscript denoting that the return difference is evaluated at the output of the plant), and the **sensitivity** function

$$S(s)=F_o^{-1}(s) \tag{1.13}$$

then we can write

$$y(s)=S(s)d(s)+G_c(s)r(s)-T(s)m(s) \tag{1.14}$$

where

$$T(s) = S(s)G(s)K(s) \tag{1.15}$$

and

$$G_c(s) = T(s)P(s) \tag{1.16}$$

We shall call $T(s)$ the **closed-loop transfer function.**

It should be noted that $S(s)$ and $T(s)$ depend only on those transfer functions which are inside the feedback loop, while the pre-filter $P(s)$ affects only $G_c(s)$. Consequently, we can consider first the problem of designing $K(s)$ to obtain desired $S(s)$ and $T(s)$, and subsequently design $P(s)$ to give a suitable $G_c(s)$. This second step is a matter of open-loop series compensation, and so is not strictly a feedback design problem. In many cases the ideal solution would be to set $P(s) = T^{-1}(s)$ (because of equation (1.16)), but we cannot do this because of the same obstacles that prevent perfect feedforward control from being realized.

In many (perhaps most) applications the pre-filter is fixed, most commonly as the unit matrix. In this case $G_c(s)$ and $T(s)$ cannot be adjusted independently of each other, and in the most common case are identical. We then have a 'one degree of freedom' design problem, whereas the general problem represented by Figure 1.8 is known as the 'two degrees of freedom' problem. We shall consider the 'one degree of freedom' problem, namely the design of $K(s)$ only, almost exclusively. (In many cases a pre-filter is effectively supplied by a plant operator, or a pilot, who learns to compensate for $T(s)$ by giving the reference signal a different profile from the one he wishes the output to follow.)

From equation (1.14) we see that, if the feedback loop is to be successful in attenuating the effects of disturbances acting on the plant, and if the output is to be insensitive to measurement errors, then we need both $S(s)$ and $T(s)$ to be small in some sense. However, from (1.12) and (1.15) we obtain

$$T(s) + S(s) = I \tag{1.17}$$

This shows that, if we succeed in making $S(s)$ nearly zero, then $T(s)$ will be nearly the identity; and conversely, if we make $T(s)$ nearly zero, then $S(s)$ will be nearly the identity. We thus have an **unavoidable trade-off** between attenuating disturbances and filtering out measurement error. This trade-off is one of the factors which make feedback design difficult.

The closed-loop transfer function $T(s)$ is often referred to as the **complementary sensitivity**, because of relation (1.17). This is useful for distinguishing $T(s)$ from other possible closed-loop transfer functions.

In process control, signals from measurement sensors are often effectively free from error because the time constants which occur are large enough to allow almost complete elimination of electrical noise by filtering. (However, measurements may be noisy for other than electrical reasons – consider measuring the level of a boiling liquid, for example.) It may appear that the trade-off is avoided in such a situation. Unfortunately this is not so, because in process control it is invariably a requirement that control signal movements should be kept reasonably small, and it turns out that this is equivalent to the requirement that $T(s)$ should be kept small (at least in the single-loop case), as we shall now show.

From Figure 1.8 we see that

$$u(s) = K(s)[P(s)r(s) - m(s) - d(s) - G(s)u(s)] \tag{1.18}$$

Hence

$$[I + K(s)G(s)]u(s) = K(s)[P(s)r(s) - m(s) - d(s)] \tag{1.19}$$

If we define the return difference at the input of the plant to be

$$F_i(s) = I + K(s)G(s) \tag{1.20}$$

then we can write

$$u(s) = F_i^{-1}(s)K(s)[P(s)r(s) - m(s) - d(s)] \tag{1.21}$$

($F_i^{-1}(s)$ is sometimes called the **input sensitivity**, in which case $S(s) = F_o^{-1}(s)$ is called the **output sensitivity**.)

For the rest of this chapter we shall confine ourselves to single-loop feedback. In this case the two return differences are the same, and we can define

$$F(s) = F_i(s) = F_o(s) \tag{1.22}$$

Let us measure the strength of variation of the control signal by $\int_0^\infty u^2(t)\,dt$. Then, by Parseval's theorem (Oppenheim *et al.*, 1983), we have

$$\int_0^\infty u^2(t)\,dt = \frac{1}{2\pi}\int_{-\infty}^\infty |u(j\omega)|^2\,d\omega \tag{1.23}$$

where we have used $u(t)$ to denote the input signal as a function of time, and $u(s)$ to denote its Laplace transform. In writing equation (1.23) we have assumed that the feedback loop is stable, and that the control signal is responding to a temporary disturbance – otherwise neither integral would exist.

If a wide class of disturbances is likely to occur, or if the disturbance is well defined but its energy spectrum extends over a wide frequency range, then we can see from equation (1.21) that the only way of ensuring that $\int |u(j\omega)|^2 \, d\omega$ is small is to keep $F^{-1}(j\omega)K(j\omega)$ small. But

$$F^{-1}(j\omega)K(j\omega) = \frac{K(j\omega)}{1 + K(j\omega)G(j\omega)} \tag{1.24}$$

$$= \frac{T(j\omega)}{G(j\omega)} \tag{1.25}$$

So we see that $|T(j\omega)|$ must be kept small, except possibly at those frequencies at which $|G(j\omega)|$ is sufficiently large, if any such frequencies exist.

1.5   The 'shape' of the solution

In practice, the conflict between keeping $T(s)$ small and keeping $S(s)$ small is resolved by making one small at some frequencies, and the other small at other frequencies. Usually the spectra of reference signals and disturbances are concentrated at low frequencies, while the spectrum of measurement errors extends over a much wider frequency range. The solution which is almost universally adopted is therefore to make $|S(j\omega)|$ small at low frequencies (for $0 \leqslant \omega < \omega_0$, say) and $|T(j\omega)|$ small at high frequencies (for $\omega > \omega_b$, say). Even if no measurement noise is present, equation (1.25) shows that $|T(j\omega)|$ should be made as small as possible at frequencies above those at which disturbances (and reference signals, if $P(s) = 1$) occur.

From equation (1.12) we see that, to make $S(s)$ small, we must make $|1 + G(s)K(s)|$ large, and hence $|G(s)K(s)|$ must be large. If $|G(j\omega)|$ is not itself sufficiently large at low frequencies, then $|K(j\omega)|$ must be made large there. We also see that, to make $T(s)$ small, and hence $S(s)$ close to unity (see equation (1.17)), we need to make $|G(j\omega)K(j\omega)|$ close to zero. Now, all real plants have the characteristic that $|G(j\omega)| \to 0$ as $\omega \to \infty$, so it may be enough to require $|K(j\omega)|$ not to increase at high frequencies, but if the rate of gain reduction with frequency of the plant is not fast enough, then $|K(j\omega)|$ is required to be small at high frequencies.

EXAMPLE 1.1

A regulator ($r = 0$) is subject to an output disturbance which consists of a sine wave of amplitude A and frequency ω_0, superimposed on occasional steps (that is, a piecewise-constant function) of unpredictable magnitude.

Its output is to remain within the bounds $\pm A/10$ (ignoring any transients from the occasional steps), and this is to be done with minimal use of control energy.

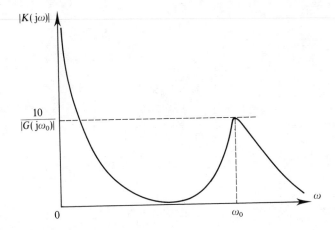

Figure 1.11 Compensator gain characteristic to cope with single-frequency disturbance.

We can meet this requirement by ensuring that

(1) $S(0)=0$ (by the final value theorem for Laplace transforms):

$$|K(0)|=\infty, \quad \text{if} \quad |G(0)|<\infty$$

(2) $|S(j\omega_0)|\leqslant 1/10$:

$$|K(j\omega_0)G(j\omega_0)+1|\geqslant 10$$

or $|K(j\omega_0)G(j\omega_0)|\geqslant 10$ (approximately)

(3) $|K(j\omega)|$ is as small as possible, consistent with 1 and 2; thus $|K(j\omega)|$ has the 'shape' shown in Figure 1.11.

Usually, the output disturbance is not concentrated at one frequency, but is known to have significant content *up to* some frequency ω_0. To get acceptable performance, we then require that

$$|S(j\omega)|<\varepsilon, \quad 0\leqslant\omega\leqslant\omega_0$$

which implies that

$$|K(j\omega)G(j\omega)|>1/\varepsilon \text{ (approximately)}, \quad 0\leqslant\omega\leqslant\omega_0$$

frequently with $|K(0)G(0)|=\infty$ as an additional constraint; Figure 1.12 shows a typical plot of the 'open-loop gain' $|K(j\omega)G(j\omega)|$. Recall that

$$T(s)=1-S(s)$$

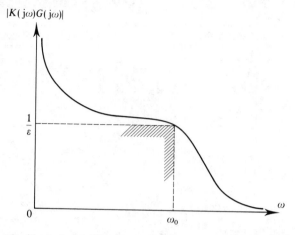

Figure 1.12 Typical specification and characteristic of compensator gain.

so corresponding to the above, we have

$$|T(j\omega)| = |1 - S(j\omega)| \approx 1, \quad \text{for } 0 \leqslant \omega \leqslant \omega_o \text{ if } \varepsilon \ll 1$$

and

$$|T(0)| = 1$$

Figure 1.13 shows the typical 'shapes' of $|S(j\omega)|$ and $|T(j\omega)|$ which result.

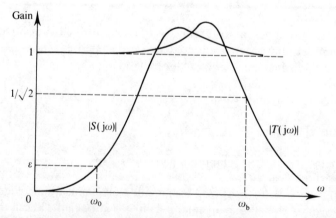

Figure 1.13 Typical gain characteristics of sensitivity and complementary sensitivity.

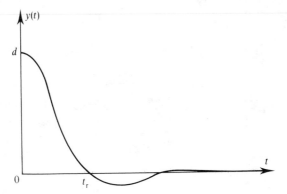

Figure 1.14   Closed-loop response to step disturbance on output.

We call ω_b the **bandwidth** of the feedback loop and define it, rather arbitrarily, as the lowest frequency such that

$$|T(j\omega_b)| = |T(0)|/\sqrt{2}$$

Note that ω_b may be considerably higher than ω_o. Clearly the requirement that $|T(j\omega)|$ should fall as quickly as possible for $\omega > \omega_o$ translates into the requirement that ω_b should be as small as possible.

Now, it is known that (roughly speaking) the bandwidth of the loop is inversely proportional to the response time of $y(t)$, in response to a step disturbance – see Figure 1.14. So keeping the bandwidth small implies keeping its response relatively 'sluggish'. Note that if $P(s) = 1$, the response to a step on the disturbance is a 'mirror image' of the response to a step on the reference input, so the 'tracking' step response is also relatively slow.

The loop bandwidth ω_b is usually very close to the '0 dB cross-over' frequency ω_c, at which $|G(j\omega_c)K(j\omega_c)| = 1$. Figure 1.15 shows a Nyquist locus. At the cross-over frequency ω_c the locus intersects the origin-centred unit circle, while at the loop bandwidth frequency ω_b it intersects the $M = 1/\sqrt{2}$ circle. If the locus were to lie entirely along the negative imaginary axis, these two intersections would occur at the same frequency, and hence we would have $\omega_b = \omega_c$. If the locus lay entirely along the negative real axis, then the rate of change of gain with frequency would be quadratic (40 dB/decade), as can be seen by considering the example of two integrators in series. Since the $M = 1/\sqrt{2}$ circle crosses the negative real axis at -0.414, the loop bandwidth would in this case be given by $\omega_b = 1.55\omega_c$. In practice Nyquist loci usually lie between these two extremes in the region shown in Figure 1.15. As we shall see in Section 1.7.1, a limit on the rate of change of gain with frequency can be obtained from the phase angle, and this enables us to obtain a third estimate. Suppose that the Nyquist locus is as shown in Figure 1.15, entering the $M = 1/\sqrt{2}$ circle at $-0.414 + j0$, and let us take its phase, between

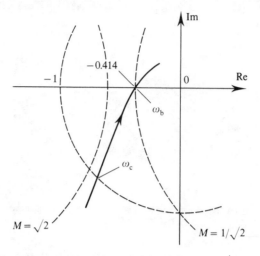

Figure 1.15 Typical Nyquist locus, showing the $M=1/\sqrt{2}$, $M=\sqrt{2}$ and $|S|=1$ circles.

the cross-over frequency and the loop bandwidth, to be $-135°$. This corresponds to the gain changing as $\omega^{-1.5}$ ($-30\,\mathrm{dB/decade}$), at most. Hence in this case $\omega_b = 1.8\omega_c$. These calculations are approximate, of course, but they do illustrate that for typical, acceptable designs we can estimate the loop bandwidth in terms of the cross-over frequency by

$$\omega_c \leqslant \omega_b \leqslant 2\omega_c \tag{1.26}$$

In general, we have a second 'degree of freedom' because $P(s)$ need not be just unity. In this case it is useful to distinguish between the 'loop bandwidth' and the 'transmission bandwidth'.

The loop bandwidth is the bandwidth of

$$T(s) = \frac{G(s)K(s)}{1+G(s)K(s)} \tag{1.27}$$

and indicates the 'speed of elimination' of a disturbance.

The transmission bandwidth is the bandwidth of

$$G_c(s) = \frac{G(s)K(s)P(s)}{1+G(s)K(s)} \tag{1.28}$$

and indicates the 'speed of response' to the reference signal. $P(s)$ is usually used to extend the transmission bandwidth beyond the loop bandwidth, but of course this demands increased control energy.

EXAMPLE 1.2

Suppose all three signals acting on the system (r, d and m) are stochastic processes with spectral densities $\Phi_r(\omega)$, $\Phi_d(\omega)$ and $\Phi_m(\omega)$, respectively. Define the tracking error $e(t)$ by

$$e(t) = r(t) - y(t)$$

Suppose we wish to minimize the variance of the error, $E\{e^2(t)\}$. From equation (1.10) we see that

$$e(s) = [1 - T(s)]d(s) + [G_c(s) - 1]r(s) - T(s)m(s) \qquad (1.29)$$

If we assume that d, r and m are mutually uncorrelated, and let the spectral density of e be $\Phi_e(\omega)$, then we obtain

$$\Phi_e(\omega) = |1 - T(j\omega)|^2 \Phi_d(\omega) + |1 - G_c(j\omega)|^2 \Phi_r(\omega)$$
$$+ |T(j\omega)|^2 \Phi_m(\omega) \qquad (1.30)$$

But

$$E\{e^2(t)\} = \frac{1}{2\pi} \int_{-\infty}^{\infty} \Phi_e(\omega)\,d\omega \qquad (1.31)$$

so it is clear that, to keep the error small, we must keep $|1 - T(j\omega)|$ small when $\Phi_d(\omega)$ is large, $|T(j\omega)|$ small when $\Phi_m(\omega)$ is large and $|1 - G_c(j\omega)|$ small when $\Phi_r(\omega)$ is large.

The quality of control is limited by the extent to which the spectra of d and m are separated (and their relative magnitudes).

Once again, the 'shape' of a solution becomes clear: $|K(j\omega)|$ large enough to keep $|S(j\omega)|$ small for $0 \leqslant \omega \leqslant \omega_0$, and $|K(j\omega)|$ as small as possible for $\omega > \omega_0$, for some frequency ω_0.

If $P(s) = 1$ then $G_c(s) = T(s)$, and the transmission and loop bandwidths coincide. Thus the better the system responds to the reference signal, the more it will respond to the (misleading) measurement noise. If there is significant overlap between $\Phi_r(\omega)$ and $\Phi_m(\omega)$, the loop bandwidth should be kept low, and $P(s)$ ($\neq 1$) should be used to increase the transmission bandwidth.

The loop bandwidth should be just large enough to give sufficient attenuation of disturbances, including those representing uncertainty about the plant model. It follows that, if there were *no* disturbances, or in other words if there were *no uncertainty* about the behaviour of the plant, then the loop bandwidth should be zero; that is, in this case we should have *no feedback*, but should use open-loop series compensation alone. Of course, such an ideal situation never occurs, and is very rarely even approached, but it serves to illustrate that *the purpose of using feedback is to combat uncertainty*, and is not just to modify a system's input/output behaviour.

1.6   Two approaches to design

From equation (1.15) we see that

$$K(s) = G^{-1}(s)F(s)T(s) \tag{1.32}$$

which suggests an obvious approach to design: we just need to choose either $F(s)$ or $T(s)$ to satisfy the requirements described in Section 1.5, and stability requirements, and then obtain $K(s)$ from equation (1.32). The feedback loop will be stable if $F(s)$ has all its zeros in the open left half-plane, or equivalently, if $T(s)$ has all its poles there.

The fundamental objection to this 'closed-loop' approach is that it forces the designer to pick one particular $F(s)$ (or $T(s)$) from the infinitely many which would be equally suitable. This choice is essentially arbitrary, in so far as it ignores the characteristics of the plant $G(s)$. It is probable that choosing a different $F(s)$ would yield a $K(s)$ which would be preferable, even though both may meet closed-loop requirements. For example, one choice of $F(s)$ may result in $K(s)$ being unstable, whereas another may not. Stable controllers are preferred to unstable ones, primarily because they are easier to set up and test before inserting them into the feedback loop. It is virtually impossible to determine the most appropriate choice of $F(s)$ directly – that is, without examining the open-loop characteristics of the plant.

If the objective of design is very narrowly defined – 'minimize the variance of the error', for example – then this approach is more attractive, since a unique $F(s)$ can be found which meets this objective. The 1950s approach to optimal design, based on Wiener filter theory, took this path (Newton *et al.*, 1957). But even in such a case the mathematically expressed objective is likely to be somewhat artificial, and the designer may well prefer a solution which is not quite 'optimal' because it is better in a sense which defies mathematical description. A more detailed assessment of this 'closed-loop' approach to design has been given by Horowitz (1963).

There are also less fundamental, more technical objections to the 'closed-loop' approach. In equation (1.32), $K(s)$ depends on $G^{-1}(s)$. If $G(s)$ has poles or zeros in the right half-plane, unstabilizable (because uncontrollable or unobservable) modes will be built in to the feedback loop. If $F(s)$ or $T(s)$ is chosen so that $|F(j\omega)T(j\omega)|$ approaches zero at high frequency less quickly than $|G(j\omega)|$ does, then $K(s)$ will be unrealizable. These objections are not insuperable because it is possible to choose the closed-loop specifications so as to overcome them. For example, if $F(s)$ has poles at those points at which $G(s)$ has unstable poles, then these poles will not be cancelled by $K(s)$. (See Franklin and Powell (1980) for a fuller discussion.)

However, if $F(s)$ is obtained as the solution to some optimization problem, then the designer does not have the opportunity of ensuring that $F(s)$ will have the special features required to give a realistic design. It is, nevertheless, possible to proceed with a 'closed-loop' approach to design, by working in a rather more elaborate framework. The so-called 'Youla parametrization' yields a means of solving general feedback-optimization problems (most commonly optimizing some closed-loop properties), with the assurance that the solution is a realizable controller which does not attempt to cancel right half-plane poles or zeros. This parametrization will be described in detail in Chapter 6.

By far the most common approach to feedback design is to work directly with the 'open-loop' return ratio $G(s)K(s)$. As we have already seen, the designer starts by knowing roughly the characteristics that $|G(j\omega)K(j\omega)|$ must have – large up to some frequency, as small as possible beyond that. He probably has some idea of the frequency region in which the transition from high gain (large $|GK|$) to low gain should occur, and he knows, from the Nyquist stability theorem, how the graph of $G(j\omega)K(j\omega)$ should behave in relation to the point $-1 + j0$. He may have much more precise information about the open-loop requirements, which are derived from closed-loop specifications. These requirements are usually stated in the following form:

(1) $|G(j\omega)K(j\omega)| > L \,(\gg 1)$, for $0 \leqslant \omega < \omega_1$;

(2) $|G(j\omega)K(j\omega)| < \varepsilon \,(\ll 1)$, for $\omega > \omega_h$;

(3)   Gain margin $> \mu$, phase margin $> \phi$;

(4)   The graph of $G(j\omega)K(j\omega)$ is to remain outside a neighbourhood of the point $-1 + j0$, defined by a particular M-circle (most commonly $M = \sqrt{2}$).

To some extent (3) and (4) are alternatives.

These requirements may vary. Commonly met additional requirements are that $|G(0)K(0)| = \infty$, which ensures zero steady-state error in the face of step disturbances, or that

$$\lim_{\omega \to 0} \omega|G(j\omega)K(j\omega)| = \infty$$

which ensures zero steady-state error in the face of ramp disturbances. These two are the most common examples of the application of a general principle which states that the long-term error in the face of a persistent disturbance (that is, one which does not decay to zero with time) can be zero only if the poles of (the transform of) the disturbance are included among the poles of the return ratio $G(s)K(s)$. In the past fifteen years this principle has been shown to hold in very general situations, and has been called the 'internal-model principle' (Åström and Wittenmark, 1984; Wonham, 1974), although its essence was known to the pioneers of feedback design methods.

Another departure from the standard form of specifications which can occur is that the required gain–frequency behaviour may not be monotonic decreasing. For example, it may be necessary to reduce the loop gain in the region of a resonance, but to keep it large at both lower and higher frequencies. As another example, it may be necessary to keep the loop gain almost constant over a range of frequencies if the plant has a zero in the right half-plane. (In such cases terms like 'loop bandwidth' should be used with care.)

The attraction of working with the open-loop rather than the closed-loop transfer functions is that it is easy to see what features $K(j\omega)$ should have if $G(j\omega)K(j\omega)$ is to satisfy the requirements, and it is usually relatively easy to find a transfer function $K(s)$ which has these features. Furthermore, the design can be pursued in stages: for example, meeting high-frequency gain reduction requirements first, then stability margins, and finally low-frequency gain enhancement. Usually such stages 'interfere' with one another, and it may be better to think of each stage as giving a better approximation to the required characteristics of $G(s)K(s)$. (Breaking the design into stages corresponds to obtaining $K(s)$ as a product, each factor being the transfer function obtained at each stage.)

Graphical displays of $G(j\omega)K(j\omega)$, in the form of Nyquist, Bode or Nichols plots, enable the designer to make a rapid visual assessment of the degree to which the requirements have been met, and, with a little practice, to judge the seriousness of any violations of the requirements.

1.7   Limitations on performance

The question arises of whether it is always possible to meet specifications for $G(j\omega)K(j\omega)$ by using some $K(s)$. It turns out that it is not possible, because the gain–frequency behaviour and phase–frequency behaviour are not independent of each other.

1.7.1   Gain–phase relationships

Suppose that $Q(s)$ is a transfer function which has all its poles and zeros in the open left half-plane. Such a transfer function is called **minimum-phase** (for reasons which will become clear in Section 1.7.2). Also suppose that $Q(0) > 0$ (and, of course, that $Q(0)$ is real), and let

$$L(\zeta) = \log |Q(j\zeta)| \qquad (1.33)$$

$$\phi(\zeta) = \arg Q(j\zeta) \qquad (1.34)$$

$$\lambda = \log(\zeta/\omega) \qquad (1.35)$$

It was shown by Bode (1945) that

$$\phi(\omega) = \frac{1}{\pi} \int_{-\infty}^{\infty} \frac{\mathrm{d}L(\zeta)}{\mathrm{d}\lambda} \log \coth |\lambda/2| \, \mathrm{d}\lambda \qquad (1.36)$$

Noting that the function $\log \coth |\lambda/2|$ behaves approximately like the impulse $\frac{1}{2}\pi^2 \delta(\lambda)$, we obtain

$$\phi(\omega) \approx \frac{\pi}{2} \frac{\mathrm{d}L}{\mathrm{d}\lambda} \bigg|_{\zeta = \omega} \qquad \text{(very approximately)} \qquad (1.37)$$

From this, the usual 'Bode plot' rule of thumb follows: if the gain falls at a sustained rate of 20 dB/decade of frequency (which corresponds to $\mathrm{d}L/\mathrm{d}\lambda = -1$), then the phase at those frequencies will be $-\pi/2$ rad. If the gain falls at 40 dB/decade, the phase will be $-\pi$ rad, and so on. Our assumption that $Q(0) > 0$ may be incorrect, in which case equation (1.36) gives the phase *change* which occurs between frequencies 0 and ω. If the transfer function has zeros in the right half-plane, then the phase will be more negative than is indicated by equation (1.36).

We have seen that the solution to the archetypal feedback design problem is of the form 'use high loop gain over some frequency range, then decrease the loop gain as rapidly as possible'. If only gain mattered, we could get arbitrarily fast cut-off of loop gain just by cascading large numbers of low-pass filters. But we need to maintain the stability of the loop. To do this we cannot have excessive phase lag at the frequency at which $|K(j\omega)G(j\omega)| = 1$ because of Nyquist's stability theorem. But we know from the gain–phase relationships that this restricts the magnitude roll-off rate: there is a phase lag of at least 90° for each 20 dB/decade of roll-off rate.

For an open-loop stable, minimum-phase plant, we can say that the roll-off rate in the region of the gain cross-over frequency (that is, where the loop gain is unity) must not exceed about 40 dB/decade. For a reasonable stability margin it must be smaller than this.

Note that this is a 'local' and rather approximate result, which can usually be used only to tell whether a specification is easy or difficult to meet. It is not usually possible to say that a specification is impossible to meet without doing a detailed design study. Consider the following (extreme) examples:

EXAMPLE 1.3

loop gain ≥ 33, $\omega \leq 1$

loop gain $\leq 1/33$, $\omega \geq 10$

The average slope required is -60 dB/decade, which implies a phase lag of

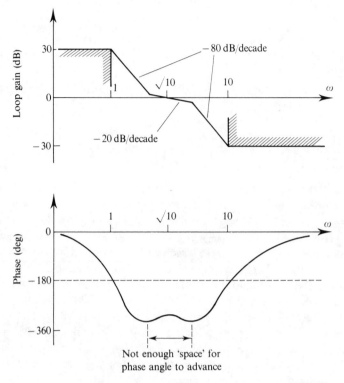

Figure 1.16 Bode plot with unattainable gain specification.

270° (average) over one decade of frequency. We could try shaping the loop gain as shown in Figure 1.16, but we find that the region over which the gain falls at only 20 dB/decade is too small to allow the phase to increase significantly.

EXAMPLE 1.4

loop gain $\geqslant 33 \times 10^6$ for $\omega \leqslant 0.001$

loop gain $\leqslant 1/33 \times 10^{-6}$ for $\omega \geqslant 100$

The average slope required is -60 dB/decade, as before, but now this is spread over five decades, and the previous strategy has a chance of success (because the -20 dB/decade section can be made to last for, say, three decades). But note that such a design could only be conditionally stable: a reduction of the loop gain could make the loop unstable.

1.7.2 Right half-plane zeros

Suppose that $G(s)$ is stable, and has one real, right half-plane zero, at $s = a$. We shall show that the achievable performance of a feedback loop is then limited, in the following sense. Suppose that $|K(j\omega)G(j\omega)| \geqslant \gamma > 1$ ($0 \leqslant \omega \leqslant \omega_0$). Then the open-loop **gain–bandwidth product** $\gamma\omega_0$ is, in practice, limited. Furthermore, an approximate practical *upper bound* on ω_0 is $\omega_0 \leqslant a$.

Note that these limitations occur even without constraints on $u(t)$, and without the presence of measurement noise; also, ω_0 is an open-loop bandwidth, but the upper bound on ω_0 in effect puts a similar bound on ω_b, the loop bandwidth.

We begin by writing

$$G(s) = G_o(s) A(s) \tag{1.38}$$

where $A(s) = (s - a)/(s + a)$ is an **all-pass** function (that is, $|(j\omega - a)/(j\omega + a)| = 1$, while its phase decreases). Its Nyquist diagram is shown in Figure 1.17, while Figure 1.18 shows how the phase of $A(s)$ varies with frequency. Suppose that the gain cross-over frequency is $\omega_c = a$, and let us make the unrealistic assumption that we could tolerate zero phase margin. Then since

$$\arg A(ja) - \arg A(0) = -\tfrac{1}{2}\pi$$

we could allow $\arg K(j\omega)G_0(j\omega)$ to change by $-\tfrac{1}{2}\pi$ at most between $\omega = 0$ and $\omega = a$. From the gain–phase relations we know this to mean that $|K(j\omega)G_0(j\omega)|$ must be falling at a rate of 20 dB/decade *at most* in the region of $\omega = a$.

Figure 1.19 shows some possible (γ, ω) specifications. It will be seen that for $a/10 < \omega_0 < a$, roughly, the gain–bandwidth product is limited to

$$\gamma\omega_0 \leqslant a \tag{1.39}$$

Of course, allowing for a realistic phase margin would reduce this further.

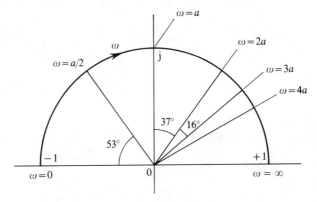

Figure 1.17   Nyquist diagram of the all-pass transfer function $(s - a)/(s + a)$.

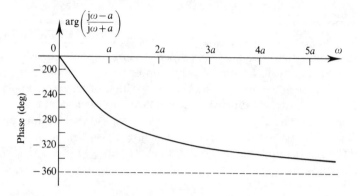

Figure 1.18 Phase–frequency characteristic of $(s-a)/(s+a)$.

If the cross-over frequency is reduced to $\omega_c = a/2$, then $A(s)$ contributes only 53° of phase lag there, so the slope of $|KG_0|$ can be increased to about 28 dB/decade (still with zero phase margin) and hence we can achieve $\gamma\omega_0 \leqslant 1.25a$.

In fact, $\gamma\omega_0$ can be increased indefinitely by pushing ω_c to a sufficiently low value. In other words, right half-plane zeros do not cause difficulties if they lie well outside the desired (closed-loop) bandwidth.

Now suppose that we raise the cross-over frequency to $\omega_c = 2a$. $A(j\omega_c)$ contributes 127° of phase lag, which leaves 53° for $\arg(KG_0)$. In practice most of this would be needed for the phase margin, leaving a very shallow slope for $|KG_0|$. However, for comparability with previous figures we assume again that the phase margin is zero. Then $|KG_0|$ can fall at 12 dB/decade, allowing

$$\gamma\omega_0 \leqslant 0.4a$$

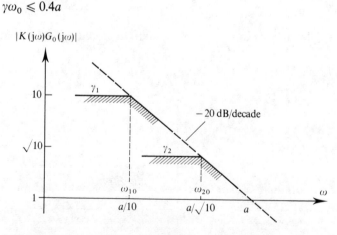

Figure 1.19 Some possible (open-loop) gain and bandwidth specifications when a zero is located at a in the right half-plane.

Table 1.1 Effects of right half-plane zero located at $a > 0$.

ω_c	$\arg A(0) - \arg A(j\omega_c)$	$-\arg[K(j\omega_c)G_0(j\omega_c)]$	$\dfrac{d}{d\omega}\lvert KG_0 \rvert$	$(\gamma\omega_0)_{max}$
$\frac{1}{2}a$	$53°$	$87°$	$-19\,\text{dB/decade}$	$0.93a$
a	$90°$	$50°$	$-11\,\text{dB/decade}$	$0.36a$
$2a$	$127°$	$13°$	$-2.9\,\text{dB/decade}$	$0.14a$
$4a$	$152°$	impossible	—	—

With a phase margin of $40°$ the figures are as shown in Table 1.1. It is seen that ω_c is limited to approximately $\omega_c \leqslant a$.

When there is more than one zero in the right half-plane, the 0 dB cross-over frequency is similarly limited. (For a full discussion see Horowitz (1963).)

As we said earlier, the loop bandwidth is usually very close to the 0 dB cross-over frequency. But this need not be the case if the plant has right half-plane zeros, so these zeros do not limit the loop bandwidth directly. For example, if we have a plant with transfer function

$$G(s) = \frac{s-1}{s+2} \tag{1.40}$$

and compensator

$$K(s) = \frac{-20(s+2)}{s(s+41)} \tag{1.41}$$

then we obtain a cross-over frequency of approximately $0.7\,\text{rad s}^{-1}$. However, the closed-loop transfer function $T(s)$ is

$$T(s) = \frac{20(1-s)}{(s+20)(1+s)} \tag{1.42}$$

so that

$$\lvert T(j\omega) \rvert = \left\lvert \frac{20}{j\omega + 20} \right\rvert \tag{1.43}$$

and we have a loop bandwidth of $20\,\text{rad s}^{-1}$. The sensitivity function $S(s)$ for this design, however, is

$$S(s) = \frac{s(s+41)}{(s+1)(s+20)} \tag{1.44}$$

so that $|S(j\omega)| > 1$ for $\omega \geqslant 0.35 \, \text{rad s}^{-1}$, approximately. Between the frequencies 0.35 and 20 rad s^{-1}, then, this design amplifies disturbances without attenuating measurement noise, and is therefore extremely inefficient. If a two-degree-of-freedom structure is permissible, then a much better design is obtained by limiting the loop bandwidth to about 1 rad s^{-1}, since no benefit is obtained by going beyond this, and using a pre-filter to extend the reference signal transmission bandwidth to 20 rad s^{-1}, if that is really required.

We see, then, that right half-plane zeros in fact limit the loop bandwidth for engineering rather than mathematical reasons.

1.7.3 Right half-plane poles

Rather surprisingly, the presence of unstable poles imposes a *lower bound* on the cross-over frequency. Suppose that a system has one real unstable pole located at $s = a$, and factorize the transfer function as

$$G(s) = G_0(s)A^{-1}(s) \tag{1.45}$$

where $A(s) = (s-a)/(s+a)$, as before. Since $G_0(s)$ has all its poles in the left half-plane, and has at least as many poles as zeros, it is impossible for its Nyquist locus to make any anticlockwise encirclements of the point $-1 + j0$ (or of any other point). But we need one such encirclement if the closed loop is to be stable (by Nyquist's theorem), and clearly we can only get this from the phase characteristic of $A^{-1}(j\omega)$.

If we further suppose that $|G(j\omega)|$, and hence $|G_0(j\omega)|$, is monotonic decreasing with respect to ω, then $\arg G_0(j\omega) - \arg G_0(0)$ is always negative (by equation (1.36)). Consider Figure 1.20, which shows a possible Nyquist locus of $G(s)$, with a phase margin ϕ. If ω_c denotes the 0 dB cross-over frequency, then

$$\arg G(j\omega_c) - \arg G(0) = \phi \tag{1.46}$$

But $\arg G(j\omega) = \arg G_0(j\omega) + \arg A^{-1}(j\omega)$, so

$$\arg G(j\omega_c) - \arg G(0) = \arg G_0(j\omega_c) - \arg G_0(0)$$
$$+ \arg A^{-1}(j\omega_c) - \arg A^{-1}(0)$$
$$< \arg A^{-1}(j\omega_c) - \pi$$

Hence

$$\arg A^{-1}(j\omega_c) > \phi + \pi \tag{1.47}$$

Thus from Figure 1.17 (which shows the Nyquist diagram of $A(s)$), we see that $\phi = \pi/4$ implies that $\omega_c > a/2$ (approximately), $\phi = \pi/2$ implies that $\omega_c > a$, and so on.

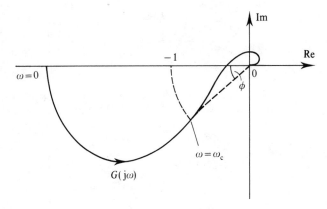

Figure 1.20 Possible Nyquist locus of open-loop unstable system.

Thus we conclude that the bandwidth of the closed-loop system must be approximately a, at least, if the $|G(j\omega)|$ characteristic has the usual shape, namely a monotonic decreasing function of frequency. As is to be expected, this lower bound for the bandwidth is increased by the presence of other right half-plane poles.

1.7.4 Bode's integral theorem

Figure 1.21 shows the Nyquist locus of some system. We see that, for $\omega_1 < \omega < \omega_2$, $|S(j\omega)| > 1$, in other words there is a frequency region in which feedback would increase the sensitivity of the system rather than decrease it. It is easy to see that for any system which is open-loop stable, and

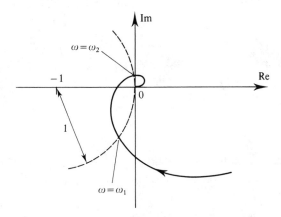

Figure 1.21 Nyquist locus of a system, with part of $|S| = 1$ circle.

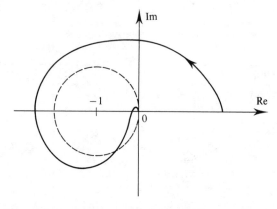

Figure 1.22 Nyquist locus of system which has two unstable open-loop poles, but is closed-loop stable.

which has a high-frequency roll-off steeper than $-20\,\mathrm{dB/decade}$, such a region must always exist since the gain–phase relations tell us that the Nyquist plot of such a system must be heading towards the origin at a phase angle more negative than $-\tfrac{1}{2}\pi$; and Nyquist's stability theorem says that, if this phase angle is more negative than $-\pi$, then the locus must cross the negative real axis between -1 and 0. Thus the locus is bound to enter the unit circle which is centred on -1.

If the system is open-loop unstable, the same is true. Figure 1.22 shows a possible locus for a system with two open-loop unstable poles (and roll-off steeper than $-20\,\mathrm{dB/decade}$). Again, the gain–phase relations are such that the locus is bound to penetrate the circle.

The reader may think that it may be possible to devise a system whose Nyquist locus is shaped in such a way as to avoid penetrating the circle, but this was shown by Bode (1945) to be untrue. The following theorem was first stated by Bode for open-loop stable systems, and extended to the form given here by Freudenberg and Looze (1985) (see also Freudenberg and Looze, 1987):

Theorem 1.2: Suppose that $G(s)K(s)$ has the set of open right half-plane poles $\{p_i\colon i = 1,\ldots,N\}$ (including multiplicities), and that

$$\lim_{\substack{R\to\infty \\ \substack{|s|>R \\ \mathrm{Re}(s)>0}}} \sup R|G(s)K(s)| = 0 \quad \text{('roll-off' condition)} \tag{1.48}$$

and let $S(s) = [1 + G(s)K(s)]^{-1}$, as before. If the closed-loop is stable, then

$$\int_0^\infty \ln|S(j\omega)|\,d\omega = \pi\sum_{i=1}^N \mathrm{Re}(p_i) \tag{1.49}$$

This theorem has the following consequences:

(1)   For open-loop stable systems,

$$\int_0^\infty \ln|S(j\omega)|\,d\omega = 0 \tag{1.50}$$

That is, in a sense there is as much sensitivity increase as there is decrease. If the sensitivity (in decibels) is plotted against frequency (linear scale), then the positive area under the graph will equal the negative area.

(2)   For open-loop unstable systems the position is worse. There is now more positive than negative area. Fast (open-loop) unstable poles are more harmful than slow ones. Intuitively, one can perhaps think of some of the feedback as being 'used up' in shifting poles into the stable region, and only part of it being 'available' for the reduction of sensitivity. This gives a theoretical justification for preferring (open-loop) stable controllers – but there are also powerful practical reasons for this preference.

(3)   If one pushes the sensitivity lower at some low frequencies, the penalty is a higher peak value of the sensitivity at some higher frequency. This is not apparent from the theorem, since one might think that the corresponding increase required in the positive area could be obtained by making an arbitrarily small increase in $S(j\omega)$, spread over an arbitrarily large frequency range. Freudenberg and Looze (1985) show that this is not so. In fact, if

$$|G(j\omega)K(j\omega)| < \frac{M}{\omega^{1+k}} < \varepsilon, \quad \omega > \omega_c \tag{1.51}$$

then

$$\int_{\omega_c}^\infty \ln|S(j\omega)|\,d\omega < \frac{\omega_c}{k} \ln\frac{1}{1-\varepsilon} \tag{1.52}$$

which tends to zero as ε tends to zero, so that most of the required positive area must be acquired at frequencies below ω_c, if ε is very small.

SUMMARY

In this chapter we have reviewed the analysis and design of single-loop feedback systems. We have considered the possibilities of both 'closed-loop' design, in which a desired closed-loop transfer function is somehow established, and a controller is then synthesized automatically, and 'open-loop' design, in which the plant characteristics are taken account of explicitly.

We have shown that there are fundamental trade-offs between the conflicting objectives of reducing sensitivity to disturbances and parameter uncertainty on the one hand, and filtering out any internally generated noise on the other. It is the existence of these trade-offs which makes feedback design difficult. The standard approach to reconciling the conflicting objectives relies on making the loop gain high at low frequencies, then reducing it as quickly as possible at high frequencies. However, the rate of gain reduction is limited by the need to maintain stability of the feedback loop, since a large rate of gain reduction is possible only with a large associated phase lag, and the allowable phase lag is limited by Nyquist's stability theorem.

The location of any zeros or poles in the right half-plane restricts the range of frequencies over which the use of feedback can be beneficial: right half-plane zeros impose an upper bound on the range of frequencies over which the sensitivity can be reduced, while right half-plane poles impose a lower bound on the range of frequencies over which noise must be passed without attenuation (the loop bandwidth).

We shall see that all of these considerations continue to hold for multivariable feedback systems.

EXERCISES

1.1 The peak phase change given by a simple first-order phase-lead or phase-lag compensator is given by equation (1.7). Plot a graph of this peak phase change for $1 \leqslant \alpha \leqslant \infty$.

In practice, various considerations usually limit the range of α to $0.1 \leqslant \alpha \leqslant 10$. What is the practical peak phase change obtainable with such a compensator?

(*Solution*: 55°)

1.2 A plant has transfer function $1/s^2$. Design phase-lead compensators that will give a 40° phase margin, and a closed-loop bandwidth of (a) 5 Hz, (b) 0.02 Hz. Compare the closed-loop transient responses of the two designs.

1.3 Let

$$T(s) = \frac{G(s)}{1 + G(s)}$$

and let ω_c denote the cross-over frequency, at which $|G(j\omega_c)| = 1$. If ϕ is the phase margin, show that

$$|T(j\omega_c)| = \frac{1}{2\sin(\frac{1}{2}\phi)}$$

Hence show that a specification framed in terms of M-circles implicitly specifies minimum gain and phase margins. What gain and phase margins are guaranteed if the Nyquist locus remains outside the $M = \sqrt{2}$ circle?

(*Solution*: 4.6 dB, 41°)

1.4   Suppose that

$$G(s) = \frac{s-3}{s+1} \quad \text{and} \quad T(s) = \frac{5}{s+5}$$

Find the corresponding compensator $K(s)$, if

$$T(s) = \frac{G(s)K(s)}{1 + G(s)K(s)}$$

Obtain the corresponding Nyquist locus. Does it predict stability? Show that the closed loop is in fact unstable, because of a hidden unstable mode. (*Hint*: this is most easily done by using state-space models.)

Suggest a $T(s)$ which leads to a stable feedback loop, but still has a bandwidth of 5 rad s^{-1}. Examine the corresponding sensitivity function, and discuss the advisability of using a two-degrees-of-freedom design.

1.5   A plant has a zero at $s = 3$ and a transport delay of 1.0 s. Estimate the highest achievable 0 dB cross-over frequency, assuming that the open-loop gain–frequency characteristic is monotonic decreasing.

(*Solution*: 2 rad s^{-1})

1.6   Give an example of a plant for which it is impossible to find a feedback design which is 'good' in the sense of giving stability, sensitivity reduction at low frequencies and noise attenuation at high frequencies.

References

Åström K.J. and Wittenmark B. (1984). *Computer-Controlled Systems*. Englewood Cliffs NJ: Prentice-Hall.

Bode H.W. (1945). *Network Analysis and Feedback Amplifier Design*. New York: Van Nostrand.

Chen C.T. (1984). *Linear System Theory and Design*. New York: Holt, Rinehart & Winston.

Franklin G.F. and Powell J.D. (1980). *Digital Control of Dynamic Systems*. Reading MA: Addison-Wesley.

Freudenberg J.S. and Looze D.P. (1985). Right half plane poles and zeros and design tradeoffs in feedback systems. *IEEE Transactions on Automatic Control*, **AC-30**, 555–65.

Freudenberg J.S. and Looze D.P. (1987). *Frequency Domain Properties of Scalar and Multivariable Feedback Systems*. Berlin: Springer-Verlag.

Horowitz I. (1963). *Synthesis of Feedback Systems*. New York: Academic Press.

Kailath T. (1980). *Linear Systems*. Englewood Cliffs NJ: Prentice-Hall.

Newton G.C., Gould L.A. and Kaiser J.F. (1957). *Analytical Design of Linear Feedback Controls*. New York: Wiley.

Oppenheim A.V., Willsky A.S. and Young I.T. (1983). *Signals and Systems*. Englewood Cliffs NJ: Prentice-Hall.

Wonham M. (1974). *Linear Multivariable Systems*. Berlin: Springer-Verlag.

CHAPTER 2

Poles, Zeros and Stability of Multivariable Feedback Systems

2.1 Introduction
2.2 The Smith–McMillan form of a transfer-function matrix
2.3 Poles and zeros of a transfer-function matrix
2.4 Matrix-fraction description (MFD) of a transfer-function
2.5 State-space realization from a transfer-function matrix
2.6 How many zeros?

2.7 Internal stability
2.8 The generalized Nyquist stability criterion
2.9 The generalized inverse Nyquist stability criterion
2.10 Nyquist arrays and Gershgorin bands
2.11 Generalized stability
Summary
Exercises
References

2.1 Introduction

We shall take the basic description of a linear multivariable system to be a **transfer-function matrix**. This is simply a matrix $G(s)$ of transfer functions, in which the (i, j) element $g_{ij}(s)$ is the transfer function relating the ith output to the jth input. If each element $g_{ij}(s)$ is rational, and **proper**, namely if

$$|g_{ij}(s)| \to |d_{ij}| < \infty, \quad \text{as } |s| \to \infty \tag{2.1}$$

then one can find a state-space model

$$\dot{x} = Ax + Bu, \qquad y = Cx + Du \tag{2.2}$$

37

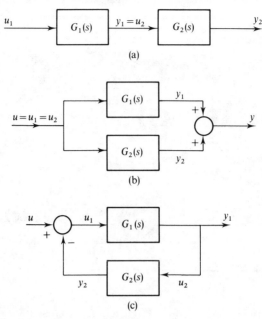

Figure 2.1   Connections of two systems: (a) in series, (b) in parallel, (c) negative feedback.

in which u is the vector of system inputs, y is the vector of outputs and x is a **state vector**. A, B, C and D are real matrices of appropriate dimensions, and are related to the transfer function $G(s)$ by

$$G(s) = C(sI - A)^{-1}B + D \tag{2.3}$$

The 4-tuple (A, B, C, D) is said to be a **realization** of $G(s)$, and we shall sometimes write $G(A, B, C, D)$ to denote this. Note that a realization is not unique, because $(\tilde{A}, \tilde{B}, \tilde{C}, \tilde{D})$ is also a realization if T is any invertible matrix and

$$\tilde{A} = T^{-1}AT, \qquad \tilde{B} = T^{-1}B, \qquad \tilde{C} = CT \tag{2.4}$$

Transmission paths through the system often contain time delays; the elements of $G(s)$ are then transcendental functions, and no state-space realization exists. This is inconvenient, because many of the procedures for analysing and designing multivariable systems are best implemented by algorithms which operate on state-space models, but it is not disastrous – many of the procedures can be implemented by operating directly on frequency-response data, namely on matrices such as $G(j\omega)$. Alternatively, transfer functions of time delays (e^{-sd}, where d is the delay duration) can be approximated by rational functions.

Consider two systems, defined by transfer-function matrices $G_i(s)$
$(i = 1, 2)$, their input signals being the vectors u_i $(i = 1, 2)$ and output signals
the vectors y_i $(i = 1, 2)$, so that

$$y_i(s) = G_i(s)u_i(s), \quad i = 1, 2 \tag{2.5}$$

If the dimensions of y_1 and u_2 are the same, then we can form the **series
connection** as shown in Figure 2.1(a), and obtain

$$y_2(s) = G_2(s)G_1(s)u_1(s) \tag{2.6}$$

(Note that the order of multiplication is important because matrix multiplication is not commutative.)

If both systems have the same number of input signals and the same
number of output signals (and hence both transfer-function matrices have the
same dimensions) then we can form the **parallel connection** as shown in
Figure 2.1(b), and obtain the relation

$$y(s) = [G_1(s) + G_2(s)]u(s) \tag{2.7}$$

If the dimensions of y_1 and u_2 are the same *and* the dimensions of u_1
and y_2 are the same, then we can form the **feedback connection** shown in
Figure 2.1(c). In this case we have

$$y_1(s) = G_1(s)[u(s) - G_2(s)y_1(s)] \tag{2.8}$$

or

$$[I + G_1(s)G_2(s)]y_1(s) = G_1(s)u(s) \tag{2.9}$$

so that

$$y_1(s) = [I + G_1(s)G_2(s)]^{-1}G_1(s)u(s) \tag{2.10}$$

Alternatively, by starting with

$$u_1(s) = u(s) - G_2(s)G_1(s)u_1(s) \tag{2.11}$$

we can derive

$$y_1(s) = G_1(s)[I + G_2(s)G_1(s)]^{-1}u(s) \tag{2.12}$$

Note that (2.10) and (2.12) are alternative expressions for the same closed-
loop transfer function. If we have a negative-feedback connection as
in Figure 2.1(c), then we call the products $-G_2(s)G_1(s)$ and $-G_1(s)G_2(s)$

return ratios, and the matrices $I + G_2(s)G_1(s)$ and $I + G_1(s)G_2(s)$ are called **return differences** (MacFarlane, 1970). Note that their dimensions are not all the same, unless both $G_1(s)$ and $G_2(s)$ are square. (If a positive-feedback convention is used, then the return ratios are G_2G_1 and G_1G_2, and the return differences are $I - G_2G_1$ and $I - G_1G_2$.)

2.2 The Smith–McMillan form of a transfer-function matrix

In order to generalize the ideas and results of Chapter 1 to multivariable feedback, we shall need to find appropriate generalizations of poles and zeros. We shall do this by showing that every rational transfer-function matrix can be reduced to a diagonal canonical form known as the **Smith–McMillan** form, and then defining poles and zeros in terms of this Smith–McMillan form.

Every rational transfer-function matrix can be expressed as a polynomial matrix, divided by a common denominator polynomial. We therefore begin by showing that every polynomial matrix can be reduced to a canonical form known as the **Smith form**.

> *Definition 2.1*:   A polynomial matrix $U(s)$ is called **unimodular** if it has an inverse which is also a polynomial matrix.

> *Theorem 2.1*: $U(s)$ is a unimodular matrix if and only if det $U(s)$ is a constant (independent of s).

> *Proof*:   If $U(s)$ is unimodular, then it has a polynomial matrix inverse $U(s)^{-1}$ such that $U(s)U(s)^{-1} = I$. This implies that
>
> $$\det [U(s)] \det [U(s)^{-1}] = 1$$
>
> Since both det $[U(s)]$ and det $[U(s)^{-1}]$ are polynomial expressions, they must be constants. Conversely, suppose that det $[U(s)] = c$ is a constant. Then
>
> $$U(s)^{-1} = \frac{1}{c}[\operatorname{adj} U(s)]$$
>
> where adj $U(s)$ denotes the adjoint of $U(s)$. Since the adjoint matrix is a polynomial matrix, then so is $U(s)^{-1}$.   ∎

There are three elementary operations which can be performed on polynomial matrices:

- interchange of two rows or columns
- multiplication of one row or column by a constant
- addition of a polynomial multiple of one row or column to another

Each of these elementary operations can be represented by multiplying a polynomial matrix by a suitable matrix, called an **elementary matrix**. It is easy to show that all elementary matrices are unimodular.

We say that two (polynomial or rational) matrices $P(s)$ and $Q(s)$ are **equivalent** (symbolized $P(s) \sim Q(s)$) if there exist sequences of left and right elementary matrices $\{L_1(s),\ldots,L_l(s)\}$ and $\{R_1(s),\ldots,R_r(s)\}$ such that

$$P(s) = L_l(s)\ldots L_1(s)Q(s)R_1(s)\ldots R_r(s) \tag{2.13}$$

The next result states that every polynomial matrix is equivalent to a diagonal polynomial matrix known as the Smith form.

> **Theorem 2.2:**   Let $P(s)$ be a polynomial matrix of normal rank r (i.e. of rank r for almost all s). Then $P(s)$ may be transformed by a sequence of elementary row and column operations into a pseudo-diagonal polynomial matrix $S(s)$ having the form
>
> $$S(s) = \mathrm{diag}\,\{\varepsilon_1(s),\varepsilon_2(s),\ldots,\varepsilon_r(s),0,0,\ldots,0\} \tag{2.14}$$
>
> in which each $\varepsilon_i(s)$ $(i = 1,\ldots,r)$ is a monic polynomial (i.e. has leading coefficient 1) satisfying the divisibility property
>
> $$\varepsilon_i(s)\,|\,\varepsilon_{i+1}(s), \quad i = 1,\ldots,r-1$$
>
> (that is, $\varepsilon_i(s)$ divides $\varepsilon_{i+1}(s)$ without remainder). Moreover, if we define the **determinantal divisors**
>
> $$\left.\begin{aligned}D_0(s) &= 1 \\ D_i(s) &= \text{greatest common divisor of all } i \times i \\ &\quad \text{minors of } P(s)\end{aligned}\right\} \tag{2.15}$$
>
> where each greatest common divisor is normalized to be a monic polynomial, then
>
> $$\varepsilon_i(s) = \frac{D_i(s)}{D_{i-1}(s)}, \quad i = 1,\ldots,r \tag{2.16}$$
>
> The matrix $S(s)$ is the Smith form of $P(s)$, and the $\varepsilon_i(s)$ are called the **invariant factors** of $P(s)$.
>
> *Proof*:   The reduction to Smith form can be carried out in the following steps:
>
> (1)   From all the non-zero elements of $P(s)$, choose one which has least degree and bring it to the $(1,1)$ position by permuting the rows and columns.

(2)   Write the element $p_{21}(s)$ (if non-zero) as $p_{21}(s) = q(s)p_{11}(s) + r(s)$, where $r(s)$ is either zero or such that $\deg r(s) < \deg p_{11}(s)$. Subtracting $q(s)$ times the first row from the second row, the $(2, 1)$ element becomes $r(s)$. If $r(s)$ is zero, proceed to the next step; otherwise interchange rows 1 and 2 and repeat the present step. Since the degree of the $(1, 1)$ element must decrease by at least one every time we repeat this step, we must end up with a zero in the $(2, 1)$ position.

(3)   In a similar manner to that described in step 2, use elementary row operations to make zero every element of the first column except the $(1, 1)$ entry.

(4)   By means of column operations, reduce every element in the first row to zero except the $(1, 1)$ entry.

(5)   Step 4 may, however, have reintroduced non-zero entries below the $(1, 1)$ position in the first column. If this has happened, return to the beginning of step 2. However, on each cycle through the steps 2, 3, 4, 5 and 2, the degree of the $(1, 1)$ element drops, so the process must terminate with a matrix of the form

$$
\left[
\begin{array}{c:cccc}
a(s) & 0 & 0 & & 0 \\
\hdashline
0 & b(s) & c(s) & \cdots & d(s) \\
0 & e(s) & f(s) & \cdots & g(s) \\
\vdots & \vdots & \vdots & & \vdots \\
0 & h(s) & k(s) & \cdots & l(s)
\end{array}
\right]
\tag{2.17}
$$

(6)   If any element of columns $2, 3, \ldots$ is not divisible by $a(s)$, we add this column to the first column and then go back to the beginning of step 2. Again, on each passage through the cycle 2, 3, 4, 5, 6 and 2, there must be a reduction in the degree of the $(1, 1)$ element. So the process terminates with a matrix of the form (2.17) in which $a(s)$ divides every other non-zero element. We may assume that $a(s)$ is monic, and put $\varepsilon_1(s) = a(s)$.

Now delete the first row and first column of the matrix and repeat the above procedure on the remaining submatrix. We then get a second matrix equivalent to $P(s)$:

$$
\left[
\begin{array}{cc:ccc}
\varepsilon_1(s) & 0 & 0 & \cdots & 0 \\
0 & \varepsilon_2(s) & 0 & \cdots & 0 \\
\hdashline
0 & 0 & w(s) & \cdots & x(s) \\
\vdots & \vdots & \vdots & & \vdots \\
0 & 0 & y(s) & \cdots & z(s)
\end{array}
\right]
$$

in which $\varepsilon_1(s)|\varepsilon_2(s)|\{w(s), x(s), \ldots, y(s), z(s)\}$. It is clear that the procedure may be repeated until we get the Smith form of $P(s)$.

The fact that there are r non-zero invariant factors follows from the observation that elementary operations do not change the rank of a matrix. The second part of the theorem follows from the Binet–Cauchy theorem (Gantmacher, 1959). The $D_i(s)$ are invariant under elementary operations, and hence the determinantal divisors of $P(s)$ and $S(s)$ are the same. Hence we have that

$$D_1(s) = \varepsilon_1(s)$$

$$D_2(s) = \varepsilon_1(s)\varepsilon_2(s)$$

$$\vdots$$

$$D_r(s) = \varepsilon_1(s)\varepsilon_2(s)\ldots\varepsilon_r(s)$$

from which equation (2.16) follows.   ∎

EXAMPLE 2.1

Let

$$P(s) = \begin{bmatrix} 1 & -1 \\ s^2 + s - 4 & 2s^2 - s - 8 \\ s^2 - 4 & 2s^2 - 8 \end{bmatrix} \tag{2.18}$$

then

$$D_0(s) = 1$$

$$D_1(s) = \gcd\{1, -1, s^2 + s - 4, 2s^2 - s - 8, s^2 - 4, 2s^2 - 8\} = 1$$

(where 'gcd' means 'greatest common divisor')

$$D_2(s) = \gcd\left\{ \begin{vmatrix} 1 & -1 \\ s^2 + s - 4 & 2s^2 - s - 8 \end{vmatrix}, \begin{vmatrix} 1 & -1 \\ s^2 - 4 & 2s^2 - 8 \end{vmatrix}, \right.$$

$$\left. \begin{vmatrix} s^2 + s - 4 & 2s^2 - s - 8 \\ s^2 - 4 & 2s^2 - 8 \end{vmatrix} \right\}$$

$$= (s + 2)(s - 2)$$

so that

$$\varepsilon_1(s) = 1 \quad \text{and} \quad \varepsilon_2(s) = (s + 2)(s - 2)$$

It follows that

$$P(s) \sim S(s) = \begin{bmatrix} 1 & 0 \\ 0 & (s+2)(s-2) \\ 0 & 0 \end{bmatrix} \qquad (2.19)$$

It is clear that the Smith form of a polynomial matrix is uniquely defined by equation (2.16), and that two equivalent polynomial matrices have the same Smith form. The Smith form is thus a canonical form for a set of equivalent polynomial matrices.

Now we extend the idea of a canonical form to rational matrices.

Theorem 2.3 **(Smith–McMillan form):**   Let $G(s)$ be a rational matrix of normal rank r. Then $G(s)$ may be transformed by a series of elementary row and column operations into a pseudo-diagonal rational matrix $M(s)$ of the form

$$M(s) = \text{diag}\left\{ \frac{\varepsilon_1(s)}{\psi_1(s)}, \frac{\varepsilon_2(s)}{\psi_2(s)}, \ldots, \frac{\varepsilon_r(s)}{\psi_r(s)}, 0, \ldots, 0 \right\} \quad (2.20)$$

in which the monic polynomials $\{\varepsilon_i(s), \psi_i(s)\}$ are coprime for each i (i.e. they have no common factors) and satisfy the divisibility properties

$$\left. \begin{array}{c} \varepsilon_i(s) \mid \varepsilon_{i+1}(s) \\ \psi_{i+1}(s) \mid \psi_i(s) \end{array} \right\} \quad i = 1, \ldots, r-1 \qquad (2.21)$$

$M(s)$ is the Smith–McMillan form of $G(s)$.

Proof: Let $d(s)$ be the least common multiple of all the denominators of the elements $g_{ij}(s)$ of $G(s)$. Then $G(s)$ can be written as

$$G(s) = \frac{1}{d(s)} P(s) \qquad (2.22)$$

where $P(s)$ is a polynomial matrix. By Theorem 2.2, $P(s)$ is equivalent to its Smith form:

$$P(s) \sim S(s) = \text{diag}\{\varepsilon_1'(s), \ldots, \varepsilon_r'(s), 0, \ldots, 0\} \qquad (2.23)$$

It follows that

$$G(s) \sim \frac{1}{d(s)} S(s) = M(s) \qquad (2.24)$$

where each diagonal element $\varepsilon_i(s)/\psi_i(s)$ of $M(s)$ is obtained by performing all possible cancellations between the numerator and denominator of $\varepsilon_i'(s)/d(s)$.

The divisibility property (2.21) is a direct consequence of the fact that

$$\varepsilon_i'(s)\,|\,\varepsilon_{i+1}'(s), \quad i = 1, \ldots, r-1 \qquad \blacksquare$$

EXAMPLE 2.2

Let

$$G(s) = \begin{bmatrix} \dfrac{1}{s^2+3s+2} & \dfrac{-1}{s^2+3s+2} \\[2mm] \dfrac{s^2+s-4}{s^2+3s+2} & \dfrac{2s^2-s-8}{s^2+3s+2} \\[2mm] \dfrac{s-2}{s+1} & \dfrac{2s-4}{s+1} \end{bmatrix}$$

Then

$$G(s) = \frac{1}{s^2+3s+2}\,P(s) \tag{2.25}$$

where $P(s)$ is the polynomial matrix of Example 2.1, with Smith form given by (2.19). Proceeding as in the proof of Theorem 2.3, we find that the Smith–McMillan form of $G(s)$ is given by

$$G(s) \sim \frac{1}{d(s)}\,S(s) = \begin{bmatrix} \dfrac{1}{(s+1)(s+2)} & 0 \\[2mm] 0 & \dfrac{s-2}{s+1} \\[2mm] 0 & 0 \end{bmatrix} \tag{2.26}$$

2.3 Poles and zeros of a transfer-function matrix

In single-input/single-output (SISO) systems, the poles and zeros of the scalar transfer function

$$g(s) = \frac{n(s)}{d(s)} \tag{2.27}$$

(with $n(s)$ and $d(s)$ coprime polynomials) are given by the roots of $d(s)$ and

$n(s)$, respectively. We now define the poles and zeros of a transfer-function matrix by means of the Smith–McMillan form.[1]

> **Definition 2.2:**   Let $G(s)$ be a rational transfer-function matrix with Smith–McMillan form $M(s)$, as in (2.20), and define the **pole polynomial** and **zero polynomial**
>
> $$p(s) = \psi_1(s)\ldots\psi_r(s) \tag{2.28}$$
>
> $$z(s) = \varepsilon_1(s)\ldots\varepsilon_r(s) \tag{2.29}$$
>
> The roots of $p(s)$ and $z(s)$ are called the **poles** and **zeros** of $G(s)$, respectively.

In other words, the poles of $G(s)$ are all the roots of the denominator polynomials $\psi_i(s)$ of the Smith–McMillan form of $G(s)$. If p_0 is a pole of $G(s)$, then $(s - p_0)^v$ ($v \geq 1$) must be a factor of some $\psi_i(s)$. The number v is called the **multiplicity** of the pole, and if $v = 1$ we say that p_0 is a **simple pole**. Zeros and their multiplicity are defined similarly, in terms of the numerator polynomials $\varepsilon_i(s)$ of the Smith–McMillan form.

EXAMPLE 2.3

The pole and zero polynomials of the transfer-function matrix $G(s)$ of Example 2.2 are

$$p(s) = (s+1)^2(s+2)$$

$$z(s) = (s-2)$$

Hence $G(s)$ has poles at $\{-1, -1, -2\}$ and a zero at 2. Since $\psi_1(s) = (s+1)(s+2)$ and $\psi_2(s) = s+1$, all the poles are simple.

> **Corollary 2.1:**   If $G(s)$ is square, then
>
> $$\det G(s) = c\frac{z(s)}{p(s)}$$
>
> for some constant c. Note that although the pair of polynomials $\{\varepsilon_i(s), \psi_i(s)\}$ is coprime for each i, it is possible that there exist common factors between $z(s)$ and $p(s)$ which cancel out in forming $\det G(s)$.

Definition 2.3: The degree of the pole polynomial $p(s)$ is the **McMillan degree** of $G(s)$.

Zeros defined via the Smith–McMillan form are often called **transmission zeros**, in order to distinguish them from other kinds of zero which have been defined. It is clear from equation (2.20) that the rank of $G(s)$ drops below its normal value whenever $s = z_0$, if z_0 is a zero of $G(s)$. Hence there exists a non-zero vector u_0 such that $G(z_0)u_0 = 0$. If the input-signal vector has transform

$$u(s) = \frac{u_0}{s - z_0} \tag{2.30}$$

then the output is given by

$$y(s) = G(s)u(s) + \text{initial-condition response} \tag{2.31}$$

$$= \frac{G(z_0)u_0}{s - z_0} + \sum_i \frac{R_i u(p_i)}{s - p_i} + \text{initial-condition response} \tag{2.32}$$

$$= 0$$

if the initial conditions are chosen so as to cancel out the second term in (2.32), in which p_i denotes a pole of $G(s)$, R_i denotes the residue of $G(s)$ at $s = p_i$ and all the poles are assumed to be simple. Hence transmission zeros have a **transmission-blocking** property. (For further details of this interpretation of zeros, and an analogous interpretation of poles, see Desoer and Shulman (1974).)

Note that the locations of poles can be determined simply by examining the elements of the transfer-function matrix – poles occur only at those points at which the elements have poles. Their multiplicities, and even their numbers, cannot be determined so easily, however. It is sometimes important to establish their numbers – for example, if there exist right half-plane poles, it is necessary to know how many of them there are before applying Nyquist-like stability theorems. The situation is worse with zeros: one cannot establish even the location of zeros by inspection of the transfer-function matrix elements.

Although the proofs of Theorems 2.2 and 2.3 provide an algorithm for finding poles and zeros, it is not a practical one for use with real systems. Reliable and practical algorithms require a state-space realization as a starting point; such algorithms are described in Chapter 8.

The significance of poles may be summarized very simply. Each pole p_i of a transfer-function matrix $G(s)$ must also appear as a pole of at least one of its elements. It is therefore possible to write $G(s)$ in 'partial fractions' as

$$G(s) = \sum_{i=1}^{v} \frac{G_i}{(s - p_i)^{k_i}} + G_0$$

(assuming that $G(s)$ is proper), where G_i and G_0 are constant matrices, and k_i is some positive integer. Hence the impulse-response matrix of the system, which is obtained as the inverse Laplace transform of $G(s)$, is

$$\bar{G}(t) = \sum_{i=1}^{v} G_i t^{k_i - 1} e^{p_i t} + G_0 \delta(t)$$

The relation of pole locations to system stability is therefore the same as for SISO systems: a system is asymptotically stable if $\mathrm{Re}\{p_i\} < 0$ for each i, and it is stable if $\mathrm{Re}\{p_i\} \leqslant 0$ and $k_i = 1$ whenever $\mathrm{Re}\{p_i\} = 0$.

We have considered only rational transfer functions; in practice, the relation of pole locations to system stability is the same even when elements of the transfer-function matrix are irrational (Callier and Desoer, 1982; Rosenbrock, 1974).

2.4   Matrix-fraction description (MFD) of a transfer function

Throughout this section, $G(s)$ will denote an $m \times l$ strictly proper, rational transfer-function matrix, namely one such that $G(\infty) = 0$. Let $L(s)^{-1}$ and $R(s)^{-1}$ be the unimodular matrices that take $G(s)$ to its Smith–McMillan form $M(s)$:

$$G(s) = L(s)M(s)R(s) = L(s)\,\mathrm{diag}\left\{\frac{\varepsilon_1(s)}{\psi_1(s)}, \ldots, \frac{\varepsilon_r(s)}{\psi_r(s)}, 0, \ldots, 0\right\}R(s) \tag{2.33}$$

We may write $M(s)$ as

$$M(s) = N'(s)D'(s)^{-1} \tag{2.34}$$

where $N'(s)$ and $D'(s)$ are polynomial matrices defined as

$$\left.\begin{array}{l} N'(s) = \mathrm{diag}\{\varepsilon_1(s), \ldots, \varepsilon_r(s), 0, \ldots, 0\} \\ D'(s) = \mathrm{diag}\{\psi_1(s), \ldots, \psi_r(s), 1, \ldots, 1\} \end{array}\right\} \tag{2.35}$$

and $D'(s)$ is a square matrix of dimension $l \times l$.

Substituting (2.34) into (2.33) yields

$$G(s) = L(s)N'(s)D'(s)^{-1}R(s)$$

$$= [L(s)N'(s)][R(s)^{-1}D'(s)]^{-1}$$

$$= N(s)D(s)^{-1} \tag{2.36}$$

where

$$N(s) = L(s)N'(s) \quad \text{and} \quad D(s) = R(s)^{-1}D'(s) \tag{2.37}$$

Note that since $R(s)$ is unimodular, $R(s)^{-1}$ is a polynomial matrix and so both $N(s)$ and $D(s)$ are polynomial matrices. We have therefore expressed $G(s)$ as a 'fraction' of two polynomial matrices. The representation (2.36) for $G(s)$ is called a **right matrix-fraction description**, and $N(s)$ and $D(s)$ are called the **numerator matrix** and the **denominator matrix**, respectively, of the MFD.

Note that in the MFD (2.36),

(1) z is a zero of $G(s)$ if and only if $N(z)$ loses rank

(2) p is a pole of $G(s)$ if and only if $D(p)$ loses rank

Another way of stating 2 is that the pole polynomial is given by the determinant of the denominator matrix.

Evidently, an MFD representation is not unique. For example, if $X(s)$ is invertible, then

$$G(s) = [N(s)X(s)][D(s)X(s)]^{-1} \tag{2.38}$$

$$= \tilde{N}(s)\tilde{D}(s)^{-1} \tag{2.39}$$

so that $(\tilde{N}(s), \tilde{D}(s))$ is also an MFD of $G(s)$. We are frequently interested in removing any unnecessary common factors, such as $X(s)$, and we therefore make the following definition:

Definition 2.4: Let $N(s)$ and $D(s)$ be polynomial matrices with the same number of columns. If there exist $\tilde{N}(s)$ and $\tilde{D}(s)$ such that

$$N(s) = \tilde{N}(s)U(s) \quad \text{and} \quad D(s) = \tilde{D}(s)U(s)$$

only for unimodular $U(s)$, then $N(s)$ and $D(s)$ are said to be **right coprime**.

If $N(s)$ and $D(s)$ have the same number of rows, and there exist $\tilde{N}(s)$ and $\tilde{D}(s)$ such that

$$N(s) = U(s)\tilde{N}(s) \quad \text{and} \quad D(s) = U(s)\tilde{D}(s)$$

only for unimodular $U(s)$, then $N(s)$ and $D(s)$ are said to be **left coprime**.

An MFD $G(s) = N(s)D^{-1}(s)$ (or $D^{-1}(s)N(s)$) is said to be **irreducible** if $N(s)$ and $D(s)$ are coprime (left or right, as appropriate); otherwise it is **reducible**.

With the aid of this definition we can state the following theorem:

Theorem 2.4: If $G(s) = N(s)D^{-1}(s)$ (or $D^{-1}(s)N(s)$), and $N(s)$ and $D(s)$ are coprime, then

(1) z is a (transmission) zero of $G(s)$ if and only if $N(s)$ loses rank at $s = z$

(2) p is a pole of $G(s)$ if and only if $D(p)$ is singular

Proof: See Kailath (1980), Chapter 6.

Corollary 2.2: If $G(s) = N(s)D^{-1}(s)$ (or $D^{-1}(s)N(s)$), and $N(s)$ and $D(s)$ are coprime, then the pole polynomial of $G(s)$ is given by

$$p(s) = \det D(s) \tag{2.40}$$

2.5   State-space realization from a transfer-function matrix

The formula (2.3) can be rewritten as

$$G(s) = \frac{C\,\mathrm{adj}\,(sI - A)B}{\det(sI - A)} + D$$

Since $C\,\mathrm{adj}\,(sI - A)B$ is a polynomial matrix, it is clear that every pole of $G(s)$ must be a zero of $\det(sI - A)$, namely an eigenvalue of A. The converse is not true, however, because zeros of $\det(sI - A)$ may be cancelled in equation (2.3), and hence not appear as poles of $G(s)$. Such cancellations occur precisely when the realization (A, B, C, D) is uncontrollable or unobservable. If the realization is both controllable and observable, then the poles of $G(s)$ are the eigenvalues of A – or, more precisely, the pole polynomial $p(s)$ is given by

$$p(s) = \det(sI - A)$$

This implies that the dimension of A cannot be smaller than the McMillan degree of $G(s)$; for this reason a realization which is both controllable and observable is called **minimal**.

We summarize these results in the following theorem:

Theorem 2.5:

        Let $G(s)$ have a minimal realization (A, B, C, D), and let $p(s)$ be the pole polynomial of $G(s)$ (as defined via the Smith–McMillan form). Then

(1) $\dim A = \deg p(s)$

        In other words, the McMillan degree of $G(s)$ is the minimal dimension of a realization.

(2)   The eigenvalues of A are just the poles of $G(s)$. (Note that if (A, B, C, D) is a realization which is not minimal, then the poles of $G(s)$ are a subset of the eigenvalues of A.)

There are several ways of finding a state-space realization which corresponds to a given transfer-function matrix. Perhaps the most straightforward is the following, in which a separate realization is found for each column of the transfer function – that is, for each input – and these separate realizations are then assembled. Let $g_i(s)$ be the ith column of $G(s)$, so that

$$G(s) = [g_1(s), g_2(s), \ldots, g_l(s)]$$

and let each column be written as

$$g_i(s) = \frac{n_i(s)}{d_i(s)} + \delta_i$$

where $d_i(s)$ is the (monic) common-denominator polynomial of $g_i(s)$,

$$d_i(s) = s^{k_i} + d_i^1 s^{k_i - 1} + \ldots + d_i^{k_i}$$

$n_i(s)$ is a vector of polynomials, each of degree smaller than k_i, and δ_i is a vector of constants. Let the jth element of $n_i(s)$ be the polynomial

$$v_{ji}(s) = v_{ji}^1 s^{k_i - 1} + v_{ji}^2 s^{k_i - 2} + \ldots + v_{ji}^{k_i}$$

Now let

$$A_i = \begin{bmatrix} 0 & 1 & 0 & 0 & \ldots & 0 \\ 0 & 0 & 1 & 0 & \ldots & 0 \\ \vdots & \vdots & \vdots & \vdots & & \vdots \\ 0 & 0 & 0 & 0 & \ldots & 1 \\ -d_i^{k_i} & -d_i^{k_i - 1} & -d_i^{k_i - 2} & -d_i^{k_i - 3} & \cdots & -d_i^1 \end{bmatrix}$$

$$
B_i = \begin{bmatrix} 0 \\ 0 \\ \vdots \\ 0 \\ 1 \end{bmatrix}, \qquad
C_i = \begin{bmatrix} v_{1i}^{k_i} & v_{1i}^{k_i-1} & \cdots & v_{1i}^{1} \\ v_{2i}^{k_i} & v_{2i}^{k_i-1} & \cdots & v_{2i}^{1} \\ \vdots & \vdots & & \vdots \\ v_{mi}^{k_i} & v_{mi}^{k_i-1} & \cdots & v_{mi}^{1} \end{bmatrix}
$$

Then it is straightforward to show that $(A_i, B_i, C_i, \delta_i)$ is a realization of $g_i(s)$. A realization of $G(s)$ is then given by (A, B, C, D), where

$$
A = \mathrm{diag}\{A_1, A_2, \ldots, A_l\}, \qquad B = \mathrm{diag}\{B_1, B_2, \ldots, B_l\},
$$

$$
C = [C_1, C_2, \ldots, C_l], \qquad D = [\delta_1, \delta_2, \ldots, \delta_l]
$$

This realization is guaranteed to be controllable, but may not be observable. For many purposes this does not matter, but if it is desired to obtain a minimal realization then the algorithm given in Section 8.3.5 may be used to remove any unobservable modes.

Other methods of finding state-space realizations are given by Kailath (1980) and Chen (1984).

2.6   How many zeros?

It is easy to tell at a glance how many zeros a scalar transfer function has, but this is not true of transfer-function matrices. Transfer functions of systems with unequal numbers of inputs and outputs are not square. If one bears in mind that zeros are points in the complex plane at which a transfer function loses rank, then it is clear that, for a non-square transfer function to have a zero at some point, several of its minors must simultaneously become singular at that point. One can construct examples for which this happens, but only with carefully chosen coefficients. Generically, non-square transfer functions have *no* zeros.

EXAMPLE 2.4

$$
G(s) = \begin{bmatrix} \dfrac{s+1}{s+2} & \dfrac{s+3}{s+4} \end{bmatrix}
$$

The normal rank of $G(s)$ is 1, and there is no value of s for which both elements simultaneously become zero.

$$G(s) = \begin{bmatrix} \dfrac{s+1}{s+2} & \dfrac{s+3}{s+4} \\[2ex] \dfrac{1}{s+5} & \dfrac{s+1}{s+6} \\[2ex] \dfrac{s+3}{s+4} & \dfrac{1}{s+2} \end{bmatrix}$$

The normal rank of $G(s)$ is 2, and there is no value of s for which the second column becomes a multiple of the first.

In order to discuss the number of zeros of a square transfer-function matrix, it is useful to introduce the concept of poles and zeros 'at infinity'. First, we examine the concept as applied to scalar transfer functions. Suppose that

$$G(s) = k\,\frac{s^m + b_1 s^{m-1} + \ldots + b_m}{s^n + a_1 s^{n-1} + \ldots + a_n}$$

If $n>m$ we say that $G(s)$ has $n-m$ zeros at infinity, and if $m>n$ we say it has $m-n$ poles at infinity. This allows us to assert that every scalar transfer function has as many zeros as poles (if those at infinity are included), and that improper (unrealizable) transfer functions are those with poles at infinity. Furthermore, the number of zeros at infinity is obviously related to the high-frequency gain behaviour of $G(j\omega)$ in a simple way, and we can now say that, on a root-locus plot, *every* pole is 'attracted' to a zero as the parameter of the locus becomes unbounded.

Definition 2.5: Let $H(\lambda)=G(1/\lambda)$. The poles and zeros of $G(s)$ at $s=\infty$ are then the poles and zeros of $H(\lambda)$ at $\lambda=0$.

Theorem 2.6: If $G(s)$ is square, it has as many zeros as poles (including those at infinity).

Proof: Let $H(\lambda)=G(1/\lambda)$. Then

$$\det H(\lambda)=\det G(1/\lambda) \tag{2.41}$$

$$=k\,\frac{z(1/\lambda)}{p(1/\lambda)} \tag{2.42}$$

where $z(s)$ and $p(s)$ are the zero and pole polynomials, respectively, of $G(s)$.

Now let

$$z(s) = \prod_{i=1}^{m} (s - z_i) \quad \text{and} \quad p(s) = \prod_{i=1}^{n} (s - p_i)$$

Then

$$z(1/\lambda) = \lambda^{-m} \prod_{i=1}^{m} (1 - \lambda z_i) \tag{2.43}$$

and

$$p(1/\lambda) = \lambda^{-n} \prod_{i=1}^{n} (1 - \lambda p_i) \tag{2.44}$$

Hence

$$\det H(\lambda) = k \lambda^{n-m} \prod_{i=1}^{m} (1 - \lambda z_i) \bigg/ \prod_{i=1}^{n} (1 - \lambda p_i) \tag{2.45}$$

Now, let $H(\lambda)$ have z_∞ zeros at 0, and p_∞ poles at 0. Then, from (2.45),

$$z_\infty - p_\infty = n - m \tag{2.46}$$

But z_∞ and p_∞ are the numbers of zeros and poles, respectively, of $G(s)$ at infinity.

Also, let z_f and p_f be the numbers of finite zeros, poles respectively, of $G(s)$. Then

$$p_f - z_f = n - m \tag{2.47}$$

Hence

$$p_f - z_f = z_\infty - p_\infty \tag{2.48}$$

or

$$p_f + p_\infty = z_f + z_\infty \tag{2.49}$$

∎

Note that this theorem does not hold for non-square transfer functions.

Now suppose that $G(s)$ is square, of dimensions $m \times m$, and has a minimal state-space realization (A, B, C, D), of state dimension n. Then

$$G(s) = D + C(sI - A)^{-1} B \tag{2.50}$$

which can be expanded as[2]

$$G(s) = D + \frac{CB}{s} + \frac{CAB}{s^2} + \frac{CA^2B}{s^3} + \cdots \tag{2.51}$$

for $|s|$ large enough, so

$$H(\lambda) = D + \lambda CB + \lambda^2 CAB + \lambda^3 CA^2 B + \dots \tag{2.52}$$

for $|\lambda|$ small enough.

From this we see that $H(0) = D$. Suppose that $\text{rank}(D) = \mu$. Then the rank of $H(\lambda)$ drops from m to μ at $\lambda = 0$, and therefore $H(\lambda)$ has at least $m - \mu$ zeros at $\lambda = 0$. Thus $G(s)$ has at least $m - \mu$ zeros at infinity, and therefore at most $n - m + \mu$ finite zeros. In the common case when $\mu = m$, the rank of $G(s)$ does not fall at infinity, and we conclude that $G(s)$ has n finite zeros, and no infinite zeros.

If $D = 0$, and therefore $\mu = 0$, we see that $G(s)$ has at most $n - m$ finite zeros, and at least m zeros at infinity. The exact number of finite zeros can be established by examining the **Markov parameters** $CA^{i-1}B$ according to a rather complicated algorithm. For our purposes it will be sufficient to note that if $\text{rank}(CB) = m$, then there are exactly $n - m$ finite zeros, but that if $\text{rank}(CB) = m - d$ then there are at most $n - m - d$ finite zeros.

For a more complete treatment of the topic raised in this section see Kailath (1980), and MacFarlane and Karcanias (1976).

An algorithm for finding zeros is given in Section 8.3.7.

2.7   Internal stability

> *Definition 2.6:*   A rational transfer-function matrix is **exponentially stable** if and only if it is proper and has no poles in the closed right half-plane.

Clearly, this definition is appropriate for continuous-time systems; when considering discrete-time systems we need to replace 'closed right half-plane' by 'boundary and exterior of the origin-centred unit circle', as usual.

The reason for insisting on properness is that improper systems may give unbounded outputs with bounded inputs, even if they have no (finite) poles, as shown by the following example:

EXAMPLE 2.5

$G(s) = s$,   input: $\sin(t^2)$,   output: $2t\cos(t^2)$

Consider the feedback loop shown in Figure 2.2, and define transfer functions $H_{11}(s)$, $H_{12}(s)$, $H_{21}(s)$ and $H_{22}(s)$ such that

$$\begin{bmatrix} e_1(s) \\ e_2(s) \end{bmatrix} = \begin{bmatrix} H_{11}(s) & H_{12}(s) \\ H_{21}(s) & H_{22}(s) \end{bmatrix} \begin{bmatrix} u_1(s) \\ u_2(s) \end{bmatrix} \tag{2.53}$$

Note that in Figure 2.2 a *positive-feedback* convention has been used.

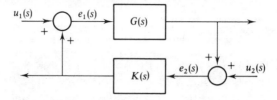

Figure 2.2   Feedback configuration for investigation of internal stability.

Definition 2.7 (see Note 3):   The feedback system shown in Figure 2.2 is **internally stable** if and only if the transfer-function matrix

$$\begin{bmatrix} H_{11}(s) & H_{12}(s) \\ H_{21}(s) & H_{22}(s) \end{bmatrix}$$

is exponentially stable.

The purpose of this definition is to exclude right half-plane pole–zero cancellations taking place between two systems $G(s)$ and $K(s)$, which cannot be detected by Nyquist-like tests, since these are applied to the return ratio $G(s)K(s)$ (or $K(s)G(s)$). The definition states, in effect, that *each* of the four transfer functions $H_{ij}(s)$ must be exponentially stable.

If $K(s)$ and $G(s)$ are both unstable, then it is necessary to check all four of these transfer functions to ensure internal stability. However, if at least one of them is stable (as is usually the case) then there is less work to do:

Theorem 2.7:   If $K(s)$ is exponentially stable, then the feedback system shown in Figure 2.2 is internally stable if and only if $H_{21}(s) = [I - G(s)K(s)]^{-1}G(s)$ is exponentially stable.

Proof: 'Only if': Immediate.
'If':

$$I + K(s)H_{21}(s) = I + K(s)[I - G(s)K(s)]^{-1}G(s) \qquad (2.54)$$

$$= I + K(s)G(s)[I - K(s)G(s)]^{-1} \qquad (2.55)$$

$$= [I - K(s)G(s)]^{-1} \qquad (2.56)$$

$$= H_{11}(s) \qquad (2.57)$$

So if $K(s)$ and $H_{21}(s)$ are exponentially stable, then so is $H_{11}(s)$. But this also implies that $H_{11}(s)K(s) = H_{12}(s)$ is exponentially stable.

Finally

$$I + H_{21}(s)K(s) = I + [I - G(s)K(s)]^{-1}G(s)K(s) \qquad \textbf{(2.58)}$$

$$= [I - G(s)K(s)]^{-1} \qquad \textbf{(2.59)}$$

$$= H_{22}(s) \qquad \textbf{(2.60)}$$

which shows $H_{22}(s)$ to be exponentially stable.
Hence the feedback system is exponentially stable. ∎

We can move towards a Nyquist-like criterion via the following:

Theorem 2.8: If $K(s)$ is exponentially stable, then $H_{21}(s)$ is exponentially stable if and only if

(1)   det $[I - G(s)K(s)]$ has no zeros in the closed right half-plane (including infinity), and

(2) $[I - G(s)K(s)]^{-1}G(s)$ is analytic (i.e. has no poles) at every closed right half-plane pole of $G(s)$ (including infinity)

Proof: In the following proof, 'closed right half-plane' is abbreviated as CRHP.
'Only if':

(1)   Let $G(s)K(s) = N(s)D^{-1}(s)$ be a coprime MFD. Then

$$H_{21}(s) = [I - G(s)K(s)]^{-1}G(s) \qquad \textbf{(2.61)}$$

$$= D(s)[D(s) - N(s)]^{-1}G(s) \qquad \textbf{(2.62)}$$

so $H_{21}(s)$ exponentially stable $\Rightarrow D(s) - N(s)$ has no zeros in CRHP since $D(s) - N(s)$ can have no zeros in common with $N(s)$ or $D(s)$ (by coprimeness) and every CRHP zero of $G(s)$ is also a zero of $N(s)$, because of the stability of $K(s)$.
Hence, $H_{21}(s)$ exponentially stable

$\Rightarrow \det[D(s) - N(s)] \neq 0$ in CRHP

$\Rightarrow \det[I - N(s)D^{-1}(s)] \neq 0$ in CRHP.

(2)   Immediate.
'If':

det $[I - G(s)K(s)] \neq 0$ in CRHP
$\Rightarrow$ any CRHP poles of $H_{21}(s)$ must be CRHP poles of $G(s)(*)$

But $H_{21}(s)$ is analytic at CRHP poles of $G(s)$

$\Rightarrow$ no CRHP poles of $H_{21}(s)$ coincide with CRHP poles of $G(s)$

$\Rightarrow H_{21}(s)$ has no CRHP poles by (*).   ∎

Note that part (2) of this theorem is needed to pick up pole–zero cancellations, and that it is needed even with SISO systems:

EXAMPLE 2.6

$$K(s) = \frac{s-1}{s+2}, \qquad G(s) = \frac{1}{s-1}$$

$$\det[I - G(s)K(s)] = \frac{s+1}{s+2} \quad \text{(looks OK)}$$

but

$$[I - G(s)K(s)]^{-1}G(s) = \frac{s+2}{(s+1)(s-1)}$$

Of course, each of the last two theorems can be restated for exponentially stable $G(s)$.

Theorems 2.7 and 2.8 together give the basis for a generalization of the Nyquist stability criterion, which will be developed in the next section.

If we revert to a negative-feedback convention, instead of the positive feedback used in Figure 2.2, we can obtain all the corresponding results simply by changing the sign of $K(s)$ (or $G(s)$). So in Theorems 2.7 and 2.8 we merely have to replace $[I - GK]$ by $[I + GK]$, and we obtain the results for negative-feedback loops.

The conditions which need to be satisfied in order to achieve internal stability apply whenever two systems are connected together in a feedback loop, and do not depend on the position of those systems in the loop. In other words, they apply not only when one system is in the forward path and the other in the feedback path, as shown in Figure 2.2, but also to the conventional configuration when both plant and compensator are in the forward path (as in Figure 1.8). Why, then, has feedback design been successfully based on the Nyquist stability theorem, which checks, in effect, only part 1 of Theorem 2.8? The reason is that all the additional complexity which has appeared in this section has been concerned with detecting the presence of right half-plane pole–zero cancellations, and thus of hidden unstable modes. When designing a compensator 'by hand', using the traditional techniques, one has complete and explicit control over the location of poles and zeros. Thus, as long as the designer obeys the basic injunction: 'do not introduce right half-plane pole–zero cancellations', it is enough to use only part 1 of Theorem 2.8. In this case the system $G(s)$ usually represents the series connection of plant and compensator, while $K(s)$ usually represents a constant gain, which is therefore exponentially stable.

This situation continues to hold when one uses the semi-automatic methods for multivariable design which we shall introduce in Chapter 4. But when one adopts the fully automated synthesis techniques described in Chapters 5 to 7, one needs to be sure that unstable hidden modes have not been introduced by the computer. In the linear quadratic Gaussian and H_∞ techniques examined in Chapters 5 and 6, the theory of internal stability is built into the synthesis theory, so that these techniques produce only internally stable designs. But the parameter-optimization techniques described in Chapter 7 offer no such guarantee, and either the internal stability of the resulting designs needs to be checked, or conditions such as those of Theorem 2.8 need to be added as constraints during the optimization.

2.8   The generalized Nyquist stability criterion

Let $G(s)$ be a square, rational transfer matrix. As we said in the previous section, this $G(s)$ will often represent the series connection of a plant with a compensator, and we must assume that there are no hidden unstable modes in the system represented by this transfer function. We wish to examine the stability of the negative-feedback loop created by inserting the compensator kI into the loop (i.e. an equal gain k into each loop), for various real values of k. To do this we apply part 1 of Theorem 2.8, with $K(s) = -kI$.

Let $\det[I + kG(s)]$ have P_O poles and P_c zeros in the closed right half-plane. Then, just as with SISO systems, we have, by the principle of the argument, that

$$\Delta \arg \det [I + kG(s)] = -2\pi(P_\mathrm{c} - P_\mathrm{O}) \tag{2.63}$$

where $\Delta\arg$ denotes the change in the argument as s traverses the usual **Nyquist contour** once – up the imaginary axis from the origin to some arbitrarily large distance, then along a semicircular arc in the right half-plane until it meets the negative imaginary axis, and finally up towards the origin; if any poles of $G(s)$ are encountered on the imaginary axis the contour is indented so as to include these poles. But for closed-loop stability we need P_c to equal zero, and the poles of $\det[I + kG(s)]$ are just the poles of $G(s)$. Hence we see that the condition for closed-loop stability is that the image of $\det[I + kG(s)]$ (as s goes once round the Nyquist contour) encircle the origin P_O times anticlockwise.

Note that the argument so far has exactly paralleled that for SISO feedback. However, if we stopped here we would have to draw the Nyquist locus of $\det[I + kG(s)]$ for each value of k in which we were interested, whereas the great virtue of the classical Nyquist criterion is that we draw a locus only once, and can then infer stability properties for all values of k.

It is easy to show that, if $\lambda_i(s)$ is an eigenvalue of $G(s)$, then $k\lambda_i(s)$ is an eigenvalue of $kG(s)$, and $1 + k\lambda_i(s)$ is an eigenvalue of $I + kG(s)$. Consequently (since the determinant is the product of the eigenvalues) we have

$$\det[I + kG(s)] = \prod_i [1 + k\lambda_i(s)] \tag{2.64}$$

and hence

$$\Delta \arg \det[I + kG(s)] = \sum_i \Delta \arg[1 + k\lambda_i(s)] \tag{2.65}$$

so we see that we can infer closed-loop stability by counting the total number of encirclements of the origin made by the graphs of $1 + k\lambda_i(s)$, or equivalently, by counting the total number of encirclements of -1 made by the graphs of $k\lambda_i(s)$. In this form, the criterion has the same powerful character as the classical criterion, since it is enough to draw the graphs once, usually for $k = 1$.

The graphs of $\lambda_i(s)$ (as s goes once round the Nyquist contour) are called the **characteristic loci**.

The above argument is not complete. Is it really true that the encirclements which we are supposed to count are well defined? There is a potential problem, as we can see if we consider the transfer-function matrix

$$G(s) = \begin{bmatrix} 0 & 1 \\ \dfrac{s-1}{s+1} & 0 \end{bmatrix} \tag{2.66}$$

This matrix has eigenvalues $\lambda_i(s) = \pm\sqrt{[(s-1)/(s+1)]}$. Neither $\lambda_1(j\omega)$ nor $\lambda_2(j\omega)$ forms a closed curve, so we see that we may have problems counting encirclements. This problem arises because the eigenvalues of a rational matrix are usually irrational functions.

It turns out that in practice there is no problem, but this is not easy to prove. We shall outline the idea of a proof. First we note that eigenvalues are continuous functions of the elements of a matrix. Consequently, if $G(s)$ has dimensions $m \times m$, we can distinguish m eigenvalue functions $\lambda_i(j\omega)$, each of which is continuous in ω (but the indexing is arbitrary, of course). Furthermore, eigenvalues of $G(j\omega)$ coincide at a finite number of values of ω, at most. So the graphs of $\lambda_i(j\omega)$ are indeed m 'well-behaved' graphs, with (possibly) isolated points of intersection.

Also, if *both* characteristic loci are drawn for the above example (Figure 2.3), it is seen that *together* they form a closed curve, since $\lambda_1(j\infty) = \lambda_2(-j\infty)$ and $\lambda_1(-j\infty) = \lambda_2(j\infty)$. This is always the case, basically because $G(j\infty) = G(-j\infty)$ (and hence the sets of eigenvalues at $j\infty$ and $-j\infty$ are the same). So, provided we consider all the characteristic loci together, we

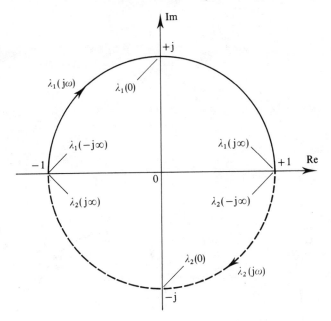

Figure 2.3   The two characteristic loci for the system defined by equation (2.66).

get a closed curve (or possibly a set of distinct closed curves) and can therefore count the encirclements of any point by this curve unambiguously.

> ***Theorem 2.9*** **(Generalized Nyquist theorem[4])**:   If $G(s)$ (as defined above) has P_0 unstable (Smith–McMillan) poles, then the closed-loop system with return ratio $-kG(s)$ is stable if and only if the characteristic loci of $kG(s)$, taken together, encircle the point -1 P_0 times anticlockwise, assuming that there are no hidden unstable modes.

EXAMPLE 2.7

$$G(s) = \frac{1}{1.25(s+1)(s+2)} \begin{bmatrix} s-1 & s \\ -6 & s-2 \end{bmatrix} \qquad (2.67)$$

The characteristic loci of this transfer function are shown in Figure 2.4. Since $G(s)$ has no unstable poles, we will have closed-loop stability if these loci give zero net encirclements of $-1/k$ when a negative feedback kI is applied. From Figure 2.4 it can be seen that for $-\infty < -1/k < -0.8$, for $-0.4 < -1/k < 0$ and for $0.53 < -1/k < \infty$ there will be no encirclements, and hence closed-loop stability will be obtained. For $-0.8 < -1/k < -0.4$

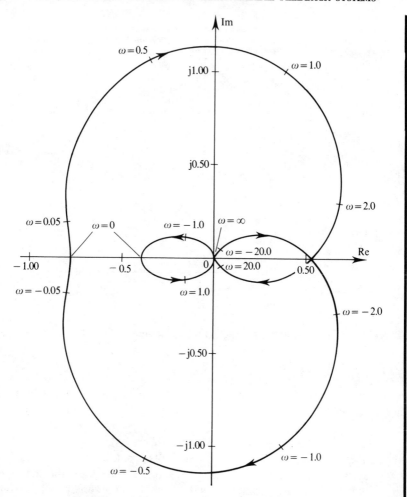

Figure 2.4 The two characteristic loci for the system defined by equation (2.67).

there is one anticlockwise encirclement, and hence closed-loop instability, while for $0 < -1/k < 0.53$ there are two clockwise encirclements, and therefore closed-loop instability again.

2.9   The generalized inverse Nyquist stability criterion

It is possible to restate the generalized Nyquist stability criterion in terms of the characteristic loci of the inverse return ratio $G^{-1}(s)$ rather than $G(s)$:

Theorem 2.10 (**Generalized inverse Nyquist theorem**[5]): If $G(s)$ (as defined in Section 2.8) has Z_O (transmission) zeros in the closed right half-plane, then the closed-loop system is stable if and only if the characteristic loci of $G^{-1}(s)/k$, taken together, encircle the point -1 Z_O times anticlockwise, assuming that there are no hidden unstable modes.

Proof: Since no pole–zero cancellations can occur between $G(s)$ and kI, it is sufficient to check part 1 of Theorem 2.8, with $F(s) = -kI$.
Now,

$$I + kG(s) = [G^{-1}(s)/k + I]kG(s) \tag{2.68}$$

so

$$\det[I + kG(s)] = k^m \det[G^{-1}(s)/k + I]\det G(s) \tag{2.69}$$

$$= k^m \det[G^{-1}(s)/k + I]\frac{z(s)}{p(s)} \tag{2.70}$$

where $z(s)$ and $p(s)$ are the zero and pole polynomials, respectively, of $G(s)$.

Now let $z_c(s)$ and $p_c(s)$ be the zero and pole polynomials of $I + kG(s)$. Then $p_c(s) = p(s)$, since the poles of $I + kG(s)$ are just those of $G(s)$. Hence

$$z_c(s) = k^m \det[G^{-1}(s)/k + I]z(s) \tag{2.71}$$

or

$$k^m \det[G^{-1}(s)/k + I] = \frac{z_c(s)}{z(s)} \tag{2.72}$$

So if $z_c(s)$ has Z_c roots in the CRHP, and $z(s)$ has Z_O roots there, then

$$\Delta \arg \det[G^{-1}(s)/k + I] = -2\pi(Z_c - Z_O) \tag{2.73}$$

But for stability we need $Z_c = 0$ – that is,

$$\Delta \arg \det[G^{-1}(s)/k + I] = 2\pi Z_O \tag{2.74}$$

The restatement in terms of encirclements by the characteristic loci of $G^{-1}(s)/k$ follows as in Theorem 2.9.   ∎

Note that if $\{\lambda_i(s)\}$ is the set of eigenvalues of $G(s)$, then $\{\lambda_i^{-1}(s)\}$ is the set of eigenvalues of $G^{-1}(s)$.

If the inverse Nyquist criterion is used, then the behaviour of the inverse characteristic loci on the semicircular portion of the Nyquist contour must be examined carefully. The inverse loci usually become large as s becomes large, so encirclements may be introduced as s moves around this part of the contour.

2.10   Nyquist arrays and Gershgorin bands

The **Nyquist array** of $G(s)$ is an array of graphs (not necessarily square), the (i, j)th graph being the Nyquist locus of $g_{ij}(s)$. Use is also made of the **inverse Nyquist array**, which is the array of graphs of Nyquist loci of the elements of $G^{-1}(s)$. (Of course, the inverse array is defined only when $G(s)$ is square.)

The key to Nyquist array methods is:

Theorem 2.11 (**Gershgorin's theorem**):   Let Z be a complex matrix of dimensions $m \times m$. Then the eigenvalues of Z lie in the union of the m circles, each with centre z_{ii} and radius

$$\sum_{\substack{j=1 \\ j \neq i}}^{m} |z_{ij}|, \quad i = 1, \ldots, m$$

They also lie in the union of the circles, each with centre z_{ii} and radius

$$\sum_{\substack{j=1 \\ j \neq i}}^{m} |z_{ji}|, \quad i = 1, \ldots, m$$

Proof: See Kreyszig (1972) or Rosenbrock (1974).   ∎

Consider the Nyquist array of some square $G(s)$. On the loci of $g_{ii}(j\omega)$ superimpose, at each point, a circle of radius

$$\sum_{\substack{j=1 \\ j \neq i}}^{m} |g_{ij}(j\omega)| \quad \text{or} \quad \sum_{\substack{j=1 \\ j \neq i}}^{m} |g_{ji}(j\omega)|$$

(making the same choice for all the diagonal elements at each frequency). The 'bands' obtained in this way are called **Gershgorin bands**; each is composed of **Gershgorin circles**. Gershgorin bands are illustrated in Figure 2.5.

By Gershgorin's theorem we know that the union of the Gershgorin bands 'traps' the union of the characteristic loci. Furthermore, it can be shown that if the Gershgorin bands occupy distinct regions, then as many characteristic loci are trapped in a region as the number of Gershgorin bands occupying it. So, *if all the Gershgorin bands exclude the point* -1, then we

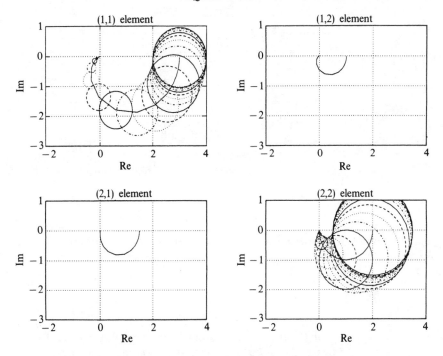

Figure 2.5 A Nyquist array, with Gershgorin bands.

can assess closed-loop stability by counting the encirclements of -1 by the Gershgorin bands, since this tells us the number of encirclements made by the characteristic loci.[6]

If the Gershgorin bands of $G(s)$ exclude the origin, then we say that $G(s)$ is **diagonally dominant** (row dominant or column dominant, if applicable). Note that to assess stability we need $[I + G(s)]$ to be diagonally dominant.

The greater the degree of dominance (of $G(s)$ or of $I + G(s)$) – that is, the narrower the Gershgorin bands – the more closely does $G(s)$ resemble m non-interacting SISO transfer functions.

Clearly, from the generalized inverse Nyquist theorem we can obtain a stability criterion in terms of the Gershgorin bands of $G^{-1}(s)$: if all the Gershgorin bands of $G^{-1}(s)$ exclude the point -1, we can assess closed-loop stability by counting the number of encirclements of -1 made by these bands. In this case we require $I + G^{-1}(s)$ to be diagonally dominant on the Nyquist contour (including the semicircular part of the contour).

These Nyquist-array-based tests check sufficient, but not necessary, conditions for stability (or instability). If any Gershgorin band *does* overlap the point -1, in other words if $I + G$ (or $I + G^{-1}$) is *not* diagonally dominant,

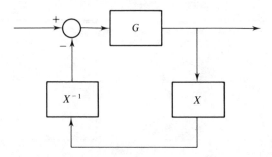

Figure 2.6 Feedback loop containing the system defined by equation (2.75).

then we cannot infer whether the system is stable or not. But suppose that we replace G by

$$\tilde{G} = XGX^{-1} \tag{2.75}$$

This is a similarity transformation, so the characteristic loci of $\tilde{G}$ must be the same as those of G, and hence we can check stability by displaying Gershgorin bands of $\tilde{G}$ (or $\tilde{G}^{-1}$) just as well as we can with those of G. Physically, equation (2.75) corresponds to inserting a system X at the output of G, and X^{-1} at the input of G, as shown in Figure 2.6. The point of doing this is that $\tilde{G}$ may be diagonally dominant, even if G is not, for some choices of X.

A method for finding a suitable X has been given by Mees (1981). For any matrix M with elements m_{ij}, we define

$$\text{abs } M = [|m_{ij}|] \tag{2.76}$$

that is, the elements of abs M are $|m_{ij}|$. Such a matrix is called **positive**. If M is square, let the eigenvalues of abs M be $\{\lambda_1, \lambda_2, \ldots, \lambda_n\}$, and let them be ordered so that $|\lambda_1| \geqslant |\lambda_2| \geqslant \ldots$. M is called **primitive** if $(\text{abs } M)^r$ has positive entries only for some integer r. For primitive positive matrices λ_1 is real and $\lambda_1 > |\lambda_i|$ for $i \neq 1$; λ_1 is called the **Perron–Frobenius eigenvalue** of M, and we shall denote it by $\lambda_P(M)$. The corresponding left and right eigenvectors of abs M are called the **Perron–Frobenius eigenvectors** of M (and are also real and positive).

Now, let

$$M_{\text{diag}} = \text{diag}\{m_{11}, m_{22}, \ldots, m_{nn}\} \tag{2.77}$$

Mees proves the following theorem:

Theorem 2.12 (Mees, 1981):   If G is square and primitive, then there exists a diagonal matrix X such that

$$\tilde{G} = XGX^{-1}$$

is diagonally dominant, if and only if

$$\lambda_{\mathrm{P}}(GG_{\mathrm{diag}}^{-1}) < 2 \tag{2.78}$$

If (2.78) is satisfied, and the Perron–Frobenius left eigenvector of GG_{diag}^{-1} is $(x_1, x_2, \ldots, x_n)^{\mathrm{T}}$, then an X which achieves diagonal dominance is

$$X = \operatorname{diag}\{x_1, x_2, \ldots, x_n\} \tag{2.79}$$

Note that the X found by this theorem can be interpreted as a scaling of the outputs of G, and X^{-1} as a scaling of the inputs of G, since X is diagonal and real. A different X is required at each frequency, of course; this is no problem if we are using X only for analysis, since X does not correspond to a system that has to be built. Later we shall use this theory for design, and we shall then have to build systems whose gain behaviour approximates that of X as frequency varies.

Theorem 2.12 requires G to be primitive, but in practice this is not a severe restriction. If all the elements of G are non-zero then G is certainly primitive. A problem occurs only if some elements of G are fixed at zero as frequency varies. If this happens then G is usually block-diagonal, or at least block-triangular, or can be made so by reordering inputs or outputs. The characteristic loci of G are then just the characteristic loci of the (primitive) blocks that occur on the principal diagonal of G, and Theorem 2.12 can be applied separately to each of these blocks.

Mees also shows that the X produced by Theorem 2.12 is optimal, in the sense that it gives the narrowest possible Gershgorin bands obtainable with any diagonal similarity transformation.

Note that Theorem 2.12 should be applied to $I + G$, rather than G, if we do not want the Gershgorin bands to cover the point -1. Since $X(I+G)X^{-1} = I + XGX^{-1}$, we can still infer stability from the Gershgorin bands of XGX^{-1}.

There is one form of stability theorem which is expressed in terms of Gershgorin bands but which we have not presented yet, and which can be very useful for checking the robustness of particular designs as gains or other real parameters vary, as we shall see in Chapter 3. This form is relevant when the feedback loop contains a square transfer function $G(s)$ connected in series with a diagonal gain matrix

$$K = \operatorname{diag}\{k_1, k_2, \ldots, k_m\} \tag{2.80}$$

and each k_i is a constant (usually real). The difference between this and the feedback loops considered so far in Sections 2.8 and 2.9, and this section, is that the gain (k_i) in each loop may now be different.

Theorem 2.13 (Rosenbrock, 1970):   Suppose that $G(s)$ is square, that $K = \text{diag}\{k_1, \ldots, k_m\}$ and that

$$\left| g_{ii}(s) + \frac{1}{k_i} \right| > \sum_{j \neq i} |g_{ij}(s)| \tag{2.81}$$

for each i and for all s on the Nyquist contour; and let the ith Gershgorin band of $G(s)$ encircle the point $-1/k_i$, N_i times anticlockwise. Then the negative feedback system with return ratio $-G(s)K$ is stable if and only if

$$\sum_i N_i = P_O \tag{2.82}$$

where P_O is the number of unstable poles of $G(s)$, and there are no hidden unstable modes.

Proof: We know from Section 2.8 that a necessary and sufficient condition for stability is that

$$N = P_O$$

where N is the number of anticlockwise encirclements of the origin made by $\det[I + G(s)K]$, as s moves once around the Nyquist contour. So, to prove the theorem we need to prove that

$$N = \sum_i N_i \tag{2.83}$$

To do this, note that the ith Gershgorin band of $G(s) + K^{-1}$ encircles the origin N_i times anticlockwise (the condition (2.81) ensuring that it does not cover the origin). But the union of these bands contains the characteristic loci of $G(s) + K^{-1}$; call these characteristic loci $\mu_i(s)$ $(i = 1, \ldots, m)$. So we have, as in Section 2.8, that

$$\Delta \arg \det[G(s) + K^{-1}] = \sum_i \Delta \arg \mu_i(s) \tag{2.84}$$

$$= \sum_i N_i \tag{2.85}$$

But

$$\Delta \arg \det [I + G(s)K] = \Delta \arg \det [G(s) + K^{-1}] + \Delta \arg \det (K)$$
(2.86)

$$= \Delta \arg \det [G(s) + K^{-1}]$$
(2.87)

Hence

$$N = \sum_i N_i$$ ■

Theorem 2.13 can clearly be restated in terms of column dominance instead of row dominance, by replacing (2.81) by

$$\left| g_{ii}(s) + \frac{1}{k_i} \right| > \sum_{j \neq i} |g_{ji}(s)|$$
(2.88)

and indeed the theorem holds if *either* (2.81) *or* (2.88) holds at each point on the Nyquist contour. A corresponding theorem can also be stated in terms of the inverse transfer function $G^{-1}(s)$ (see Exercise 2.11).

2.11   Generalized stability

So far we have used the term 'stability' in its familiar and usual sense, and have tested for its existence by checking whether any poles lie in the right

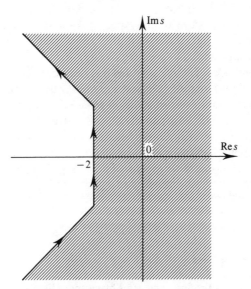

Figure 2.7   Generalized stability: closed-loop poles may not lie in the hatched area (for example).

half-plane. We can obviously transfer all the results to discrete-time systems by replacing the left half-plane by the interior of the unit circle. But we can also replace the left half-plane by any other region around which we can draw a single, closed boundary. (If the left half-plane does not appear to meet this requirement, consider the points $+j\infty$ and $-j\infty$ to be the same.)

Suppose, for example, that we require of a system not merely that it be stable, but also that all its poles and hidden modes decay at least as fast as e^{-2t}, and to have damping factors no smaller than $1/\sqrt{2}$; in other words, they should all be located to the left of the hatched region shown in Figure 2.7. To check whether this condition is satisfied, we can use *all* the results of Sections 2.7 to 2.10, replacing 'right half-plane' by the hatched region shown in Figure 2.7, and 'imaginary axis' by the boundary of this region, as shown by the arrows in Figure 2.7.

In this way we can generalize the notion of stability without needing any new theory.

SUMMARY

In this chapter we have introduced transfer function matrices as descriptions of multivariable, linear dynamic systems, and related them to state-space models. Matrix fraction descriptions were also introduced as alternative representations which are often useful for resolving theoretical questions.

We have generalized the notions of poles and zeros to multivariable systems by means of the Smith–McMillan form of a transfer function matrix. The poles determine system stability, just as is the case for SISO systems; in the next chapter we shall see that the zeros may limit the achievable performance of multivariable feedback systems.

The stability of feedback systems has been considered. The notion of 'internal stability' was introduced, and a generalization of Nyquist's stability theorem to multivariable systems was developed. This theorem was stated in terms of characteristic loci, namely graphs of the eigenvalue functions of a multivariable return ratio. Sufficient, but not necessary, conditions for closed-loop stability or instability have been given, stated in terms of Gershgorin bands, along with means of reducing their conservativeness.

Notes

1. The definition of the poles and zeros by means of the Smith–McMillan form is taken from Rosenbrock (1970). Most of Sections 2.2 to 2.5 can also be traced back to Rosenbrock's work.

2. This is the Laurent series expansion of $G(s)$ about $s = \infty$.

3.   The definition of internal stability, and much of the rest of Section 2.7, follows Desoer and Chan (1975).

4.   This theorem was first conjectured by MacFarlane in about 1970. It has been proved several times, in various degrees of generality and correctness, and by several different approaches. The argument outlined in the text follows Desoer and Wang (1980). This paper also proves the theorem for the case when $G(s)$ is irrational, and relies heavily on the use of graph theory. A very different approach, using the theories of Riemann surfaces and algebraic functions, is taken by MacFarlane and Postlethwaite (1977), Postlethwaite and MacFarlane (1979), and Smith (1984).

5.   This statement of the generalized inverse Nyquist theorem is taken from Postlethwaite (1977).

6.   Stability theorems phrased in terms of Gershgorin bands were first developed by Rosenbrock (1970). His proofs make no use of characteristic loci.

EXERCISES

2.1   If (A, B, C, D) is a realization of $G(s)$, and $\tilde{A} = T^{-1}AT$, $\tilde{B} = T^{-1}B$ and $\tilde{C} = CT$, verify that $(\tilde{A}, \tilde{B}, \tilde{C}, D)$ is also a realization of $G(s)$. Also show that $G(\infty) = D$.

2.2   If the discrete-time state-space model

$$x_{k+1} = Ax_k + Bu_k, \qquad y_k = Cx_k + Du_k$$

has transfer function $G(z)$, show that $G(z)$ is related to the realization (A, B, C, D) by the same formula that holds for continuous-time systems. (As a consequence of this result, everything in this chapter holds for discrete-time systems.)

2.3   Show how the elementary operations on polynomial matrices can be represented by multiplication with suitable matrices (the elementary matrices). Verify that such matrices are unimodular.

2.4   Find the Smith–McMillan form of

$$\frac{1}{(s+1)(s+3)} \begin{bmatrix} 1 & 0 \\ -1 & 2(s+1)^2 \end{bmatrix}$$

Hence find its poles and zeros.

2.5   Find MFDs of

(a)

$$\frac{1}{s^2 + 2s + 2} \begin{bmatrix} s^2 & 2 \\ -2 & s^2 \end{bmatrix}$$

(b)

$$\frac{1}{(s+1)(s+2)} \begin{bmatrix} 1 & -1 \\ s^2+s-4 & 2s^2-s-8 \\ s^2-4 & 2s^2-8 \end{bmatrix}$$

Hence find their poles and zeros.

2.6 By considering a coprime MFD of $G(s)$, show that the poles of $\det[I+kG(s)]$ are the same as the poles of $G(s)$, if k is a scalar constant.

2.7 Show that the zeros of a transfer function in the forward path of a feedback loop remain as zeros of the closed-loop transfer function, unless they are deliberately cancelled. Show by giving an example that this is not true of the zeros of the elements of the transfer function.

2.8 (a) Find a state-space realization of

$$G(s) = \begin{bmatrix} \dfrac{2+5}{s^2+2s+3} & \dfrac{1}{s+1} \\ \dfrac{1}{s+4} & \dfrac{1}{s+2} \end{bmatrix}$$

Check the minimality of your realization (preferably using suitable software), and hence find the McMillan degree of $G(s)$. What are the poles of $G(s)$?

(*Solution*: McMillan degree = 5, poles = $\{-1, -2, -4, -1\pm j\sqrt{2}\}$)

(b) Repeat (a), but with the $(1, 1)$ element of $G(s)$ replaced by $(s+5)/(s^2+3s+2)$.

(*Solution*: McMillan degree = 4, poles = $\{-1, -2, -2, -4\}$)

2.9 If $G(s)$ is square and not identically singular, show that the zeros and poles of $G^{-1}(s)$, including those at infinity, are just the poles and zeros, respectively, of $G(s)$.

2.10 (a) Show that the transfer functions H_{11}, H_{12}, H_{21} and H_{22}, which appear in equation (2.53), are given in terms of K and G (which appear in Figure 2.2) by

$$H_{11}=(I-KG)^{-1}, \qquad H_{12}=(I-KG)^{-1}K$$
$$H_{21}=(I-GK)^{-1}G, \qquad H_{22}=(I-GK)^{-1}$$

(b) Hence show that

$$\begin{bmatrix} H_{11} & H_{12} \\ H_{21} & H_{22} \end{bmatrix} = \begin{bmatrix} I & -K \\ -G & I \end{bmatrix}^{-1}$$

(*Hint:* $X[I+YX]^{-1}=[I+XY]^{-1}X$)

(c) Deduce that a necessary condition for internal stability is that $\det[I-GK]$ has no zeros in the CRHP. Why is it not sufficient?

(*Hint:* $\det\begin{bmatrix} W & X \\ Y & Z \end{bmatrix} = \det W \det[Z-YW^{-1}X]$, known as *Schur's formula*)

2.11 Let $G(s)$ be a square transfer function, let $\hat{g}_{ij}$ be the (i,j) element of the inverse $\hat{G}(s)=G^{-1}(s)$, and $K=\text{diag}\{k_1,k_2,\ldots,k_m\}$. Suppose that

$$|\hat{g}_{ii}(s)+k_i| > \sum_{j\neq i}|\hat{g}_{ij}(s)|$$

for each i and for all s on the Nyquist contour. Let the ith Gershgorin band of $\hat{G}(s)$ encircle the point $-k_i$, N_i times anticlockwise. Show that the negative feedback system with return ratio $-G(s)K$ is stable if and only if

$$\sum_i N_i = Z_0$$

if $G(s)$ has Z_0 transmission zeros in the right half-plane.

2.12 (This exercise is based on Section 3.5 of Owens (1978).) Suppose that a strictly proper system with transfer function $G(s)$ has m inputs, outputs and states, and a minimal state-space realization $(A,B,C,0_{mm})$, and that B^{-1} and C^{-1} exist. Show that

$$G^{-1}(s)=sA_0+A_1$$

and find the constant matrices A_0 and A_1 in terms of A,B and C. Where are the zeros of $G(s)$ located?

Suppose that a compensator of the form

$$K(s)=A_0\,\text{diag}\left\{\alpha_j+\beta_j+\frac{\alpha_j\beta_j}{s}\right\}-A_1$$

is used, where α_j and β_j are real parameters ($j=1,\ldots,m$). By writing $I+GK=G(G^{-1}+K)$, show that the closed loop is stable for any positive values of α_j and β_j, and that all the closed-loop poles are real.

(Note that this holds even if $G(s)$ is unstable. Owens calls $G(s)$ a **multivariable first-order system**; process plant may sometimes be modelled by such systems. The compensator $K(s)$ has a fixed structure, determined by A_0 and A_1, and parameters α_j and β_j which can be tuned (possibly on-line) to give acceptable dynamic characteristics.)

References

Callier F.M. and Desoer C.A. (1982). *Multivariable Feedback Systems.* Berlin: Springer-Verlag.

Chen C.T. (1984). *Linear System Theory and Design.* New York: Holt, Rinehart & Winston.

Desoer C.A. and Shulman J.D. (1974). Zeros and poles of matrix transfer-functions and their dynamical interpretation. *IEEE Transactions on Circuits and Systems,* **CAS-21**, 3–8.

Desoer C.A. and Chan W.S. (1975). The feedback interconnection of linear time-invariant systems. *Journal of the Franklin Institute,* **300**, 335–51.

Desoer C.A. and Wang, Y.T. (1980). On the generalized Nyquist stability criterion. *IEEE Transactions on Automatic Control,* **AC-25**, 187–96.

Gantmacher F.R. (1959). *Theory of Matrices* Vol. I. New York: Chelsea.

Kailath T. (1980). *Linear Systems.* Englewood Cliffs NJ: Prentice-Hall.

Kreyszig E. (1972). *Advanced Engineering Mathematics.* New York: Wiley.

MacFarlane A.G.J. (1970). The return-difference and return-ratio matrices and their use in the analysis and design of multivariable feedback control systems. *Proceedings of the Institution of Electrical Engineers,* **117**, 2037–49.

MacFarlane A.G.J. and Karcanias N. (1976). Poles and zeros of linear multivariable systems: A survey of the algebraic, geometric and complex variable theory. *International Journal of Control,* **24**, 33–74.

MacFarlane A.G.J. and Postlethwaite I. (1977). The generalized Nyquist stability criterion and multivariable root loci. *International Journal of Control,* **25**, 81–127.

Mees A.I. (1981). Achieving diagonal dominance. *System and Control Letters,* **1**, 155–8.

Owens D.H. (1978). *Feedback and Multivariable Systems.* Stevenage: Peter Peregrinus.

Postlethwaite I. (1977). A generalized inverse Nyquist stability criterion. *International Journal of Control,* **26**, 325–40.

Postlethwaite I. and MacFarlane A.G.J. (1979). *A Complex Variable Approach to the Analysis of Linear Multivariable Feedback Systems.* Berlin: Springer-Verlag.

Rosenbrock H.H. (1970). *State-Space and Multivariable Theory.* London: Nelson.

Rosenbrock H.H. (1974). *Computer-Aided Control System Design.* New York: Academic Press.

Smith M.C. (1984). Applications of algebraic function theory in multivariable control. In *Multivariable Control: New Concepts and Tools* (Tzafestas S.G., ed.), pp. 3–26. Dordrecht: Reidel.

CHAPTER 3

Performance and Robustness of Multivariable Feedback Systems

3.1 Introduction
3.2 Principal gains (singular values)
3.3 The use of principal gains for assessing performance
3.4 Relations between closed-loop and open-loop principal gains
3.5 Principal gains and characteristic loci
3.6 Limitations on performance
3.7 Transmission of stochastic signals

3.8 The operator norms $\| G \|_2$ and $\| G \|_\infty$
3.9 The use of operator norms to specify performance
3.10 Representations of uncertainty
3.11 Stability robustness
3.12 Performance robustness
Summary
Exercises
References

3.1 Introduction

The main purpose of using feedback is to reduce the effects of uncertainty. Some uncertainty is always present, both in the environment of the system – we do not know in advance exactly what disturbance and noise signals the system will be subjected to, and in the behaviour of the system itself – we know our models are not perfect, and the behaviour of the system may change in unpredictable ways. A further reason for using feedback is to stabilize an unstable system.

We shall to some extent separate out the assessment of how well a feedback system deals with unwanted or unexpected signals – its **performance** – from the assessment of how resilient it is in the face of internal changes in behaviour – its **robustness**. But we shall see that the same tools are used to evaluate these two aspects of a feedback system's behaviour, and towards the end of the chapter we shall formulate a design problem in which performance goals and robustness goals become mathematically indistinguishable.

A feedback system's ability to follow a reference input is usually considered to be an important part of its performance. This can also be assessed in the same way as the system's ability to recover from external disturbances, but it should be recognized that this is not, strictly speaking, a part of *feedback* design. As we saw in Chapter 1, the transfer function between a reference input and a system output can be modified by use of a pre-filter which is outside the feedback loop. We can therefore picture a two-stage design procedure in which the feedback loop is designed first, in order to reduce the effects of uncertainty, and the overall closed-loop transfer function is then adjusted by inserting a pre-filter, which requires only open-loop design. This distinction becomes blurred, however, when no pre-filter is to be included – either because there is no reference input, as in a regulator, or because practical considerations exclude it. In this case the response to a reference signal is essentially the same as the response to a disturbance at the plant output, provided there are no dynamics in the feedback path. The distinction is also blurred when an optimization technique is used which designs both the feedback compensator and the pre-filter at the same time.

3.2   Principal gains (singular values)

In a SISO system the performance of a feedback loop is determined by the variation of the loop gain with frequency; disturbance rejection, noise transmission and differential sensitivity to parameter variations all depend only on the gain (assuming that stability is achieved). If the open-loop transfer function (return ratio) has no right half-plane zeros, then stability margins and the closed-loop transient response are also determined by the open-loop gain characteristic.

In attempting to extend this correlation to multivariable feedback, the main problem is that a matrix does not have a unique gain: the norm $\|G(s)u(s)\|$ depends on the direction of the vector $u(s)$. However, we can bound the ratios

$$\frac{\|G(s)u(s)\|}{\|u(s)\|} \quad \text{and} \quad \frac{\|G^{-1}(s)y(s)\|}{\|y(s)\|}$$

using **matrix norms**. (We shall usually assume that $G(s)$ is square and invertible, but neither of these assumptions is necessary.) Thus we are led to replace the idea of a single gain by the notion of a range of gains, this range being bounded below and above.

If $\|x\|$ denotes any vector norm, then an **induced** (or **subordinate**) **matrix norm** is defined by

$$\|G\| = \sup_{x \neq 0} \frac{\|Gx\|}{\|x\|} \tag{3.1}$$

In particular, if we use the Euclidean vector norm (for *complex* vectors)

$$\|x\| = \sqrt{(x^H x)} \tag{3.2}$$

then the induced matrix norm is the **Hilbert** or **spectral norm**:

$$\|G\|_s = \bar{\sigma}$$

where $\bar{\sigma}^2$ is the maximum eigenvalue of $G^H G$ (or of GG^H). Here x^H denotes $\bar{x}^T$, and similarly for G^H. Now, if G has m rows and l columns, and $m \geq l$, then the positive square roots of the eigenvalues of $G^H G$ are called the **singular values** of G. (If $m \leq l$ then the square roots of the eigenvalues of GG^H are the singular values of G.)

If instead of G we have $G(s)$, and set $s = j\omega$ ($0 \leq \omega < \infty$), then the singular values of $G(j\omega)$ are functions of ω, and they are then called the **principal gains**[1] of $G(s)$. We shall denote them by $\{\sigma_i(\omega)\}$ when we wish to emphasize their dependence on frequency, or by $\{\sigma_i(G)\}$ when we wish to distinguish the principal gains of G from those of some other system. We shall also adopt the ordering $\sigma_1 \geq \sigma_2 \geq \ldots \geq \sigma_m$, when necessary, and denote σ_1, σ_m by $\bar{\sigma}$ and $\underline{\sigma}$ respectively, when we wish to emphasize the use of the largest or smallest principal gain.

Note that

$$\bar{\sigma}(G(j\omega)) = \|G(j\omega)\|_s \tag{3.3}$$

but that this is a norm on the matrix $G(j\omega)$, which changes with ω. Later (in Section 3.8) we shall introduce $\|G\|_2$ and $\|G\|_\infty$, norms on the transfer function G that are independent of frequency.

The singular-value decomposition

Let $\Sigma = \text{diag}\{\sigma_1, \sigma_2, \ldots, \sigma_m\}$, and let G be a complex matrix. Then, G can always be written as

$$G = Y\Sigma U^H \tag{3.4}$$

where

if $m \geqslant l$:	if $m \leqslant l$:
$Y \in C^{m \times l}$	$Y \in C^{m \times m}$
$\Sigma \in R^{l \times l}$	$\Sigma \in R^{m \times m}$
$U^H \in C^{l \times l}$	$U^H \in C^{m \times l}$
$Y^H Y = I_l$	$Y^H Y = YY^H = I_m$
$U^H U = UU^H = I_l$	$U^H U = I_m$

and $C^{m \times l}$ denotes the set of complex matrices with m rows and l columns, and $R^{m \times l}$ denotes the set of real matrices with these dimensions.

This is known as the **singular-value decomposition**. For further details see Golub and van Loan (1983) or Stewart (1973). Note that this decomposition is not unique: $G = Y' \Sigma U'^H$, where $Y' = Ye^{j\theta}$ and $U' = Ue^{-j\theta}$, for any θ, also gives a singular-value decomposition. However, the σ_i *are* unique.

We shall now derive some properties of the singular-value decomposition.

$$\begin{aligned} GG^H &= (Y\Sigma U^H)(Y\Sigma U^H)^H \\ &= (Y\Sigma U^H)(U\Sigma Y^H) \\ &= Y\Sigma^2 Y^H \end{aligned} \tag{3.5}$$

If $m \leqslant l$ then $Y^H = Y^{-1}$, which shows that in this case Y is the matrix of eigenvectors of GG^H, and $\{\sigma_i^2\}$ are its eigenvalues.

$$\begin{aligned} G^H G &= (U\Sigma Y^H)(Y\Sigma U^H) \\ &= U\Sigma^2 U^H \end{aligned} \tag{3.6}$$

If $m \geqslant l$ then $U^H = U^{-1}$, so in this case U is the matrix of eigenvectors of $G^H G$, and $\{\sigma_i^2\}$ are its eigenvalues.

Note that $\text{rank}(G) = \text{rank}(\Sigma)$, since U and Y are unitary. So if $\text{rank}(G) = r$, then only the first r singular values of G are positive; the remainder are zero.

Consider

$$H = U\Sigma^{-1} Y^H \tag{3.7}$$

Then

$$\begin{aligned} HGH &= (U\Sigma^{-1} Y^H)(Y\Sigma U^H)(U\Sigma^{-1} Y^H) \\ &= H \end{aligned} \tag{3.8}$$

and

$$GHG = (Y\Sigma U^H)(U\Sigma^{-1} Y^H)(Y\Sigma U)$$

$$= G \tag{3.9}$$

which shows that H is a pseudo-inverse of G. Thus we have

$$G^\dagger = U\Sigma^{-1} Y^H \tag{3.10}$$

where $G^\dagger$ denotes the pseudo-inverse of G. We have assumed here that $\mathrm{rank}(G) = \min(l, m)$. If $r < \min(l, m)$ then instead of Σ^{-1} we take

$$\begin{bmatrix} \Sigma_r^{-1} & 0 \\ 0 & 0 \end{bmatrix}$$

where $\Sigma_r = \mathrm{diag}\{\sigma_1, \ldots, \sigma_r\}$.

If G is square and non-singular, then

$$G^{-1} = U\Sigma^{-1} Y^H \tag{3.11}$$

from which

$$\| G^{-1}(j\omega) \|_s = \frac{1}{\underline{\sigma}(\omega)} \tag{3.12}$$

Hence we have that

$$\frac{\| G^{-1}(j\omega) y(j\omega) \|}{\| y(j\omega) \|} \leqslant \frac{1}{\underline{\sigma}(\omega)}$$

and

$$\frac{\| G(j\omega) u(j\omega) \|}{\| u(j\omega) \|} \leqslant \bar{\sigma}(\omega)$$

Putting $y(j\omega) = G(j\omega) u(j\omega)$, we obtain

$$\underline{\sigma}(\omega) \leqslant \frac{\| G(j\omega) u(j\omega) \|}{\| u(j\omega) \|} \leqslant \bar{\sigma}(\omega) \tag{3.13}$$

which shows that the gain of a multivariable system is sandwiched between the smallest and largest principal gains.

With each principal gain we can associate a pair of **principal directions,** as follows. Assume that $m \geqslant l$, and let the rows of Y^H be $y_1^H, y_2^H, \ldots, y_l^H$, and the columns of U be $u_1, u_2, \ldots, u_l$. Then by using (3.4) we can write

$$y = Gu \tag{3.14}$$

$$= \sum_{k=1}^{l} \sigma_k y_k u_k^H u \tag{3.15}$$

Since $\|u_k\| = 1$, we have

$$|u_k^H u| \leqslant \|u_k\| \, \|u\| = \|u\| \tag{3.16}$$

the equality holding only if $u = \alpha u_k$ for some scalar α. Suppose, then, that $u = \alpha u_i$, with $|\alpha| = 1$, so that $\|u\| = 1$. Then $u_k^H u = 0$ for $k \neq i$, and hence the resulting output is

$$y = \sigma_i y_i \alpha \tag{3.17}$$

so that

$$\|y\| = \sigma_i \tag{3.18}$$

This shows that the gain of the system is precisely σ_i if the input signal is in the direction of u_i. The set $\{u_1, u_2, \ldots, u_l\}$ is called the set of **input principal directions** of G. In particular, the greatest possible gain $\bar{\sigma} = \sigma_1$ occurs if the input signal is in the direction of u_1, and the smallest possible gain $\underline{\sigma} = \sigma_l$ occurs if it is in the direction of u_l. Note that the principal directions are orthogonal to each other (that is, $u_i^H u_j = 0$ if $i \neq j$), since $U^H U = I$.

If the input vector is in the direction u_i, then (3.17) shows that the output vector is in the direction y_i. The set $\{y_1, y_2, \ldots, y_l\}$ is called the set of **output principal directions**. Again, these are orthogonal to each other.

A useful characteristic of a system is its **condition number**, which is defined as

$$\text{cond}(G) = \frac{\bar{\sigma}(G)}{\underline{\sigma}(G)} \tag{3.19}$$

and which depends, of course, on frequency. In numerical analysis the condition number measures the difficulty of inverting a matrix (see Section 8.2.2). It has been argued that it has a control-theoretic significance, in that it measures the inherent difficulty of controlling a given plant. We shall examine this point later.

3.3   The use of principal gains for assessing performance

From Chapter 1, we have, for the system shown in Figure 3.1, the following fundamental relations:

$$y(s) = S(s)d(s) + [I - S(s)]P(s)r(s) - [I - S(s)]m(s) \qquad (3.20)$$

where

$$S(s) = [I + G(s)K(s)]^{-1} \qquad (3.21)$$

is the sensitivity, or alternatively

$$y(s) = [I - T(s)]d(s) + T(s)P(s)r(s) - T(s)m(s) \qquad (3.22)$$

where

$$T(s) = S(s)G(s)K(s) \qquad (3.23)$$

is the closed-loop transfer function.

Obviously, what we should do to assess the disturbance-rejection (sensitivity) properties of the loop is to examine the principal gains of S (see Figure 3.2). If the region lying between $\bar{\sigma}(S)$ and $\underline{\sigma}(S)$ is very narrow, then we are almost back with the SISO single gain, and we can describe the loop's sensitivity properties very accurately.

Usually the region will *not* be narrow – in this case, since we are always interested in keeping sensitivity as small as possible, it is the upper boundary of the region (that is, $\bar{\sigma}$) which will be important.

To assess the propagation of measurement noise we look at the principal gains of T (see Figure 3.2). Again, the upper boundary is important, since we wish to minimize the propagation of measurement noise. Note that

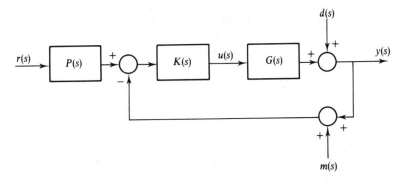

Figure 3.1 Standard feedback configuration.

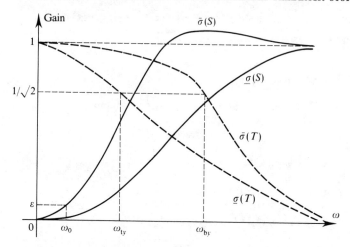

Figure 3.2   Smallest and largest principal gains of multivariable sensitivity function S (solid curves) and complementary sensitivity function T (dashed curves).

there are now two *loop* bandwidths: ω_{by}, the bandwidth at the output, and ω_{bu}, the bandwidth at the input.

As with SISO feedback, there is a conflict between keeping $\bar{\sigma}(S)$ small and keeping $\bar{\sigma}(T)$ small, because of:

Theorem 3.1:

$$|1-\bar{\sigma}(S)| \leqslant \bar{\sigma}(T) \leqslant 1+\bar{\sigma}(S) \tag{3.24}$$

and

$$|1-\bar{\sigma}(T)| \leqslant \bar{\sigma}(S) \leqslant 1+\bar{\sigma}(T) \tag{3.25}$$

Proof: $\bar{\sigma}$ is a matrix norm, which satisfies the triangle inequality ($\|A+B\| \leqslant \|A\| + \|B\|$). Therefore

$$\bar{\sigma}(T)=\bar{\sigma}(I-S)=\bar{\sigma}(I+(-S))$$

$$\leqslant 1+\bar{\sigma}(S) \tag{3.26}$$

Now,

$$\bar{\sigma}(S)=\sup_{x \neq 0} \frac{\|Sx\|}{\|x\|}$$

Let w be a vector for which this supremum is achieved – that is,

$$\frac{\|Sw\|}{\|w\|}=\bar{\sigma}(S)$$

Then

$$\| Tw \| = \|(I-S)w\| = \| w - Sw \|$$

Now,

$$\| w - Sw \| \geqslant \| w \| - \| Sw \| \quad \text{(from triangle inequality)}$$

and

$$\| w - Sw \| \geqslant \| Sw \| - \| w \| \quad \text{(from triangle inequality)}$$

Hence

$$\| Tw \| \geqslant | \, \| w \| - \| Sw \| \, |$$

So

$$\frac{\| Tw \|}{\| w \|} \geqslant \left| 1 - \frac{\| Sw \|}{\| w \|} \right|$$

$$= |1 - \bar{\sigma}(S)|$$

Therefore

$$\bar{\sigma}(T) \geqslant |1 - \bar{\sigma}(S)| \tag{3.27}$$

Relations (3.26) and (3.27) together prove (3.24).
Clearly the results for $\bar{\sigma}(S)$ in terms of $\bar{\sigma}(T)$ can be proved in the same way. ∎

If we have only one degree of design freedom (i.e if $P(s)=I$), then the reference-signal transmission properties are again determined by the principal gains of $T(s)$. If we define the transmission bandwidth ω_{ty} to be that frequency at which $\underline{\sigma}(T(j\omega_{ty})) = \underline{\sigma}(T(0))/\sqrt{2}$, then it is clear that *even when there is just one degree of freedom*,

$$\omega_{ty} < \omega_{by} \tag{3.28}$$

except in very special circumstances (see Figure 3.2). In the frequency range

$$\omega_{ty} \leqslant \omega \leqslant \omega_{by} \tag{3.29}$$

sensor noise is being propagated without reference signals being followed particularly well, so one of the design aims could be to

$$\text{minimize } (\omega_{by} - \omega_{ty}) \tag{3.30}$$

(Of course, the exact definitions of these frequencies are rather arbitrary, and may be changed to suit particular problems, but the principle will always hold.) Note that this design aim appears only for multivariable systems, since $\omega_{by} = \omega_{ty}$ for SISO loops (and for one degree of freedom).

If $P(s) \neq I$, so that we have a two degrees of freedom design, then the transmission of reference signals is determined by $T(s)P(s)$, and therefore the relevant principal gain is $\underline{\sigma}(TP)$. Now ω_{ty} is defined by $\underline{\sigma}(TP)$ rather than $\underline{\sigma}(T)$, and we can therefore have a more ambitious design aim:

$$\text{maximize } (\omega_{ty} - \omega_{by}) \tag{3.31}$$

the point being that $\omega_{ty} > \omega_{by}$ is now possible.

Another fundamental relation for the feedback loop is

$$u(s) = F_i^{-1}(s) K(s) P(s) r(s) - F_i^{-1}(s) K(s)[m(s) + d(s)] \tag{3.32}$$

where

$$F_i(s) = I + K(s)G(s) \tag{3.33}$$

If we are concerned to keep the control signals small, then clearly we must keep $\bar{\sigma}(F_i^{-1}K)$ small. Since

$$\bar{\sigma}(F_i^{-1}K) \leqslant \bar{\sigma}(F_i^{-1})\bar{\sigma}(K)$$

$$= \frac{\bar{\sigma}(K)}{\underline{\sigma}(F_i)}$$

then to ensure small $\bar{\sigma}(F_i^{-1}K)$, we need to make $\underline{\sigma}(F_i) \gg \bar{\sigma}(K)$. But

$$\underline{\sigma}(F_i) = \underline{\sigma}(I + KG) \leqslant \bar{\sigma}(I + KG)$$

$$\leqslant 1 + \bar{\sigma}(KG)$$

$$\leqslant 1 + \bar{\sigma}(K)\bar{\sigma}(G) \tag{3.34}$$

So if $\underline{\sigma}(F_i) \gg \bar{\sigma}(K)$, then $1 + \bar{\sigma}(K)\bar{\sigma}(G) \gg \bar{\sigma}(K)$, which implies that

$$\frac{1}{\bar{\sigma}(K)} + \bar{\sigma}(G) \gg 1 \tag{3.35}$$

This is only possible if $\bar{\sigma}(G) \gg 1$ or $\bar{\sigma}(K) \ll 1$. So, at frequencies at which $\bar{\sigma}(G)$ is not much larger than 1, control signals can be minimized only by keeping $\bar{\sigma}(K)$ as small as possible.

If we use the largest or smallest principal gains to assess performance, we are in effect assessing the worst-case performance, in the sense that we are examining the effects of least favourable signal directions. But in many multivariable systems we know in advance that certain signal directions can never occur, or are so unlikely that they need not be considered. In such cases the assessment may be unnecessarily pessimistic if the least favourable signal directions are among those which cannot occur in practice.

Consider, for example, simultaneous control of the yaw and roll angles of an aircraft. If we take the positive directions of yaw and roll to correspond to yawing of the nose to starboard, and a clockwise roll when looking towards the nose, then output disturbances – namely disturbances of yaw and roll – of the same sign are relatively easy to correct: the control-surface movements required to correct for each of these disturbances are consistent with each other. Disturbances having opposite signs, however, are much harder to correct: correction of yaw requires one set of rudder and aileron deflections, while correction of roll requires approximately the opposite set of control-surface deflections. The largest principal gain of the sensitivity function S will therefore correspond to disturbances of this kind. If we knew, however, that such disturbances could never occur, then it would be clearly inappropriate, or at least unnecessarily conservative, to use this principal gain as an indicator of performance.

We see, therefore, the need to modify the use of principal gains if the possible signal directions are restricted. If the signals are restricted to a subspace of the signal space then the modification is rather obvious. Suppose, for example, that the disturbance is confined to a p-dimensional subspace. Then we can write it as

$$d = Q\delta \tag{3.36}$$

where δ is a p-dimensional vector, and Q a matrix whose columns span the required subspace. If the columns of Q are chosen to be orthonormal,[2] so that $Q^H Q = I$, then $\|d\| = \|\delta\|$. Now let $r = 0$ and $m = 0$ in (3.20), so that

$$y = SQ\delta \tag{3.37}$$

We obtain an accurate measure of performance in the face of realistic disturbances by examining $\bar{\sigma}(SQ)$, the largest principal gain of the **modified sensitivity** SQ. Once we use such modified sensitivities, however, results such as Theorem 3.1 cease to apply.

Returning to the example of control of an aircraft, suppose that a typical disturbance results in a change in yaw of α units and a change in roll of β units. This can be represented by taking

$$Q = \frac{1}{\sqrt{(\alpha^2 + \beta^2)}} \begin{bmatrix} \alpha \\ \beta \end{bmatrix} \tag{3.38}$$

and δ is just a scalar in this case. Here Q is written as a constant, but it can just as well be a transfer function. In the present example we may prefer to take

$$Q(s) = \frac{1}{s\sqrt{(\alpha^2 + \beta^2)}} \begin{bmatrix} \alpha \\ \beta \end{bmatrix} \tag{3.39}$$

for instance.

Perhaps a more realistic representation of possible disturbances is to confine the disturbance signal to a region which is not a subspace, such as

$$d = \alpha d_1 + \beta d_2, \quad \alpha\beta > 0 \tag{3.40}$$

(where α and β are real scalars). In this case d is confined to a convex cone, and it is appropriate to find the largest value of $\|Sd\|/\|d\|$ in this cone. This value will occur either when the signal direction coincides with one of the system's input principal directions – if such a direction lies within the cone, or somewhere on the boundary of the cone.

A third possible representation of realistic disturbances is obtained by using equation (3.36), but with Q a square matrix, and δ of the same dimension as d, so that the disturbance is not confined to a subspace. Now, however, the columns of Q should have unequal norms, so that some directions are emphasized more than others. The choice of weightings is usually rather subjective, unless some specific information, such as the probability distribution of possible disturbances, is available. For example, if it is known that the disturbances have a multivariable Gaussian distribution with covariance V, then Q can be chosen as a square root of V:

$$Q = V^{\frac{1}{2}} \tag{3.41}$$

In such a case, however, when a description of the disturbances as a stochastic process is available, one can obtain a statistical measure of performance by examining all the principal gains of SQ rather than just the greatest one, as described in Section 3.7.

As an example of this method of representing disturbances, and referring back to the example of yaw and roll control, suppose that disturbances in the direction $d_1 = [3, 1]^T$ are considered to be most typical, while those in the orthogonal direction $d_2 = [1, -3]^T$ are least typical, and no direction is to be totally excluded. Then we take

$$Q = \begin{bmatrix} 3\alpha & 1 \\ \alpha & -3 \end{bmatrix}, \quad \alpha > 1 \tag{3.42}$$

probably making a rather arbitrary choice for α, such as $\alpha = 10$. This choice of Q would also be appropriate if we knew that the disturbance was a Gaussian random variable with covariance

$$V = QQ^T \tag{3.43}$$

$$= \begin{bmatrix} 9\alpha^2 + 1 & 3\alpha^2 - 3 \\ 3\alpha^2 - 3 & \alpha^2 + 9 \end{bmatrix} \tag{3.44}$$

3.4 Relations between closed-loop and open-loop principal gains

For SISO feedback, closed-loop performance specifications can be easily converted into requirements on the open-loop gain (that is, the modulus of the return ratio). We now seek to express closed-loop requirements in terms of open-loop principal gains, for multivariable systems.

First, we summarize our closed-loop requirements:

(1)   Sensitivity: keep $\bar{\sigma}[(I+GK)^{-1}]$ as small as possible.

(2)   Noise propagation: keep $\bar{\sigma}[I-(I+GK)^{-1}]$ as small as possible. (This conflicts with 1.)

(3)   Tracking of reference signal (for one degree of freedom): keep $\underline{\sigma}[I-(I+GK)^{-1}] \approx 1$ and $\bar{\sigma}[I-(I+GK)^{-1}] \approx 1$. (This conflicts with 2, but not with 1.)

(4)   Minimization of control energy: keep $\bar{\sigma}(K)$ as small as possible. (We expect this to conflict with 1 and 3.)

We shall need the following lemma:

Lemma 3.1:

$$\max(0, \bar{\sigma}(Q)-1) \leqslant \bar{\sigma}(Q+I) \leqslant \bar{\sigma}(Q)+1 \tag{3.45}$$

Proof: The right-hand-side (RHS) inequality follows immediately from the triangle inequality since $\bar{\sigma}(.)$ is a norm. To prove the left-hand-side (LHS) inequality, consider the identity

$$Q = I + Q - I$$

Therefore

$$\bar{\sigma}(Q) \leqslant \bar{\sigma}(I+Q) + 1 \quad \text{(triangle inequality)}$$

But

$$\bar{\sigma}(Q+I) \geqslant 0$$

so

$$\bar{\sigma}(Q+I) \geqslant \max(0, \bar{\sigma}(Q)-1) \qquad \blacksquare$$

We shall also need:

Lemma 3.2:

$$\max(0, \underline{\sigma}(Q)-1) \leqslant \underline{\sigma}(Q+I) \leqslant \underline{\sigma}(Q)+1 \qquad \textbf{(3.46)}$$

Proof: LHS inequality: Let w be a vector such that

$$\frac{\|(Q+I)w\|}{\|w\|} = \underline{\sigma}(Q+I)$$

Then

$$\|(Q+I)w\| = \|Qw+w\| \geqslant \|Qw\| - \|w\| \quad \text{(triangle inequality)}$$

Therefore

$$\|Qw\| - \|w\| \leqslant \underline{\sigma}(Q+I)\|w\|$$

Hence

$$\underline{\sigma}(Q)-1 \leqslant \frac{\|Qw\|}{\|w\|} - 1 \leqslant \underline{\sigma}(Q+I)$$

and since $\underline{\sigma}(Q+I) \geqslant 0$, the LHS inequality is obtained.

RHS inequality: Let v be a vector such that $\dfrac{\|Qv\|}{\|v\|} = \underline{\sigma}(Q)$. Then

$$\underline{\sigma}(I+Q) \leqslant \frac{\|(I+Q)v\|}{\|v\|} = \frac{\|Qv+v\|}{\|v\|}$$

$$\leqslant \frac{\|Qv\|}{\|v\|} + 1 \quad \text{(triangle inequality)}$$

$$= \underline{\sigma}(Q)+1 \qquad \blacksquare$$

We can now restate our requirements in terms of open-loop principal gains:

(1) *Sensitivity*

$$\bar{\sigma}[(I+GK)^{-1}] = \frac{1}{\underline{\sigma}(I+GK)}$$

$$\leqslant \frac{1}{\underline{\sigma}(GK)-1} \quad \text{(by Lemma 3.2, if } \underline{\sigma}(GK) > 1)$$

$$\approx \frac{1}{\underline{\sigma}(GK)} \quad \text{(if } \underline{\sigma}(GK) \gg 1)$$

that is, low sensitivity is ensured by making $\underline{\sigma}(GK)$ large. Note how similar the required calculation is to that for SISO feedback: for example,

$$\text{required sensitivity} \leqslant 0.01 \Rightarrow \underline{\sigma}(GK) \geqslant 100 \quad \text{(approximately)}$$

(2) *Noise propagation*

$$\bar{\sigma}[I-(I+GK)^{-1}] = \bar{\sigma}[(I+(GK)^{-1})^{-1}] \quad \text{(matrix inversion lemma)}^3$$

$$\text{(3.47)}$$

$$= \frac{1}{\underline{\sigma}(I+(GK)^{-1})}$$

So we want $\underline{\sigma}(I+(GK)^{-1})$ as large as possible and hence, by Lemma 3.2, we want $\underline{\sigma}(GK)^{-1}$ as large as possible or $\bar{\sigma}(GK)$ as small as possible.

Again, note that if $\bar{\sigma}(GK) \ll 1$ then $\bar{\sigma}[I-(I+GK)^{-1}] \approx \bar{\sigma}(GK)$ (by Lemma 3.2), so that conversion from closed-loop to open-loop specifications is very easy.

(3) *Tracking of reference signal*

(For one degree of freedom.) We require that

$$I-(I+GK)^{-1} \approx I \quad \text{or} \quad (I+GK)^{-1} \approx 0$$

More precisely, we require that

$$\frac{\|(I+GK)^{-1}x\|}{\|x\|} \ll 1 \quad \text{for any } x$$

so

$$\bar{\sigma}(I+GK)^{-1} \ll 1$$

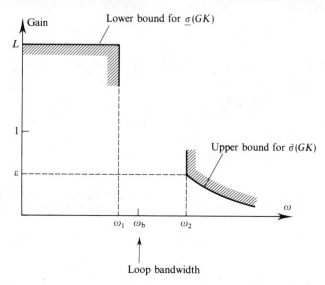

Figure 3.3 Typical specification for a multivariable return ratio.

or

$$\underline{\sigma}(I + GK) \gg 1$$

or

$$\underline{\sigma}(GK) \gg 1 \quad \text{(by Lemma 3.2)}$$

Conflicts

Summary of open-loop requirements:

(1) $\underline{\sigma}(GK)$ large
(2) $\bar{\sigma}(GK)$ small
(3) $\underline{\sigma}(GK)$ large
(4) $\bar{\sigma}(K)$ small

Since $\bar{\sigma}(GK) \geqslant \underline{\sigma}(GK)$ there are obviously conflicts between (1 and 3) and 2. Furthermore, since $\bar{\sigma}(GK) \leqslant \bar{\sigma}(G)\bar{\sigma}(K)$, there are also conflicts between (1 and 3) and 4.

We can see from this that the conflicts that arise are exactly the same as those that arise in the design of SISO systems, and we try to resolve them in the same way – by making $\underline{\sigma}(GK)$ large at low frequencies, and $\bar{\sigma}(GK)$ small at high frequencies.

A typical specification for the principal gains of the return ratio GK is shown in Figure 3.3.

(4) *Control energy*

We know that the control energy required is determined by the matrix

$$F_i^{-1}K = (I + KG)^{-1}K$$

To assess the 'gain' of this matrix from open-loop characteristics we can use the relation

$$\underline{\sigma}(F_i^{-1}K) \geqslant \underline{\sigma}(F_i^{-1})\underline{\sigma}(K)$$

$$= \frac{\underline{\sigma}(K)}{\bar{\sigma}(F_i)} = \frac{\underline{\sigma}(K)}{\bar{\sigma}(I + KG)}$$

$$\approx \frac{\underline{\sigma}(K)}{\bar{\sigma}(KG)} \quad (\text{if } \bar{\sigma}(KG) \gg 1)$$

So at low frequencies we can use the open-loop characteristics, but note that for this we need the principal gains of KG, not those of GK. In some circumstances these can be very different.

3.5 Principal gains and characteristic loci

In Chapter 4 we shall present a design method which is based on manipulating characteristic loci. In this section we show how the characteristic loci can be related to the principal gains. Without such a relation we would not know how to manipulate the characteristic loci, which in some circumstances can be very misleading indicators of performance and robustness.

Definition 3.1: Let $Q(s)$ be a square matrix. Then $Q(s)$ is **normal** if

$$Q^H(s)Q(s) = Q(s)Q^H(s)$$

Lemma 3.3: Suppose that $Q(s)$ is normal and

$$Q(s) = W(s)\Lambda(s)W^{-1}(s)$$

where $\Lambda(s) = \text{diag}\{\lambda_i(s)\}$ and each λ_i is an eigenvalue function of $Q(s)$. Then

$$W^H(s) = W^{-1}(s)$$

(that is, the eigenvectors of a normal matrix form an orthonormal set).

Theorem 3.2:   Let $Q(s)$ be normal, with eigenvalues $\{\lambda_i(s)\}$ and principal gains $\{\sigma_i(s)\}$. Then we can order the λ_i so that

$$\sigma_i(s) = |\lambda_i(s)|, \quad i = 1, 2, \ldots, m \tag{3.48}$$

Proof: Let $\Lambda(s) = \mathrm{diag}\{\lambda_i(s)\}$ and $\Sigma(s) = \mathrm{diag}\{\sigma_i(s)\}$. Then

$$Q(s) = W(s)\Lambda(s)W^{-1}(s)$$

$$= W(s)\Lambda(s)W^H(s) \quad \text{(by Lemma 3.3)}$$

$$= \sum_{i=1}^{m} \lambda_i(s)w_i(s)w_i^H(s)$$

Now let

$$\lambda_i(s) = \rho_i(s)e^{j\theta_i(s)}, \quad \rho_i \in R, \ \rho_i \geqslant 0$$

and

$$y_i(s) = w_i(s)e^{j\theta_i(s)}$$

Then

$$Q(s) = \sum_{i=1}^{m} \rho_i(s)y_i(s)w_i^H(s)$$

$$= Y(s)\,\mathrm{diag}\{\rho_i(s)\}\,W^H(s)$$

·where

$$Y(s) = W(s)\,\mathrm{diag}\{e^{j\theta_i(s)}\}$$

Hence

$$Y^H(s)\,Y(s) = \mathrm{diag}\{e^{-j\theta_i(s)}\}\,W^H(s)W(s)\,\mathrm{diag}\{e^{j\theta_i(s)}\} = I$$
$$\text{(by Lemma 3.3)}$$

and similarly

$$Y(s)\,Y^H(s) = I$$

Hence $Q(s) = Y(s)\,\mathrm{diag}\{\rho_i(s)\}\,W^H(s)$ is a singular-value decomposition of $Q(s)$. But the singular values of $Q(s)$ are unique, so

$$\mathrm{diag}\{\rho_i(s)\} = \Sigma(s) \quad \text{(up to ordering)}$$

that is,

$$\rho_i(s) = \sigma_i(s)$$

∎

This theorem shows that we can assess closed-loop performance from the moduli of the characteristic loci, or **characteristic gains**, if the return ratio is normal.

This result is of more importance to design than to analysis. We can assume that the designer can easily obtain displays of the principal gains of any proposed design. One point of the theorem is that if he wants to achieve particular *principal* gains, one way of doing it is to design K such that GK is normal and has particular *characteristic* gains.

Another reason for normal return ratios being desirable is that the sensitivity of the eigenvalues of a matrix to perturbations of its elements is minimized if the matrix is normal (Wilkinson, 1965). Knowing that one's model of the plant is bound to be inaccurate, one hopes that the use of a controller which makes the return ratio normal also reduces the likelihood of the closed-loop being unstable when the compensator is used with the real plant, since stability is determined by the characteristic loci. This argument should be applied carefully, because it is possible for relatively small changes in the plant to produce relatively large changes in the return ratio, and consequently large departures from normality. However, this seems to occur only for plant with a high condition number, and it can be argued that such plants are in any case inherently difficult to control.

Even when the return ratio is not exactly normal, the characteristic gains approximately equal the principal gains as long as its eigenvectors are nearly orthogonal. When this is not the case, the return ratio is said to be **skew**.

The following measure of the skewness of a square matrix G has been proposed by Hung and MacFarlane (1982). First, a so-called **Schur decomposition** of G is obtained:

$$G = S(D+T)S^H \tag{3.49}$$

where D is diagonal, T is strictly triangular and S is unitary (see Section 8.2 for details of how to compute this). Then a suitable measure of skewness is

$$\mathrm{ms}\,(G) = \frac{\|T\|_F}{\|G\|_F} \tag{3.50}$$

where $\|.\|_F$ denotes the Frobenius norm

$$\|X\|_F = \sqrt{\{\mathrm{tr}\,(X^H X)\}}$$

and $\mathrm{tr}(.)$ denotes the **trace** of a matrix. This measure has the property

$$0 \leqslant \mathrm{ms}\,(G) \leqslant 1 \tag{3.51}$$

with 0 denoting complete normality and 1 denoting complete skewness. Hung and MacFarlane (1982) show that $\mathrm{ms}\,(G)$ can be related to the sensitivity of the characteristic loci of G, while Pang and MacFarlane (1986) show that it can be related to the divergence between the principal and characteristic gains as follows (with the same notation as in the proof of Theorem 3.2):

$$\text{ms}\,(G)=\sqrt{\left\{\sum_{i=1}^{m}\sigma_i^2\left(1-\frac{\rho_i^2}{\sigma_i^2}\right)\Big/\sum_{i=1}^{m}\sigma_i^2\right\}} \tag{3.52}$$

When the skewness of a matrix is appreciable, we can assert only that

$$\underline{\sigma}\leqslant|\lambda_i|\leqslant\bar{\sigma} \qquad (i=1,2,\ldots,m)$$

namely that each characteristic gain has a value lying between the smallest and largest principal gains. (This follows from (3.13) by letting u be an eigenvector.)

3.6 Limitations on performance

We have seen that the performance of SISO feedback loops is limited by the Bode gain–phase relationships: an ideal gain characteristic would be as shown in Figure 3.4. Such a gain characteristic is not realizable, and any close approximation would be accompanied by an excessive phase lag at the cross-over frequency, and so would inevitably lead to an unstable loop.

We have also seen that, if the plant to be controlled is non-minimum-phase (that is, if it has one or more zeros in the right half-plane), then the achievable gain–bandwidth product is limited and, in practice, the achievable bandwidth is usually limited.

To what extent do these limitations apply to multivariable systems?

3.6.1 Gain–phase relationships

Suppose that we have a performance specification which translates into open-loop principal-gain specifications as shown in Figure 3.3. For $\omega\leqslant\omega_1$,

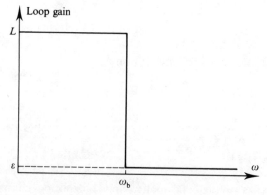

Figure 3.4 Ideal, but unrealizable, open-loop gain characteristic.

$\underline{\sigma}(GK) \geqslant L$, and usually $L > 1$. Now,

$$\underline{\sigma}(GK) \leqslant |\lambda_i(GK)|$$

where $\{\lambda_i(GK)\}$ is the set of eigenvalues of GK. So,

$$|\det(GK)| = \prod_{i=1}^{m} |\lambda_i(GK)| \geqslant L^m \tag{3.53}$$

Similarly, for $\omega \geqslant \omega_2$,

$$|\det(GK)| \leqslant \varepsilon^m \tag{3.54}$$

Now, assuming that $\det(G(s)K(s))$ is an analytic function, we can apply Bode's gain–phase relationship (1.36) to it: suppose that $\det(GK)$ is a minimum-phase function, and that $\det(G(0)K(0)) > 0$. Then, if $\log|\det(GK)|$ decreases linearly with $\log\omega$ ($\omega_1 \leqslant \omega \leqslant \omega_2$),

$$\arg\det(GK) \leqslant -\frac{\pi}{2}\frac{m\log(L/\varepsilon)}{\log(\omega_2/\omega_1)} \quad \text{(approximately)} \tag{3.55}$$

over most of this frequency range. If $\det(GK)$ is non-minimum-phase then the phase of $\det(GK)$ will be more negative than this (by the usual argument). Now,

$$\arg\det(GK) = \sum_{i=1}^{m} \arg\lambda_i(GK)$$

so

$$\sum_{i=1}^{m} \arg\lambda_i(GK) \leqslant -\frac{\pi}{2}\frac{m\log(L/\varepsilon)}{\log(\omega_2/\omega_1)} \tag{3.56}$$

Closed-loop stability is determined by the number of encirclements of the point $-1 + j0$ by the λ_i. Suppose that GK is open-loop stable, so that for stability we require *no* encirclements. To see how large L/ε can be while maintaining stability, let us make the most favourable assumption about the phases of the characteristic loci – that the greatest phase lag of any characteristic locus should be as small as possible (for each ω). Obviously this happens if all the loci have the same phase – that is, if

$$\arg\lambda_i(GK) \leqslant -\frac{\pi}{2}\frac{\log(L/\varepsilon)}{\log(\omega_2/\omega_1)} \tag{3.57}$$

for each i.

If we try to reduce the lag of any one locus, then the lag of some other characteristic locus is bound to increase, so that the total phase lag of $\det(GK)$ is maintained.

The relation (3.57) between the most favourable values of $\{\arg \lambda_i\}$, L/ε and ω_2/ω_1 is the same as the relation between the phase and gain of SISO return ratios. We can therefore deduce the same performance limitations as for SISO loops: if

$$\frac{20 \log_{10}(L/\varepsilon)}{\log_{10}(\omega_2/\omega_1)} > 40 \text{ dB/decade}$$

and $\log_{10}(\omega_2/\omega_1)$ is quite small (<2, say), then closed-loop stability will be difficult to achieve, and may be impossible. Whatever the value of $\log_{10}(\omega_2/\omega_1)$, unconditional stability will not be possible.

As for SISO loops, things are worse for open-loop unstable and/or non-minimum-phase systems.

3.6.2   Right half-plane zeros and poles

The question arises of whether individual loci obey the Bode gain–phase relationships. This has been investigated by Smith (1982), who found that if the eigenvalue function of the return ratio has no **branch points** in the right half-plane, then the individual loci *do* obey the gain–phase relationships. (A necessary, but not sufficient, condition for $s = s_0$ to be a branch point is that $\lambda_i(G(s_0)K(s_0)) = \lambda_j(G(s_0)K(s_0))$ for some pair (i, j).) What happens if there *are* right half-plane branch points is still not understood, but violations of the gain–phase relationships are not observed in practice. (And we know, of course, from Section 3.6.1 that it is impossible for all the phases of the characteristic loci simultaneously to be more positive than they would be if they obeyed the gain–phase relationships.)

Let us assume that the individual loci obey the gain–phase relationships, and that GK has a zero in the right half-plane (and that $\det(GK)$ has too). Then at least one of the eigenvalue functions has a zero in the right half-plane, since $\det(GK) = \Pi_i \lambda_i$, and will therefore be subject to the gain–bandwidth limitation which is imposed on SISO loops by right half-plane zeros.

But $\underline{\sigma}(GK) \leqslant |\lambda_i(GK)|$, so the same limitation is imposed on the open-loop gain–bandwidth product of the multivariable loop, if this is measured by the smallest principal gain $\underline{\sigma}(GK)$.

If GK has a pole in the right half-plane then it is *probably* true that at least one characteristic gain must exceed unity up to some minimum frequency, which is determined by the pole location in the same way as for SISO loops. If this is true then $\bar{\sigma}(GK)$ must also exceed unity below this frequency, and hence the loop bandwidth cannot be much lower than this frequency (by

Lemma 3.1). This has not been proved, however, and the argument cannot parallel the SISO argument completely because the eigenvalue functions of *GK* are almost always irrational.

3.7 Transmission of stochastic signals

In Section 3.3 we derived qualitative closed-loop requirements for principal gains. In Section 3.4 we found how to convert both qualitative and quantitative closed-loop specifications into open-loop specifications. We now state some results which can be used to determine closed-loop specifications quantitatively, if there are stochastic signals in the loop.

We shall assume that all signals are zero-mean, stationary stochastic processes. We shall use $\{x(t)\}$ to denote such a process, and $x(t)$ to denote a single sample function from such a process. $E\{x\}$ will denote the mean value of the process (which will always be zero, by assumption), $E\{xx^T\}$ will denote its covariance, and so on; $\Phi_{xx}(\omega)$ will denote its power spectral density, namely the Fourier transform of its autocovariance function $\phi_{xx}(\tau) = E\{x(t)x^T(t+\tau)\}$.

The performance of feedback systems which contain stochastic signals is often best measured in terms of the variances of the various signals. Typically, the variance of the error between the reference input and the output, and the variance of the plant input signal, are of interest. Since we are dealing with vectors of signals, it is mathematically convenient to assess performance in terms of the sum of the variances of the component signals; so if $\{x(t)\}$ is a vector-valued process, a measure of its 'signal strength' (more accurately, its power) is $\text{tr}(E\{xx^T\})$. Using the fact that $\text{tr}(AB) = \text{tr}(BA)$, and the properties of the expectation operator $E\{.\}$, we obtain the result that

$$\text{tr}(E\{xx^T\}) = E\{x^Tx\} \tag{3.58}$$

For SISO systems, if $y(s) = G(s)u(s)$ then the power spectral densities of y and u are related by (Papoulis, 1984; Priestley, 1981)

$$\Phi_{yy}(\omega) = |G(j\omega)|^2 \Phi_{uu}(\omega) \tag{3.59}$$

Also, we have

$$E\{y^2\} = \frac{1}{2\pi} \int_{-\infty}^{\infty} \Phi_{yy}(\omega)\, d\omega \tag{3.60}$$

Thus, by knowing $\Phi_{uu}(\omega)$ and $|G(j\omega)|$ we can compute $E\{y^2\}$, and we can use graphical displays to identify any frequency band at which most of the contributions to $E\{y^2\}$ are concentrated.

These relations can be generalized so as to apply to multivariable systems, but there are two complications. The first is that there is not just a single gain – we expect to use instead a band of gains, bounded by the smallest and largest principal gains. The second complication is that if $y(t)$ is a vector, then a suitable index of performance will be, in general, $E\{y^T Wy\}$ rather than $E\{y^T y\}$, where W is a weighting matrix which is real, symmetric (usually diagonal) and positive-semidefinite.

It can be shown that

$$E\{y^T Wy\} = \frac{1}{2\pi} \int_{-\infty}^{\infty} \mathrm{tr}\,[\,W\Phi_{yy}(\omega)\,]\,d\omega \tag{3.61}$$

and, if $y(s) = G(s)u(s)$ and $G(s)$ is stable,

$$\Phi_{yy}(\omega) = G(j\omega)\Phi_{uu}(\omega)G^T(-j\omega) \tag{3.62}$$

The weighting matrix need cause no trouble, because

$$\mathrm{tr}\,[\,W\Phi_{yy}\,] = \mathrm{tr}\,[\,W^{1/2}\Phi_{yy}\,W^{1/2}\,] \tag{3.63}$$

$$= \mathrm{tr}\,[\,W^{1/2}G(j\omega)\Phi_{uu}(\omega)G^T(-j\omega)W^{1/2}\,]$$

so

$$E\{y^T Wy\} = E\{z^T z\} \tag{3.64}$$

where

$$z(s) = W^{1/2}G(s)u(s) \tag{3.65}$$

Thus we can confine ourselves to the case $W = I$.
The following results can be shown to hold (Postlethwaite *et al.*, 1981).

> ***Theorem 3.3:*** If $y(s) = G(s)u(s)$, $G(s)$ is stable, and the power spectral density of $\{u(t)\}$ is $\Phi_{uu}(\omega)$, then
>
> $$E\{y^T y\} = \frac{1}{2\pi} \int_{-\infty}^{\infty} \sum_i \sigma_i^2 (\Phi_{uu}^{1/2}(\omega)G(j\omega))\,d\omega \tag{3.66}$$

This result is in a useful form for calculating total variances obtained with a particular design. But for guidance in design we would like a result in terms of the principal gains of $G(s)$ alone. Such a result is:

> ***Theorem 3.4:*** If $y(s) = G(s)u(s)$ and $G(s)$ is stable, then
>
> $$\underline{\sigma}^2(G(j\omega)) \leqslant \frac{\mathrm{tr}\,(\Phi_{yy}(\omega))}{\mathrm{tr}\,(\Phi_{uu}(\omega))} \leqslant \bar{\sigma}^2(G(j\omega)) \tag{3.67}$$

3.8   The operator norms $\|G\|_2$ and $\|G\|_\infty$

So far we have considered the 'gain' of a transfer function only at individual frequencies. But it is possible, and useful, to have a cruder measure of the 'gain' – a single number associated with the transfer function. We shall define two such measures, or norms, on G, and denote them by $\|G\|_2$ and $\|G\|_\infty$, respectively.

Definition 3.2:   Let $G(s)$ be a proper transfer function with no poles on the imaginary axis. Then

$$\|G\|_2 = \sqrt{\left\{ \frac{1}{2\pi} \int_{-\infty}^{\infty} \text{tr}[G(j\omega)G^{\text{T}}(-j\omega)]\,d\omega \right\}} \tag{3.68}$$

and

$$\|G\|_\infty = \sup_\omega \bar{\sigma}(G(j\omega)) \tag{3.69}$$

It can be shown that $\|G\|_2$ and $\|G\|_\infty$ both satisfy the usual properties of norms, namely:

$\|G\| \geq 0$,   with $\|G\| = 0$ if and only if $G = 0$

$\|\alpha G\| = |\alpha| \, \|G\|$,   for any scalar α

$\|G + H\| \leq \|G\| + \|H\|$

$\|G\|_\infty$ also satisfies

$\|GH\|_\infty \leq \|G\|_\infty \|H\|_\infty$

but this inequality is *not* satisfied by $\|G\|_2$. These norms are often referred to as **operator norms**, as the system represented by a transfer function is an operator which maps functions – input signals – into other functions – output signals, and these norms measure the amplification (or at least the greatest possible amplification) of this mapping. We can be more specific about this: from equations (3.61) and (3.62) we see that, if $\Phi_{uu}(\omega) = I$, $y(s) = G(s)u(s)$ and $G(s)$ is stable, then

$$E\{y^{\text{T}}y\} = \|G\|_2^2 \tag{3.70}$$

so that $\|G\|_2$ gives us precise information about the power gain of G when the input is a white stochastic process.

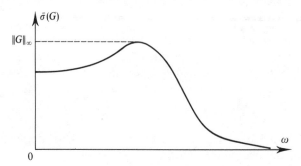

Figure 3.5   Reading off $\|G\|_\infty$ from plot of largest principal gain.

To see the meaning of $\|G\|_\infty$, suppose that the input signal $u(t)$ has finite energy, as measured by

$$\|u\|_2 = \sqrt{\left\{\int_{-\infty}^{\infty} u^{\mathrm{T}}(t)u(t)\,\mathrm{d}t\right\}} \tag{3.71}$$

(which is again a norm, this time on *signals* rather than systems), but that we have no other information about it. It is meaningful to ask what the greatest increase in energy is that can occur between the input and output for a given system. This is answered by:

Theorem 3.5   (Vidyasagar, 1985; Francis, 1987): If $\|u\|_2 < \infty$, $y(s) = G(s)u(s)$ and $G(s)$ is stable and proper and has no poles on the imaginary axis, then

$$\sup_{u} \frac{\|y\|_2}{\|u\|_2} = \|G\|_\infty \tag{3.72}$$

Proof: omitted.

Clearly both $\|G\|_2$ and $\|G\|_\infty$ can be related to the principal gains of G. For $\|G\|_2$ the relation is through the expression which appears in Theorem 3.3, whereas for $\|G\|_\infty$ the relation is very simple. Figure 3.5 shows a plot of $\bar{\sigma}(G(j\omega))$ for some transfer function G, and indicates how $\|G\|_\infty$ can be read off from the plot, since it is simply the peak value of $\bar{\sigma}(G)$. (An alternative way of finding $\|G\|_\infty$ from a state-space realization, which is more efficient if $\|G\|_\infty$ is the only quantity being computed, is given by Boyd *et al.* (1988).)

In applications these operator norms are applied to closed-loop transfer functions such as the sensitivity function $S(s)$ and its complement $T(s)$. They are generally used in connection with optimization problems; for example, the linear quadratic Gaussian (LQG) theory used in Chapter 5 is concerned with minimizing $\|Q\|_2$ for a suitably specified Q, while the problem of obtaining the best possible sensitivity function is concerned with minimizing $\|S\|_\infty$. When performing such optimizations it is of course essential that the transfer functions whose norms are being optimized remain stable, and this requirement gives rise to the following notation (which we shall not need here, but is becoming widespread in the literature). The set of all proper transfer functions G for which $\|G\|_\infty < \infty$ is denoted by L_∞ (because it is a **Lebesgue space**), while the set of all transfer functions which are members of L_∞, but which are also exponentially stable, is denoted by H_∞ (because it is a **Hardy space**). For a discussion of these spaces, and a more rigorous treatment of operator norms, see Vidyasagar (1985) or Francis (1987).

We point out that the 'peak M value' M_p which appears in classical control theory is, in the notation introduced here, the same as $\|T\|_\infty$, where T is the complementary sensitivity $I - S$.

3.9   The use of operator norms to specify performance

The kind of specifications we formulated earlier can be recast into bounds on $\|\,.\,\|_\infty$ by using suitable weighting functions. Suppose, for example, that we have the specification

$$\bar{\sigma}(S(j\omega)) \leqslant 0.2, \quad \omega \leqslant 1 \tag{3.73}$$

and

$$\bar{\sigma}(T(j\omega)) \leqslant 0.1, \quad \omega \geqslant 50 \tag{3.74}$$

We can refine this specification to that shown in Figure 3.6, in which the gain characteristics have been chosen to be those of simple transfer functions. The upper bound for the sensitivity can be approximated by the gain of the transfer function

$$w_1(s) = 0.141(1 + s) \tag{3.75}$$

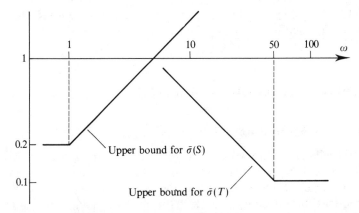

Figure 3.6   Specifications which satisfy equations (3.73) and (3.74).

(in which the steady-state gain is 0.141, rather than 0.2, to allow for the departure from the straight-line approximation at $\omega = 1$), and the upper bound for the closed-loop transfer function can be approximated by the gain of

$$w_2(s) = \frac{3.536(1 + 0.02s)}{s} \tag{3.76}$$

The original specification is then included in the new specification:

$$\| w_1^{-1} S \|_\infty \leqslant 1 \tag{3.77}$$

and

$$\| w_2^{-1} T \|_\infty \leqslant 1 \tag{3.78}$$

The choice of w_1 and w_2 is somewhat arbitrary, of course, and many other choices would also include the original specifications.

The purpose of stating specifications in this form is that a compensator can then be obtained automatically, by using the H_∞ optimization theory to be presented in Chapter 6. But problems arise which are similar to the problems discussed in Chapter 1 in connection with the idea of choosing S or T, and then deducing the required compensator. Here again we have the problem that non-essential details of the specification are being determined arbitrarily, and not by the plant's characteristics. There is also the problem of checking whether the specification is consistent. If we recall that

$$S + T = I \tag{3.79}$$

and that both S and T are determined by a return ratio which is constrained by the requirements of the generalized Nyquist stability theorem, then we realize that (3.77) and (3.78) may be inconsistent with each other.

The difficulties of using $\| . \|_\infty$ to formulate performance requirements have been explored in depth by Freudenberg and Looze (1986).

3.10   Representations of uncertainty

3.10.1   Unstructured uncertainty

Classical feedback-system design dealt with the problem of plant uncertainty by prescribing stability margins, by means of specified gain or phase margins or specified peak M values. Implicit in the use of such margins was a rather crude model of the uncertainty to which the plant was subject. If the only real concern was to maintain stability, and the prescribed phase margin was 40°, for example, then there was an implication that the plant model may be underestimating phase lag by as much as 40° at the cross-over frequency. Such a description of uncertainty is unstructured in the sense that it bounds the magnitude of possible perturbations, but does not trace the origins of the perturbations to specific elements of the plant.

In this section we are concerned with making explicit models of unstructured uncertainty. For multivariable plant the three most commonly used models are as follows: let $G_0(s)$ be a nominal transfer function, which is a best estimate, in some sense, of the true plant behaviour, and let $G(s)$ denote the true transfer function of the plant; then

$$G(s) = G_0(s) + \Delta_a(s) \tag{3.80}$$

$$G(s) = G_0(s)[I + \Delta_i(s)] \tag{3.81}$$

or

$$G(s) = [I + \Delta_o(s)]G_0(s) \tag{3.82}$$

where Δ_a represents an **additive perturbation**, Δ_i an **input multiplicative perturbation**, and Δ_o an **output multiplicative perturbation**.

The only restriction on the perturbations is on their 'size', which is measured by $\|\Delta\|_\infty$. If we want to make the size frequency-dependent then we can use $\bar{\sigma}(\Delta)$, or, equivalently, we can set $\Delta = W_1 \tilde{\Delta} W_2$, where W_1 and W_2 are minimum-phase transfer functions (no poles or zeros in the right half-plane) which serve as frequency-dependent weighting functions. In this case we can always take $\|\tilde{\Delta}\|_\infty \leqslant 1$.

The additive model (3.80) may be used to pose some robust stabiliz-ation problems which have nice solutions (see Glover, 1986), but the multi-plicative models (3.81) and (3.82) are often more realistic, since $\|\Delta_i\|_\infty$ and $\|\Delta_o\|_\infty$ represent relative rather than absolute magnitudes. For example, $\|\Delta_i\|_\infty \leqslant 0.1$ implies that the size of the perturbation is at most 10% of the 'size' of G_0, since

$$\|G - G_0\|_\infty = \|G_0 \Delta_i\|_\infty$$

$$\leqslant \|G_0\|_\infty \|\Delta_i\|_\infty$$

$$\leqslant 0.1 \|G_0\|_\infty \tag{3.83}$$

However, specifying $\|\Delta_a\|_\infty \leqslant 0.1$ implies that

$$\|G - G_0\|_\infty = \|\Delta_a\|_\infty$$

$$\leqslant 0.1 \tag{3.84}$$

We need both models (3.81) and (3.82) because multiplication of transfer-function matrices is non-commutative. In effect, (3.81) pretends that all the uncertainty occurs at the plant input, while (3.82) pretends that it all occurs at the output. Similarly, two weighting matrices (W_1 and W_2) rather than one ensure that the disturbance bound can always be normalized to

$$\| W_1^{-1} \Delta W_2^{-1} \|_\infty \leqslant 1 \tag{3.85}$$

EXAMPLE 3.1   (Skogestad and Morari, 1987)

A process plant operates with certain flow-rates in the region of 100 kmol min^{-1}. Typically, changes of up to 10 kmol min^{-1} are required. These changes are effected by servo-controlled valves which rely on measurement of the flow, but the flow measurement is in error by 1% (see Figure 3.7). So when a change of flow-rate from 100 to 110 kmol min^{-1} is required, an actual change to 111 kmol min^{-1} may occur. The error in the required change is therefore about 10%. Thus our plant model, which describes changes about some operating point, is subject to errors of up to 10% on each input channel (assuming that each input is a flow-rate, regulated as described). Since the error on each input channel is independent of the others, a suitable representation of 'model uncertainty' is by means of (3.81), with

$$\Delta_i = \text{diag}\{\delta_k\}, \quad |\delta_k| \leqslant 0.1 \tag{3.86}$$

which gives

$$\| 10\Delta_i \|_\infty \leqslant 1 \tag{3.87}$$

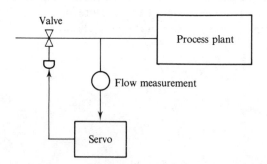

Figure 3.7   Example given by Skogestad and Morari (1987) of uncertainty about flow-rate.

In some cases it will not be appropriate to refer all the uncertainty to the input or the output. It is then necessary to use either the additive model

(3.80), or some other multiplicative model such as

$$G(s) = [I + \Delta_o(s)]G_0(s)[I + \Delta_i(s)] \tag{3.88}$$

or

$$G(s) = G_0(s) . * [I + \Delta_S(s)] \tag{3.89}$$

where '. *' represents the **Schur product**, namely element-by-element multiplication. Among other models which have been proposed are the **inverse multiplicative** model

$$G(s) = [I + \Delta_R(s)]^{-1} G_0(s) \tag{3.90}$$

and a model based on matrix fraction descriptions:

$$G(s) = N(s)D^{-1}(s) \tag{3.91}$$

where

$$N(s) = N_0(s) + \Delta_N(s) \quad \text{and} \quad D(s) = D_0(s) + \Delta_D(s) \tag{3.92}$$

Models (3.91) and (3.92) can also be extended to so-called **fractional representations**, where $N(s)$ and $D(s)$ are allowed to be stable transfer functions rather than polynomial matrices. Such fractional representations will be treated in some detail in Chapter 6.

A major advantage of using operator norms to describe model uncertainty is that it is not necessary to postulate the existence of a 'true' transfer-function model. Real plants are always non-linear, and frequently time-varying, so no 'true' transfer function description can exist. But we can interpret a model of uncertainty, such as (3.80), in the following way. Suppose that we apply an input $u(t)$ to a plant, with $\|u\|_2 = 1$, and that the resulting output is $y(t)$. Suppose also that the same input applied to a model with transfer function G_0 gives the output $y_0(t)$. Then (3.80) really says little other than

$$\|y - y_0\|_2 \leqslant \|\Delta_a\|_\infty \tag{3.93}$$

3.10.2 Structured uncertainty

In the above Example 3.1 on flow control, we had information about the *structure* of the input uncertainty: we knew that the uncertainty for each valve was independent of the uncertainty for the others. If there are two such valves, a correct description of the uncertainty is

$$\Delta_i = \begin{bmatrix} \delta_1 & 0 \\ 0 & \delta_2 \end{bmatrix}, \quad |\delta| \leqslant 0.1 \tag{3.94}$$

But when we write $\| 10\Delta_i \|_\infty \leqslant 1$ instead, we lose all the structural information, since this description also allows perturbations such as

$$\Delta_i = \begin{bmatrix} 0 & 0.1 \\ 0 & 0 \end{bmatrix} \tag{3.95}$$

and

$$\Delta_i = \frac{1}{\sqrt{2}} \begin{bmatrix} 0.1 & 0.1 \\ 0.1 & 0.1 \end{bmatrix} \tag{3.96}$$

which do not correspond to any real perturbations. The use of the unstructured description generally leads to compensator designs which are unnecessarily conservative, because they perform satisfactorily (in some sense) even in the face of perturbations which can never occur. The controller ends up having to be 'detuned' just to guard against non-existent events.

In practice both structured and unstructured information may be available about plant uncertainty. Particular parameters in a state-space model may be known to vary over known ranges, for example – which is certainly highly structured information – while at the same time it may be known that all the details of the model become highly inaccurate above a certain frequency, because of unmodelled lags, hysteresis, resonance, parasitic couplings and so on – this is unstructured information, since it would not be practical or possible to describe all these in detail. The question is how to incorporate both kinds of information simultaneously into an accurate description of the plant's uncertainty.

In practice it is always possible to represent information about uncertainty very accurately, in the following way. We consider the plant to have three sets of inputs and three sets of outputs, as shown by the block P in Figure 3.8. The first set of inputs consists of all manipulable control variables, and the second set consists of all other external signals, such as disturbances and measurement noise. The first set of outputs contains all the measured signals which are available to a feedback compensator, while the second set consists of any other outputs whose behaviour may be of interest (these need not be real signals which can be measured at some point in the plant, but may include conceptual quantities such as error signals). Figure 3.8 shows a feedback compensator K connected between the first pair of outputs and inputs, with its own external signals which represent reference or demand signals.

The third set of inputs and outputs of P is novel. Wherever there is uncertainty about some part of the plant, we can imagine representing it

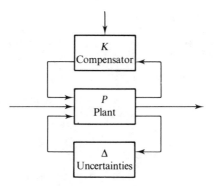

Figure 3.8 A standard representation of an inaccurately known plant under feedback control. P and K are assumed to be known accurately, and any uncertainties have been 'pulled out' into the system Δ.

'locally' by using one of the descriptions we employed in Section 3.10.1 – for example, if a parameter θ has nominal value θ_0, but an uncertainty of $\pm 20\%$, then $\theta = \theta_0(1 + \delta)$ ($|\delta| \leqslant 0.2$). We can therefore envisage a block diagram of the plant with several Δ blocks scattered about in it. Now, each of these blocks has an input (which may be a scalar or a vector) and an output. We collect all the inputs together, and these form the third set of outputs of P; similarly we collect all the outputs together to form the third set of inputs to P. In this way all the Δ blocks are taken outside the plant; in Figure 3.8 they are all contained inside the block labelled Δ. Viewed as a transfer-function matrix, this external Δ has a very special structure: it is block-diagonal, and the blocks on the diagonal are just the small Δs which have been 'pulled out' from inside the plant.[4] We write

$$\Delta(s) = \text{diag}\{\Delta_1(s), \ldots, \Delta_n(s)\} \tag{3.97}$$

if n such blocks have been 'pulled out'. Note that each Δ_j may be a scalar or a matrix.

We can make one further refinement to this model: we insert scalings in the plant so that all the uncertainties are normalized to

$$\|\Delta_j\|_\infty \leqslant 1 \tag{3.98}$$

This is always possible, and helps in formulating standard conditions for achieving robust designs, as we shall see in later sections. Note that we obtain the nominal plant model by setting $\Delta = 0$.

If the compensator K is already known, then it can be amalgamated with the 'certain' part of the plant, P, to form a single system Q, as shown in Figure 3.9. It is clear from this that any uncertainty associated with the

Compensated plant

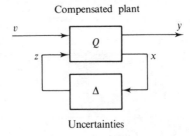

Uncertainties

Figure 3.9 As Figure 3.8, but with the plant P and compensator K combined into the single system Q.

compensator can be modelled in the same way as plant uncertainty, and results in the addition of some blocks to Δ.

EXAMPLE 3.2

Let us return to the process plant with the flow control valves which we considered in Example 3.1. For concreteness, let us suppose that the plant has two inputs and three outputs, and that its nominal model, excluding the servo-valve dynamics, is $G_0(s)$. Also suppose that each servo-valve has a nominal transfer function $V_0(s)$, and each has a 10% multiplicative uncertainty, as already described, which is present at all frequencies. But now we shall also assume that the plant has unstructured multiplicative uncertainty Δ_3 at its input, as shown in Figure 3.10, which comes into effect at high frequencies. Suppose that Δ_3 is to have almost no effect below a frequency ω_0, but that thereafter its effect is to increase so

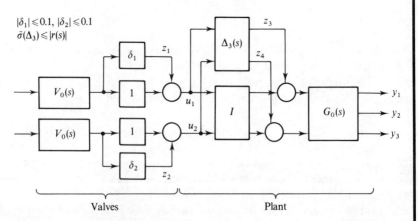

Figure 3.10 A combination of structured and unstructured uncertainty at the input of a two-input, three-output plant.

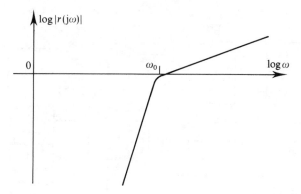

Figure 3.11   Gain characteristic of $r(s)$ defined by equation (3.101).

that $\bar{\sigma}(G_0(I + \Delta))$ is bounded by a constant gain, even though $\bar{\sigma}(G_0)$ decreases to zero at high frequencies. If

$$\bar{\sigma}(G_0(j\omega)) \leqslant \frac{c}{\omega} \qquad (3.99)$$

for large enough ω (that is, the plant exhibits at least 'first-order roll-off') then a suitable characterization of the unstructured uncertainty may be

$$\bar{\sigma}(\Delta_3) \leqslant |r(s)| \qquad (3.100)$$

where

$$r(s) = \left[1 - b_3 \left(\frac{s}{\omega_0} \right) \right] \left(1 + \frac{s}{\omega_0} \right) \qquad (3.101)$$

and $b_3(s)$ is the transfer function of a normalized third-order Butterworth filter. This gives a sharp increase from zero at frequencies just below ω_0, followed by a gradual (first-order) increase thereafter, as shown in Figure 3.11. The bounds on δ_1 and δ_2 are $|\delta_1| \leqslant 0.1$ and $|\delta_2| \leqslant 0.1$, as before. Figure 3.10 can be redrawn as Figure 3.12, which is now in the general form described above, with

$$\Delta = \left[\begin{array}{cc|cc} \tilde{\delta}_1 & 0 & 0 & 0 \\ 0 & \tilde{\delta}_2 & 0 & 0 \\ \hline 0 & 0 & & \\ 0 & 0 & & \tilde{\Delta}_3 \end{array} \right] \qquad (3.102)$$

$$|\tilde{\delta}_1| \leqslant 1, \qquad |\tilde{\delta}_2| \leqslant 1 \qquad (3.103)$$

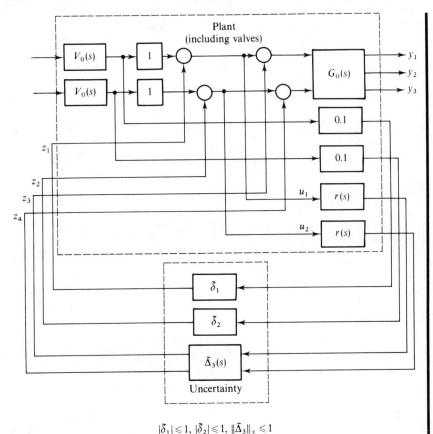

$$|\tilde{\delta}_1| \leqslant 1, \ |\tilde{\delta}_2| \leqslant 1, \ \|\tilde{\Delta}_3\|_\infty \leqslant 1$$

Figure 3.12   The system of Figure 3.10 redrawn in the form of Figure 3.8.

and

$$\bar{\sigma}(\tilde{\Delta}_3) \leqslant 1 \qquad\qquad\qquad\qquad (3.104)$$

Note that scalings have been inserted inside the plant to normalize the uncertainty bounds, and that (3.104) can be rewritten as

$$\|\tilde{\Delta}_3\|_\infty \leqslant 1 \qquad\qquad\qquad\qquad (3.105)$$

From this example it should be clear that the representation of uncertainty shown in Figures 3.8 and 3.9 is both general and accurate, in that it allows a wide variety of structured uncertainty models to be described. There is one respect in which the representation is inaccurate, however. The bounds

$$\|\Delta_j\|_\infty \leqslant 1$$

allow Δ_j to contain complex elements. In many cases, such as the valve-gain uncertainties δ_1 and δ_2 in Example 3.2, it is known that only real elements can occur. To capture this accurately, a more elaborate description of Δ_j than (3.98) is required.

3.10.3 Uncertainty templates

A straightforward and accurate representation of structured uncertainty is obtained by computing the *set* of possible frequency responses of a plant at each frequency. This set can be displayed as a region on each element of a Nyquist array. Such regions are often referred to as **templates**, because if they are displayed on Nichols charts the design of a compensator at a given frequency can be viewed as a process of sliding the uncertainty template around the Nichols chart, its shape unchanged, until every point of it satisfies the specifications at that frequency (Horowitz, 1982).[5]

In practice, of course, each region is defined by a finite set of frequency responses, this set being generated by varying some model parameters over ranges which define the extent of the uncertainty. In theory there could be considerable difficulties in determining the boundaries of the regions in this way, since these boundaries do not in general correspond to the boundaries of the expected parameter variations. (Consider templates for $[s^2 + (a-2)s + 4]^{-1}$, for example, if $0 \leqslant a \leqslant 10$.) But in practice it appears to be possible, with some effort, to determine the uncertainty templates, at least if only a few model parameters are allowed to vary. This is probably so because designers have other sources of information about particularly troublesome parameter values – for instance, values which cause a severe resonance to occur may be discovered by experience even before the mathematical model of the plant is developed.

3.11 Stability robustness

Now that we have some means of describing the extent of our uncertainty about a system model, we turn our attention to ways of checking whether a particular feedback design is robust in the face of this uncertainty. The fundamental property which must be retained for all possible perturbations of the plant (and of the compensator) is stability of the feedback system.

3.11.1 Unstructured uncertainty

Consider the additive model of unstructured uncertainty defined by (3.80)

$$G = G_0 + \Delta_a$$

with

$$\Delta_a = W_1 \tilde{\Delta}_a W_2 \quad \text{and} \quad \|\tilde{\Delta}_a\|_\infty \leqslant 1$$

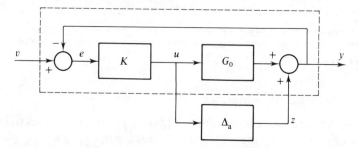

Figure 3.13 Feedback loop containing plant with additive uncertainty.

Suppose that we design a compensator K which stabilizes the nominal model G_0. Then the loop with the actual plant is as shown in Figure 3.13, and this can be redrawn in the standard form, with $\tilde{\Delta}_a$ appearing in a feedback path, as in Figure 3.14. Note that in both figures the dashed line encloses a stable system.

If we write

$$\begin{bmatrix} y \\ \tilde{u} \end{bmatrix} = Q \begin{bmatrix} v \\ \tilde{z} \end{bmatrix} = \begin{bmatrix} Q_{11} & Q_{12} \\ Q_{21} & Q_{22} \end{bmatrix} \begin{bmatrix} v \\ \tilde{z} \end{bmatrix} \tag{3.106}$$

where $\tilde{u}$, v, y and $\tilde{z}$ are as defined in Figures 3.13 and 3.14, then it is easy to show that

$$Q_{11} = G_0 K (I + G_0 K)^{-1} \tag{3.107}$$

$$Q_{12} = (I + G_0 K)^{-1} W_1 \tag{3.108}$$

$$Q_{21} = W_2 (I + K G_0)^{-1} K \tag{3.109}$$

and

$$Q_{22} = - W_2 (I + K G_0)^{-1} K W_1 \tag{3.110}$$

To check whether the feedback system will remain stable under permissible perturbations, we need to check whether the feedback combination of Q_{22} and $\tilde{\Delta}_a$ will remain stable, for all allowable $\tilde{\Delta}_a$. Now, we know that Q_{22} is stable (since K stabilizes G_0), and *we assume that $\tilde{\Delta}_a$ is stable*. Then the feedback system can become unstable only if one or more of the characteristic loci of $-Q_{22}\tilde{\Delta}_a$ encircles the point -1; we know this from the generalized Nyquist theorem (see Chapter 2). Now, if $\lambda(Q_{22}\tilde{\Delta}_a)$ is any eigenvalue of $Q_{22}\tilde{\Delta}_a$, then

$$|\lambda(Q_{22}\tilde{\Delta}_a)| \leqslant \rho(Q_{22}\tilde{\Delta}_a) \leqslant \bar{\sigma}(Q_{22}\tilde{\Delta}_a) \tag{3.111}$$

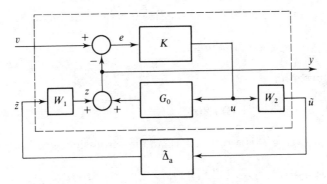

Figure 3.14 The feedback loop of Figure 3.13 redrawn in the form of Figure 3.9.

where $\rho(.)$ denotes the spectral radius, and so no encirclement of -1 can occur if

$$\bar{\sigma}(Q_{22}\tilde{\Delta}_a) < 1 \qquad (3.112)$$

at each frequency, or equivalently, if

$$\|Q_{22}\tilde{\Delta}_a\|_\infty < 1 \qquad (3.113)$$

Now,

$$\|Q_{22}\tilde{\Delta}_a\|_\infty \leqslant \|Q_{22}\|_\infty \|\tilde{\Delta}_a\|_\infty \qquad (3.114)$$

and

$$\|\tilde{\Delta}_a\|_\infty \leqslant 1$$

so

$$\|Q_{22}\|_\infty < 1 \qquad (3.115)$$

is certainly a sufficient condition to guarantee that instability cannot occur for any of the possible perturbations. For a given compensator design, this condition can be checked easily by displaying the principal gains of Q_{22}, and seeing whether the greatest one ever exceeds unity. The quantity $(\|Q_{22}\|_\infty)^{-1}$ is sometimes known as the **gain margin excess** (assuming that (3.115) is satisfied). It is often informative to display a plot of $[\bar{\sigma}(Q_{22})]^{-1}$ against frequency, in order to see the frequencies at which stability is most vulnerable.

If all possible perturbations can occur for which $\|\tilde{\Delta}_a\|_\infty \leqslant 1$ holds, then the condition (3.115) is necessary as well as sufficient for robust stability. For suppose that

$$\|Q_{22}\|_\infty = 1 + \varepsilon \qquad (3.116)$$

for some $\varepsilon \geqslant 0$, and let Q_{22} have the singular-value decomposition

$$Q_{22}(s) = Y(s)\Sigma(s)U^H(s) \tag{3.117}$$

The perturbation

$$\tilde{\Delta}_a(s) = -e^{-sD}U(s)Y^H(s) \tag{3.118}$$

may occur (where D is a real scalar), and is permissible, and leads to

$$-Q_{22}(s)\tilde{\Delta}_a(s) = e^{-sD}Y(s)\Sigma(s)Y^H(s) \tag{3.119}$$

which is a **normal** return ratio. Hence

$$|\lambda_{\max}(Q_{22}(j\omega_0)\tilde{\Delta}_a(j\omega_0))| = \bar{\sigma}(Q_{22}(j\omega_0)) \tag{3.120}$$

$$= 1 + \varepsilon$$

for some ω_0. Now, as we can choose D to be as large as we like, so as to make $\arg e^{-j\omega_0 D}$ as negative as we like, we can certainly choose D to be large enough to force at least one clockwise encirclement of -1 by the characteristic loci of $-Q_{22}\tilde{\Delta}_a$. (From (3.119) we see that the characteristic loci are given by $-e^{-j\omega D}\sigma_i(Q_{22}(j\omega))$, since $Y^H = Y^{-1}$.) In other words, we can find a destabilizing perturbation if (3.116) holds.

A crucial assumption which we made at the beginning of the derivation above is that $\tilde{\Delta}_a$ is stable. This means that the plant's unstable poles are assumed to remain fixed as it varies over its region of uncertainty. If we start instead with one of the multiplicative descriptions of uncertainty, such as (3.81) or (3.82), the whole of the preceding derivation still holds, so that we again end up with the condition (3.115), the only difference being that the transfer function Q is changed (see Exercise 3.8). But in each case we need to assume that the perturbation is stable, which amounts to assuming that unstable poles remain fixed. If this assumption is too unrealistic we can use (3.90), which allows the creation of new unstable poles, even if Δ_R is stable, or the even more exotic

$$G = G_0(I + \Delta_F G_0)^{-1} \tag{3.121}$$

which allows unstable poles to move in the plane even if Δ_F remains stable. Both (3.90) and (3.121) can be treated in exactly the same way as we treated the straightforward additive perturbation (3.80), again obtaining condition (3.115), but each time with Q_{22} a different transfer function.

For some uncertainty models we can relax the assumption about the stability of the perturbation. Consider the output multiplicative model (3.82). We shall now allow Δ_o to be unstable, but impose the alternative assumption that the nominal plant G_0 and the perturbed plant G have *the same number of unstable poles.* The argument is now slightly different from the previous one. Since the loop is stable for $\Delta_o = 0$, then by assumption, and since G and G_0 always have the same number of unstable poles, the loop will remain stable provided the number of encirclements of -1 by the characteristic loci of GK remains unchanged, which it will if no locus passes through -1 as G varies; in other words

$$\det [I + G(j\omega)K(j\omega)] \neq 0 \tag{3.122}$$

for any ω and any permissible G. (Readers who remember the derivation of the generalized Nyquist theorem will note that we could have come directly to this statement.) Now, (3.122) is the same as

$$\underline{\sigma}[I + G(j\omega)K(j\omega)] > 0 \tag{3.123}$$

for all ω, which is the same as

$$\underline{\sigma}[I + G_0 K + \Delta_o G_0 K] > 0 \tag{3.124}$$

or

$$\underline{\sigma}\{[(\Delta_o G_0 K)^{-1} + \Delta_o^{-1} + I]\Delta_o G_0 K\} > 0 \tag{3.125}$$

which holds if

$$\underline{\sigma}\{[(G_0 K)^{-1} + I]\Delta_o^{-1} + I\} > 0 \tag{3.126}$$

By Lemma 3.2, this will hold if

$$\underline{\sigma}\{[(G_0 K)^{-1} + I]\Delta_o^{-1}\} > 1 \tag{3.127}$$

which is implied by

$$\underline{\sigma}(\Delta_o^{-1})\underline{\sigma}[(G_0 K)^{-1} + I] > 1 \tag{3.128}$$

Hence

$$\bar{\sigma}(\Delta_o)\bar{\sigma}\{[(G_0 K)^{-1} + I]^{-1}\} < 1 \tag{3.129}$$

or

$$\bar{\sigma}\{G_0 K[I + G_0 K]^{-1}\} < \frac{1}{\bar{\sigma}(\Delta_o)} \tag{3.130}$$

Note that this condition is essentially the same as that obtained by using the previous approach (see Exercise 3.8), although the set of allowed perturbations is rather different. We have established the sufficiency of (3.130) for robust stability, but it can again be shown to be a necessary condition, if all permissible perturbations may actually occur.

Stability conditions such as (3.115) and (3.130) are special cases of a very general result, known as the **small gain theorem**, which states that a feedback loop composed of stable operators will certainly remain stable if the product of all the operator gains is smaller than unity. This is true for non-linear as well as linear operators.

Postlethwaite *et al.* (1981) introduce a 'small phase theorem', which may be useful when the phases of perturbations, rather than their gains, can be bounded.

3.11.2   Structured uncertainty

We now consider structured uncertainty, as represented in Figure 3.9; Δ has a block-diagonal structure, each block Δ_j satisfying $\|\Delta_j\|_\infty \leqslant 1$. We shall also assume that each Δ_j is stable. Partitioning the compensated plant Q as before, the condition found earlier, namely

$$\|Q_{22}\|_\infty < 1 \tag{3.131}$$

is certainly sufficient for robust stability. But now it is no longer a necessary condition, because most perturbations which satisfy $\|\Delta\|_\infty < 1$ are no longer permissible: the only permissible ones are those that have the appropriate block-diagonal structure. Furthermore, (3.131) can, in some cases, be so conservative as to be useless. Therefore robustness tests are needed which are better indicators of necessary conditions than is (3.131).

Let BD_δ denote the set of stable, block-diagonal perturbations with a particular structure, and with $\|\Delta_j\|_\infty \leqslant \delta$. To describe this set fully for each design problem we should really write something like $\mathrm{BD}_\delta(m_1, m_2, \ldots, m_n, k_1, k_2, \ldots, k_n)$, to denote that Δ contains $m = \sum_{i=1}^n m_i$ blocks, each block being repeated m_i times and having dimensions $k_i \times k_i$. (Repeated blocks allow some correlated perturbations to be modelled.) But we shall usually employ the abbreviated notation, with the appropriate structure understood to be fixed.[6]

A permissible perturbation destabilizes the system if and only if

$$\det[I - Q_{22}(j\omega)\Delta(j\omega)] = 0 \tag{3.132}$$

for some ω and some $\Delta \in \mathrm{BD}_1$. We therefore define

$$\mu(Q_{22}(j\omega)) = \begin{cases} 0 \quad \text{if } \det[I - Q_{22}\Delta] \neq 0 \quad \text{for any } \Delta \in \mathrm{BD}_\infty \\ \{\min_{\Delta \in \mathrm{BD}_\infty} (\bar\sigma(\Delta(j\omega))): \\ \det[I - Q_{22}(j\omega)\Delta(j\omega)] = 0\}^{-1} \quad \text{otherwise} \end{cases} \tag{3.133}$$

We also define

$$\| Q_{22} \|_\mu = \sup_\omega \mu(Q_{22}(j\omega)) \tag{3.134}$$

which is suggestive but rather misleading, since $\| . \|_\mu$ is not a norm. The function $\mu(Q)$ is called the **structured singular value** of Q, and depends on the structure of the set BD_δ as well as on Q.

Although (3.133) provides a definition of $\mu(.)$, it gives no help in computing it. Doyle (1982) has proved a number of properties of $\mu(.)$, the most important of which, for us, are as follows:

$$\mu(\alpha Q) = |\alpha| \, \mu(Q) \tag{3.135}$$

$$\mu(I) = 1 \tag{3.136}$$

$$\mu(AB) \leqslant \bar{\sigma}(A)\mu(B), \quad A, B \text{ both square} \tag{3.137}$$

$$\mu(\Delta) = \bar{\sigma}(\Delta), \quad \Delta \in BD_\delta \tag{3.138}$$

If $n = 1$ and $m_1 = 1$ (see above) then $\mu(Q) = \bar{\sigma}(Q)$ $\qquad$ (3.139)

If $n = 1$ and $k_1 = 1$ (see above) then $\mu(Q) = \rho(Q)$ $\qquad$ (3.140)

(namely the spectral radius, $\max_i |\lambda_i(Q)|$)

Let D be any real, diagonal, positive matrix with the structure

$$\text{diag}\{d_1 I_{k_1}, \ldots, d_{m_1} I_{k_1}, d_{m_1+1} I_{k_2}, \ldots, d_m I_{k_n}\} \tag{3.141}$$

and $d_i > 0$, then

$$\mu(DQD^{-1}) = \mu(Q) \tag{3.142}$$

Let U be any unitary matrix ($UU^H = I$) with the same block-diagonal structure as the set BD_δ. Then

$$\max_U \rho(UQ) \leqslant \mu(Q) \leqslant \inf_D \bar{\sigma}(DQD^{-1}) \tag{3.143}$$

Most of these results can be proved quite easily (see Exercise 3.9).

The importance of $\mu(.)$ can be seen from the following central result:

***Theorem 3.6* (Stability robustness theorem** – Doyle *et al.*, 1982):   The system shown in Figure 3.9 remains stable for all $\Delta \in BD_1$ if and only if

$$\| Q_{22} \|_\mu < 1 \qquad\qquad (3.144)$$

Proof:

$$\sup_\omega \rho(Q\Delta) \leqslant \sup_\omega \mu(Q\Delta) \qquad\qquad \text{(by (3.143))}$$

$$\leqslant \sup_\omega [\mu(Q)\bar{\sigma}(\Delta)] \qquad\qquad \text{(by (3.137))}$$

$$\leqslant \| Q \|_\mu \qquad\qquad \text{(since } \bar{\sigma}(\Delta) \leqslant 1)$$

So, if $\| Q_{22} \|_\mu < 1$, then $\rho(Q\Delta) < 1$, and $I - Q\Delta$ cannot be singular (on the imaginary axis). Hence the system remains stable. Conversely, if $\| Q_{22} \|_\mu \geqslant 1$ then $\mu(Q_{22}(j\omega_0)) \geqslant 1$ for some ω_0 and so, by definition (3.133), there exists a perturbation $\Delta_0 \in \mathrm{BD}_\infty$ such that $\det[I - Q_{22}(j\omega_0)\Delta_0] = 0$ and $\bar{\sigma}(\Delta_0) \leqslant 1$. Hence the system is unstable, and $\Delta_0 \in \mathrm{BD}_1$. This proof relies on the fact that $\| \Delta_i \|_\infty \leqslant 1$ if and only if $\| \Delta \|_\infty \leqslant 1$ (see Exercise 3.9). ∎

Theorem 3.6 does not really provide a test for stability robustness, because we do not have a means for computing $\| Q_{22} \|_\mu$. However, (3.143) provides the basis for some computable approximations to $\| Q_{22} \|_\mu$. Doyle (1982) shows that the first inequality in (3.143) can be strengthened to

$$\max_U \rho(UQ) = \mu(Q) \qquad\qquad (3.145)$$

but the maximization problem here is likely to lead to local maxima being found, and so is not reliable. In fact the second inequality in (3.143) is a more useful starting point, particularly as it is dangerous to underestimate μ, but safe to overestimate it.

For $n \leqslant 3$ and $m_i = 1$ $(i = 1, \ldots, n)$ – that is, if there are no more than three blocks in Δ, with no repetitions – Doyle (1982) has shown that

$$\mu(Q) = \inf_D \bar{\sigma}(DQD^{-1}) \qquad\qquad (3.146)$$

So, if we have up to three uncorrelated perturbations, then (3.146) holds. Note that the size of each block Δ_i is not limited here. For $n > 3$, numerical experiments indicate that the lower and upper bounds in (3.143) are usually within 5% of each other, and almost always within 15%. Thus the evidence is that $\inf \bar{\sigma}(DQD^{-1})$ is a useful estimate of μ, and it is therefore worth developing ways of computing, or at least estimating, this infimum.

The obvious approach is to use hill-climbing to find the D which minimizes $\bar{\sigma}(DQD^{-1})$ at each frequency. This is quite attractive, as $\bar{\sigma}(DQD^{-1})$ can be shown to be convex in D (Safonov and Doyle, 1984), and since the search is over only $m-1$ parameters, regardless of the dimensions of the blocks in Δ. (The search is over $d_1, d_2, \ldots, d_m$, one of which can be fixed arbitrarily.) Doyle (1982) shows how to generate descent directions for such hill-climbing; the algorithm is rather involved, and simpler alternatives are worth considering. The basis of a completely different algorithm for computing the structured singular value is given by Fan and Tits (1988).

Safonov (1982) has given a simple way of estimating $\inf \bar{\sigma}(DQD^{-1})$ for $k_i = 1$ $(i = 1, \ldots, n)$ – that is, when each perturbation is a scalar, but may be repeated. He has shown that, in this case,

$$\inf_{D} \bar{\sigma}(DQD^{-1}) \leqslant \inf_{D} \bar{\sigma}(\text{abs}(DQD^{-1})) = \lambda_{\text{P}}(Q) \tag{3.147}$$

where $[\text{abs}(X)]_{ij} = |x_{ij}|$ and $\lambda_{\text{P}}(.)$ denotes the Perron–Frobenius eigenvalue of a matrix (see Chapter 2). The D which minimizes $\bar{\sigma}(\text{abs}(DQD^{-1}))$ is given by

$$D = \text{diag}\{d_i\}, \quad i = 1, \ldots, n \tag{3.148}$$

$$d_i = \sqrt{(y_i/x_i)} \tag{3.149}$$

where $x = (x_1, \ldots, x_n)$ and $y = (y_1, \ldots, y_n)$ are the right and left Perron–Frobenius eigenvectors of Q, respectively. Note, however, that the estimate (3.147) can be obtained without finding D.

This result can be extended to the general case, which allows $k_i > 1$, as follows. Let Q be partitioned conformally with the block diagonal structure of Δ:

$$Q = \begin{bmatrix} Q_{11} & \cdots & Q_{1m} \\ \vdots & & \vdots \\ Q_{m1} & \cdots & Q_{mm} \end{bmatrix} \tag{3.150}$$

(which should not be confused with the partitioning of (3.106), as the Q there is different; Q in (3.150) corresponds to Q_{22} in (3.106)). Now define the matrix

$$\text{norm}(Q) = \begin{bmatrix} \bar{\sigma}(Q_{11}) & \cdots & \bar{\sigma}(Q_{1m}) \\ \vdots & & \vdots \\ \bar{\sigma}(Q_{m1}) & \cdots & \bar{\sigma}(Q_{mm}) \end{bmatrix} \tag{3.151}$$

Then

$$\inf_D \bar{\sigma}(DQD^{-1}) \leqslant \lambda_P(\text{norm}(Q)) \tag{3.152}$$

where the infimum is over the matrices D with the diagonal structure defined in (3.141). However, this algorithm is of limited usefulness since a better estimate of $\|Q\|_\mu$ is often provided simply by $\bar{\sigma}(Q)$ (if $k_i > 1$).

Another way of estimating $\inf \bar{\sigma}(DQD^{-1})$, but for $k_i = 1$ only, is to use the algorithm put forward by Osborne (1960), which finds

$$D^* = \arg\min_D \sum_i \sigma_i^2(DQD^{-1}) \tag{3.153}$$

and to take $\bar{\sigma}(D^*QD^{*-1})$ as an estimate.

Another useful estimate has been obtained by Kouvaritakis and Latchman (1985), but for a different class of perturbations. Suppose that Δ does not have any particular diagonal structure, but that each element of Δ is bounded by

$$|\delta_{ij}(j\omega)| \leqslant r_{ij}(\omega) \tag{3.154}$$

where $r_{ij}(\omega) > 0$. If Δ is an $m \times m$ matrix, this uncertainty description can also be represented by a purely diagonal perturbation, but of dimensions $m^2 \times m^2$; the bounds (3.154) therefore define a structured perturbation. Let R be the positive matrix

$$R = \begin{bmatrix} r_{11} & \cdots & r_{1m} \\ \vdots & & \vdots \\ r_{m1} & \cdots & r_{mm} \end{bmatrix} \tag{3.155}$$

Now let x and y be the right and left Perron–Frobenius eigenvectors of $R\,\text{abs}(Q)$, respectively, and let u and v be the right and left Perron–Frobenius eigenvectors of $\text{abs}(Q)R$, and define

$$D_1 = \text{diag}\{\sqrt{(y_i/x_i)}\}, \quad i = 1, \ldots, m \tag{3.156}$$

$$D_2 = \text{diag}\{\sqrt{(u_i/v_i)}\}, \quad i = 1, \ldots, m \tag{3.157}$$

Now, for any positive diagonal D_1 and D_2, we have (see Exercise 3.10)

$$\rho(Q\Delta) \leqslant \bar{\sigma}(D_1 R D_2)\bar{\sigma}(D_2^{-1}QD_1^{-1}) \tag{3.158}$$

so that a sufficient condition for stability of the perturbed system is

$$\bar{\sigma}(D_1 R D_2)\bar{\sigma}(D_2^{-1}QD_1^{-1}) < 1 \tag{3.159}$$

Now suppose that the uncertainty description is rewritten in terms of a diagonal perturbation $\tilde{\Delta}$, so that the corresponding compensated plant transfer function is $\tilde{Q}$, with scaling introduced so that $|\tilde{\delta}_{ij}| \leqslant 1$ (see Exercise 3.10). Then (3.159) implies that

$$\mu(\tilde{Q}) \leqslant \frac{\bar{\sigma}(D_1 R D_2)}{\underline{\sigma}(D_1 Q^{-1} D_2)} \tag{3.160}$$

and since this holds for any positive diagonal D_1 and D_2, we have

$$\mu(\tilde{Q}) \leqslant \inf_{D_1, D_2} \frac{\bar{\sigma}(D_1 R D_2)}{\underline{\sigma}(D_1 Q^{-1} D_2)} \tag{3.161}$$

(3.156) and (3.157) give D_1 and D_2 which approximately achieve the infimum. A tighter estimate of $\mu(\tilde{Q})$ can be obtained by finding D_1 and D_2 which actually do achieve it, but this requires a more elaborate algorithm based on hill-climbing. Such an algorithm is described by Kouvaritakis and Latchman (1985), and Daniel *et al.* (1986).

3.11.3 Loop failures and gain variations

An important property of a multivariable feedback system is its stability robustness in the face of one or more loop failures. This is sometimes called the **integrity** of the system. With a SISO system the position is straight-forward: if the feedback loop is broken because of some equipment failure, then stability is determined by the open-loop stability of the plant and compensator. But it is quite possible to have a multivariable plant and compensator, each of which is open-loop stable, and which together comprise a stable feedback system when everything is working correctly, but which will be unstable if one or more loops fail, perhaps as a result of a sensor or an actuator failing.

This is an important special case of the more general question of stability robustness in the face of frequency-independent gain variations anywhere in the system. Clearly, such questions can be investigated by the methods presented in Section 3.11.2, with Δ a diagonal frequency-independent matrix. These methods are likely to be unnecessarily conservative, however, since they will check for robustness in the face of complex 'gains' as well as real ones, and in the face of gain increases as well as decreases.

An exact assessment of integrity against loop failures can of course be obtained by examining the characteristic loci for each possible failure mode, but if there are m loops, and any combination of these can fail, then there are

$$\sum_{r=1}^{m-1} \binom{m}{r}$$

displays to be checked. This is feasible, although tedious, for small numbers of loops (six displays if $m = 3$, fourteen displays if $m = 4$), but gives no information about the effects of variations of gains between their nominal values and zero. Such variations may occur if equipment fails gradually rather than abruptly.

A better, albeit conservative method of assessment is obtained by using a Nyquist array with generalized Gershgorin bands. As usual, let us represent gain variations in the form of Figure 3.9, with the uncertainty described by

$$\Delta = \text{diag}\{\delta_1, \delta_2, \ldots, \delta_m\} \qquad (3.162)$$

each δ_i being real. Now consider the Nyquist array of Q_{22} (with Q partitioned as in (3.106)), with its Gershgorin bands superimposed. Since Q_{22} is itself stable, Theorem 2.13 tells us that the feedback loop will remain stable whenever the total number of encirclements of the points $\{1/\delta_i\}$, each by the ith Gershgorin band, is zero. (Note that we look at $1/\delta_i$, not $-1/\delta_i$, since in Figure 3.9 we have used a positive-feedback convention.) Usually this will be the set of values of δ_i for which none of the points $\{1/\delta_i\}$ is encircled by the ith Gershgorin band. The conservativeness of this assessment can be minimized by applying various similarity transformations to Q_{22}, including the Mees scaling described in Section 2.10. It may also be reduced by choosing another representation of the gain variations: for example, the additive representation $k_i = k_{0i} + \delta_i$ may allow a larger range of values of k_i than the multiplicative representation $k_i = k_{0i}(1 + \delta_i)$, since Q_{22} will be different in each case.

EXAMPLE 3.3

A two-input plant $G(s)$ is placed in a negative-feedback loop with a series compensator $C(s)$. The two input actuators are modelled by frequency-independent gains k_{01} and k_{02}. Is the system robustly stable against the failure of either loop? Modelling the actual actuator gains by $k_i = k_{0i}(1 + \delta_i)$ leads to the arrangement shown in Figure 3.15, with

$$Q_{22} = -\text{diag}\{k_{01}, k_{02}\} CG[I + \text{diag}\{k_{01}, k_{02}\} CG]^{-1} \qquad (3.163)$$

Nominal actuator gains correspond to $\delta_i = 0$, while complete actuator failures correspond to $\delta_i = -1$. If the generalized Gershgorin bands of Q_{22} are as shown in Figure 3.16 (drawn for positive frequencies only), we can deduce that the system remains stable whenever

$$\left(-\infty < \frac{1}{\delta_1} < -0.9 \quad or \quad -0.8 < \frac{1}{\delta_1} < -0.1 \quad or \quad \frac{1}{\delta_1} > 0.8 \right)$$

$$and \quad \left(-\infty < \frac{1}{\delta_2} < -2 \quad or \quad -0.8 < \frac{1}{\delta_2} < 0 \quad or \quad \frac{1}{\delta_2} > 0.75 \right) \qquad (3.164)$$

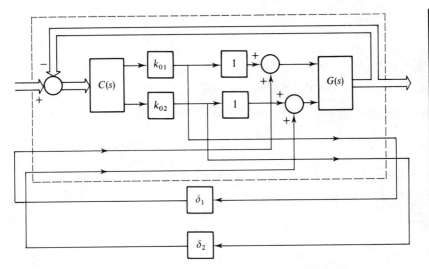

Figure 3.15   Investigating stability robustness in the face of actuator failures.

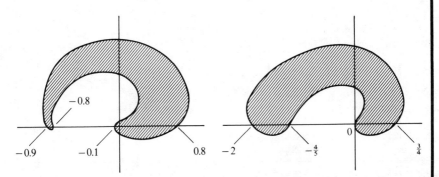

Figure 3.16   Generalized Gershgorin bands of Q_{22}, as defined by equation (3.163), for the system shown in Figure 3.15.

or, more simply, whenever

$$(-1.11 < \delta_1 < 1.25 \quad or \quad -10 < \delta_1 < -1.25)$$

and $(-0.5 < \delta_2 < 1.33 \quad or \quad \delta_2 < -1.25)$ **(3.165)**

This system therefore remains stable if actuator 1 fails, but cannot be guaranteed to remain stable if the gain of actuator 2 falls below $0.5\,k_{02}$. Note that (3.165) gives information about *simultaneous* gain variations: we can deduce, for example, that stability is maintained if $k_1 = 0$ and $k_2 = 0.6\,k_{02}$.

A different approach to robustness in the face of real parameter variations has been developed by Battacharyya (1987).

3.12   Performance robustness

In the previous section we were concerned with the robustness of stability in the face of various perturbations. In practice we are usually concerned with maintaining more than stability. We would like to maintain an acceptable level of performance, as well as stability, in the face of a prespecified class of perturbations – that is, we would like to achieve **performance robustness** as well as stability robustness. Fortunately, we have already developed all the tools we need to formulate this problem precisely and assess whether it has been achieved by a given design.

In Section 3.9 we saw that it is possible to formulate frequency-domain performance specifications in terms of the operator norm $\|.\|_\infty$. Using out standard representation of uncertainty, shown in Figure 3.9, we partition the transfer function Q so that

$$\begin{bmatrix} y \\ x \end{bmatrix} = Q \begin{bmatrix} v \\ z \end{bmatrix} = \begin{bmatrix} Q_{11} & Q_{12} \\ Q_{21} & Q_{22} \end{bmatrix} \begin{bmatrix} v \\ z \end{bmatrix} \tag{3.166}$$

and define the variables v and y such that

$$\|Q_{11}\|_\infty < 1 \tag{3.167}$$

becomes the performance specification.

EXAMPLE 3.4

Suppose that the performance specification is

$$\left\| \begin{bmatrix} W_1 S \\ W_2 T \end{bmatrix} \right\|_\infty < 1 \tag{3.168}$$

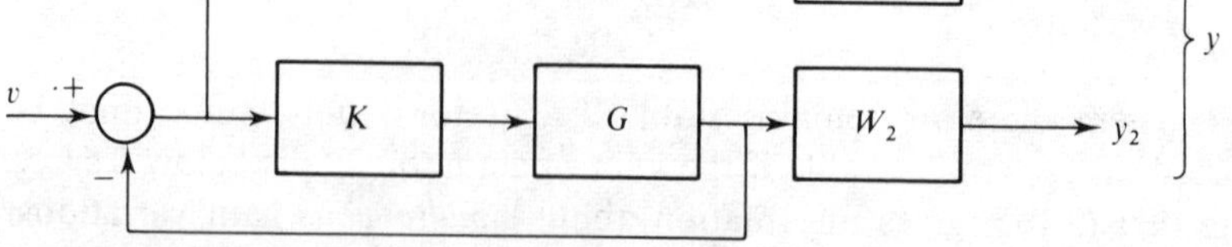

Figure 3.17   Definition of the variables v and y in equation (3.166) so that the specification (3.168) can be expressed as $\|Q_{11}\|_\infty < 1$.

where S is the sensitivity function, T is the closed-loop transfer function ($T = I - S$), and W_1 and W_2 are frequency-dependent weighting matrices. Then v and $y = [y_1 \, y_2]^T$ are defined as in Figure 3.17, in which G and K are the *perturbed* plant and compensator, respectively.

With the perturbation Δ present, as shown in Figure 3.9, the relationship between v and y becomes[7]

$$y = [Q_{11} + Q_{12}\Delta(I - Q_{22}\Delta)^{-1}Q_{21}]v \tag{3.169}$$

so that the criteria for performance robustness can be stated precisely as:

$$\| Q_{11} + Q_{12}\Delta(I - Q_{22}\Delta)^{-1}Q_{21} \|_\infty < 1 \tag{3.170}$$

and

$$\| Q_{22} \|_\mu < 1 \tag{3.171}$$

where $\Delta = \text{diag}\{\Delta_1, \ldots, \Delta_n\}$ and $\|\Delta_i\|_\infty \leqslant 1$, and (3.171) comes from Theorem 3.6.

Suppose that, instead of imposing a performance specification, we were to add an additional uncertainty block Δ_0 between y and v, with $\|\Delta_0\|_\infty \leqslant 1$, as shown in Figure 3.18, and require stability robustness in the face of any possible perturbation, including the fictitious Δ_0. That is, suppose that we required

$$\| Q \|_\mu < 1 \tag{3.172}$$

where $\| . \|_\mu$ is now computed with respect to the structure $\text{diag}\{\Delta_0, \Delta\} = \text{diag}\{\Delta_0, \Delta_1, \ldots, \Delta_n\}$. Then, since stability would be required for perturbations $\text{diag}\{\Delta_0, 0\}$ and $\text{diag}\{0, \Delta\}$, (3.172) would certainly imply that both

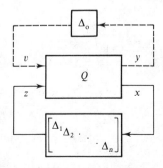

Figure 3.18 Replacement of a performance specification by a fictitious uncertainty Δ_0.

$\|Q_{11}\|_\infty < 1$ and $\|Q_{22}\|_\mu < 1$ were satisfied. Rather remarkably, it can be shown that (3.172) also implies that (3.170) is satisfied for all permissible Δ; moreover, (3.172) is actually a necessary condition for (3.170) to hold.

This result is of great importance in the analysis of feedback systems, and opens the way to systematic methods of synthesis, as we shall see in Chapter 6.

Theorem 3.7 (Performance robustness theorem – Doyle et al., 1982):
The condition (3.170) holds for all $\Delta \in BD_1$ and (3.171) is satisfied (computed with respect to the structure of Δ) if and only if

$$\|Q\|_\mu < 1$$

where

$$Q = \begin{bmatrix} Q_{11} & Q_{12} \\ Q_{21} & Q_{22} \end{bmatrix} \tag{3.173}$$

and $\|Q\|_\mu$ is computed with respect to the structure of $\text{diag}\{\Delta_o, \Delta\}$ (provided Q is stable).

Proof:

$$\|Q\|_\mu = \sup_\omega \mu(Q(j\omega)) \quad \text{(by definition)}$$

Now,

$$\mu(Q(j\omega)) < 1 \tag{3.174}$$

if and only if

$$\det(I - Q(j\omega)\tilde{\Delta}) > 0 \tag{3.175}$$

for all ω (by definition of $\mu(.)$), and for all $\tilde{\Delta} = \text{diag}\{\Delta_0, \Delta\}$, with $\|\Delta_0\|_\infty \leqslant 1$ and $\Delta \in BD_1$. Now,

$$\det(I - Q\tilde{\Delta}) = \det\begin{bmatrix} I - Q_{11}\Delta_0 & -Q_{12}\Delta \\ -Q_{21}\Delta_0 & I - Q_{22}\Delta \end{bmatrix} \tag{3.176}$$

$$= \det(I - Q_{22}\Delta)\det[(I - Q_{11}\Delta_0)$$
$$- Q_{12}\Delta(I - Q_{22}\Delta)^{-1}Q_{21}\Delta_0]$$
$$\text{(by Schur's formula)} \tag{3.177}$$

$$= \det(I - Q_{22}\Delta)\det(I - [Q_{11} + Q_{12}\Delta$$
$$\times (I - Q_{22}\Delta)^{-1}Q_{21}]\Delta_0) \tag{3.178}$$

Since (3.178) must hold for $\Delta = 0$ and $\Delta_0 = 0$, we deduce that (3.175) holds if and only if

$$\det(I - Q_{22}\Delta) > 0 \tag{3.179}$$

and

$$\det(I - [Q_{11} + Q_{12}\Delta(I - Q_{22}\Delta)^{-1}Q_{21}]\Delta_0) > 0 \tag{3.180}$$

for all permissible Δ and Δ_0. But (3.179) is equivalent to

$$\| Q_{22} \|_\mu < 1$$

and, using the same argument as in Section 3.11.1, (3.180) is equivalent to

$$\| Q_{11} + Q_{12}\Delta(I - Q_{22}\Delta)^{-1}Q_{21} \|_\infty < 1 \qquad \blacksquare$$

EXAMPLE 3.5 (Doyle *et al.*, 1982)

We are given a nominal plant

$$G_0(s) = \frac{1}{s}\begin{bmatrix} 10 & 9 \\ 9 & 8 \end{bmatrix} \tag{3.181}$$

with unstructured multiplicative input uncertainty, so that the true plant is

$$G(s) = G_0(s)(I + \Delta_i) \tag{3.182}$$

where

$$\bar{\sigma}(\Delta_i) \leqslant \left| \frac{1 + j\omega}{\alpha} \right| \tag{3.183}$$

for some constant α. The performance specification is

$$\bar{\sigma}(S(j\omega)) \leqslant \left| \frac{j\omega\alpha}{1 + j\omega} \right| \tag{3.184}$$

where $S = (I + GK)^{-1}$. Note that increasing α has the effect of making performance robustness easier to achieve.

We wish to determine the values of α for which the specification is achieved in the face of uncertainty, with the particular compensator design

$$K = \begin{bmatrix} 0.118 & 1 \\ 1 & -0.118 \end{bmatrix} \tag{3.185}$$

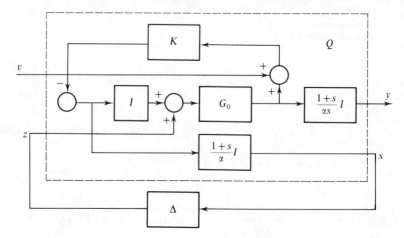

Figure 3.19   Block diagram corresponding to Example (Doyle *et al.*, 1982).

To do this we first reformulate (3.183) and (3.184) as

$$\| \Delta \|_\infty = \left\| \frac{\alpha}{1 + j\omega} \Delta_i \right\|_\infty \leqslant 1 \tag{3.186}$$

$$\left\| \frac{1 + j\omega}{j\omega\alpha} S \right\|_\infty \leqslant 1 \tag{3.187}$$

and obtain Q corresponding to Figure 3.9. The detailed block diagram for this case is shown in Figure 3.19, from which we obtain

$$Q = \begin{bmatrix} Q_{11} & Q_{12} \\ Q_{21} & Q_{22} \end{bmatrix}$$

$$= \begin{bmatrix} \dfrac{1+s}{\alpha s}(I + G_0 K)^{-1} & \dfrac{1+s}{\alpha s}(I + G_0 K)^{-1} G_0 \\[2ex] -\dfrac{1+s}{\alpha}(I + KG_0)^{-1}K & -\dfrac{1+s}{\alpha}(I + KG_0)^{-1}KG_0 \end{bmatrix}$$

$$\tag{3.188}$$

Now we have

$$\mu(Q) = \inf_D \bar{\sigma}(DQD^{-1}) \tag{3.189}$$

where

$$D = \begin{bmatrix} d_1 I_2 & 0 \\ 0 & d_2 I_2 \end{bmatrix} \tag{3.190}$$

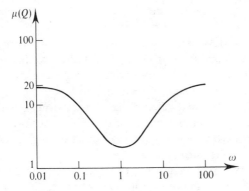

Figure 3.20   Structured singular value of Q with $\alpha = 1$ (see text).

the equality holding in (3.189) since D has only two blocks, and (3.189) simplifies, in this case, to

$$\mu(Q) = \inf_{d} \bar{\sigma}\left(\begin{bmatrix} Q_{11} & dQ_{12} \\ \frac{1}{d}Q_{21} & Q_{22} \end{bmatrix}\right) \tag{3.191}$$

where d is a positive real scalar. A plot of $\mu(Q)$ with $\alpha = 1$ is shown in Figure 3.20, from which we find $\| Q \|_{\mu} = 18.5$.

Since $\| Q \|_{\mu}$ scales inversely with α, the compensator (3.185) achieves performance robustness provided $\alpha > 18.5$.

SUMMARY

In this chapter we have seen that the all-important notion of the gain of a feedback system can be extended to multivariable systems. For multivariable systems this gain is a range of values, each corresponding to a particular input signal direction. The lower and upper boundaries of this range are given by the smallest and largest frequency-dependent singular values, or principal gains. Analysis of feedback systems often requires the inspection of graphs of these principal gains, which look just like the familiar Bode plots used for SISO design.

We have seen that Bode's gain–phase relations continue to hold for multivariable systems, and that the same broad conclusions can be drawn about gain characteristics near the cross-over frequency, and about the effects of right half-plane zeros and poles, as for SISO systems.

To examine the robustness of multivariable designs we have introduced a new representation of plant uncertainty which is powerful enough to model a wide variety of both structured and unstructured perturbations. By exploiting the flexibility of this representation, we found that we could examine a wide variety of robustness questions by borrowing only one idea from traditional analysis – that of gain margin.

The use of principal gains, or operator norms of suitably weighted transfer functions, allows the specification of both performance and robustness requirements to take the same mathematical form, and this leads to a very powerful result: performance specifications are, in a precise sense, interchangeable with uncertainty specifications, and the same tools can be used to assess the attainment of both types of specification. This begins to explain the well-known fact that systems have to be 'detuned' if their behaviour is very uncertain.

If the approach taken in Sections 3.10 and 3.12 is adopted, then all problems of feedback analysis become multivariable. There is always some performance specification, and some description of uncertainty, so that Q is at least a 2×2 matrix, even if both the plant and the compensator are SISO. This does not mean, of course, that multivariable techniques should necessarily be applied to design.

Notes

1. The term 'principal gain' was introduced by MacFarlane and Scott-Jones (1979). It has not become well-established, and principal gains are often referred to merely as 'singular values'. But because of the possible confusion with the singular values of the matrix A of a state-space realization, which are not frequency-dependent, it seems preferable to make the distinction, just as it is useful to use the term 'characteristic loci' rather than 'eigenvalues'.

2. Orthonormal columns can always be obtained by means of the QR factorization, described in Section 8.2.3.

3. The identity $I - (I + X)^{-1} = (I + X^{-1})^{-1}$ occurs frequently in the analysis of feedback systems, and it should be noted that it also equals $X(I + X)^{-1}$. It can be verified with a little algebra, but can also be obtained as a special case of the matrix-inversion lemma

    $$(A + BCD)^{-1} = A^{-1} - A^{-1}B(DA^{-1}B + C^{-1})^{-1}DA^{-1}$$

    which is useful in systems theory, optimal control and Kalman filtering theory. See Kailath (1980) – where it is called the 'modified matrices formula' – for references, and Åström and Wittenmark (1984) for a derivation.

4. The standard block-diagonal representation of structured uncertainty described in this section was introduced by Doyle, two of the earliest references being Doyle *et al.* (1982) and Doyle (1982).

5. The design technique developed by Horowitz has recently acquired the name 'quantitative feedback theory' (QFT). It is described in Chapter 4.

6. The notation here follows Doyle (1982), who also gives proofs of some of the results presented in this section. It is possible for each Δ_i to be non-square, but we do not treat this case.

7. In the notation introduced in Section 6.1, we can write the inequality (3.170) as $\| F_l(Q,\Delta) \|_\infty < 1$.

EXERCISES

3.1 Using Lemma 3.2, verify that if $\bar\sigma(Q) \ll 1$ then $\bar\sigma[I-(I+Q)^{-1}] \approx \bar\sigma(Q)$.

3.2 Suppose that G and K are both square and invertible. By writing $GK = G(KG)G^{-1}$ show that:

(a) $\bar\sigma(GK) \leqslant \bar\sigma(KG)\,\text{cond}(G)$

(b) $\underline{\sigma}(GK) \geqslant \dfrac{\underline{\sigma}(KG)}{\text{cond}(G)}$

and establish corresponding bounds on $\underline{\sigma}(KG)$ and $\bar\sigma(KG)$.

3.3 Consider a plant

$$G = \begin{bmatrix} 1 & 0 \\ g & 1 \end{bmatrix}$$

and a controller

$$K = \begin{bmatrix} k_{11} & 0 \\ k_{21} & k_{22} \end{bmatrix}$$

Show that choosing $k_{21} = -k_{11}g$ makes the return ratio GK normal (in fact, diagonal). Suppose that $k_{11} = 1$ and nominally $g = 10^3$, so that $k_{21} = -10^3$. Show that a 1% change in g can make the return ratio extremely skew. What is the condition number of G in this case?

3.4 (a) Express $\| G \|_2$ in terms of the principal gains of G.

(b) Suppose that $\{u(t)\}$ has power spectral density $\Phi_{uu}(\omega)$, that $y(s) = G(s)u(s)$ and that $G(s)$ is stable. Obtain an expression for $E\{y^TWy\}$ in terms of the norm $\|.\|_2$, assuming that $W^T = W \geqslant 0$.

3.5   A multivariable-feedback system is to attenuate all output disturbances by at least a factor of 10 at frequencies below 0.1 rad s^{-1}, and to attenuate output sensor noise by at least a factor of 10 at frequencies above 2 rad s^{-1}. Constant output disturbances should be attenuated (in the steady-state) by a factor of at least 100.

(a)   Formulate suitable specifications on the closed-loop transfer function T and sensitivity S, in terms of their principal gains.

(b)   Obtain a corresponding specification on the open-loop return ratio GK.

(c)   Reformulate the closed-loop specifications using the operator norm $\|.\|_\infty$ and suitable weighting functions.

(d)   From (b), estimate the cross-over frequency, the maximum 'phase margin' which could be exhibited by a characteristic locus, and hence a lower bound for $\|T\|_\infty$.

(*Solution:* $\omega_c \approx 0.45$ rad s^{-1}, max. 'phase margin' $\approx 36°$, $\|T\|_\infty \geqslant 1.6$ (approximately))

(e)   Is the lower bound obtained in (d) consistent with the specification you obtained in (c)?

3.6   Consider the aircraft model AIRC given in the Appendix. Suppose that $a_{13} = 1.132 \pm 5\%$ and $b_{41} = 4.419 \pm 10\%$, and that all other parameters are known exactly.

(a)   Using the controller designed in Section 4.4, obtain a representation of the closed-loop system in the form of Figure 3.9.

(b)   Examine the stability robustness of the system, by computing $\|Q_{22}\|_\infty$.

(c)   Examine the stability robustness by estimating $\|Q_{22}\|_\mu$.

(d)   Examine the stability robustness by the use of a Nyquist array.

3.7   (a)   Find the transfer-function matrix of the plant model $P(s)$ shown in Figure 3.12.

(b)   In Figure 3.12, y_1 and y_2 are the controlled variables. These are fed back through conventional proportional-plus-integral compensators, each with transfer function

$$\frac{k_I + sk_P}{s}$$

to the flow control values (see Section 3.10.1 and 3.10.2). In order to investigate the effect of a loop failure, one of the compensators is assumed to have a multiplicative uncertainty of 100% in each of its parameter values. Draw a detailed block diagram which includes this uncertainty, in the form shown in Figure 3.9.

(c)   Does (b) describe a sensible way of investigating loop failure?

3.8   (a)   Consider the feedback loop shown in Figure 3.13, with $\Delta_a = 0$. For each of the possible descriptions of plant uncertainty which appear in equations (3.81), (3.82), (3.90) and (3.121), draw the diagram corresponding to Figure 3.14, find expressions for Q_{11}, Q_{12}, Q_{21} and Q_{22} (replacing those which appear in (3.107)–(3.110)), and hence find a necessary and sufficient condition for robust stability, assuming only that $\tilde{\Delta}$ is stable and that $\|\tilde{\Delta}\|_\infty \leqslant 1$ in each case, where $\Delta = W_1 \tilde{\Delta} W_2$.

(b)   Repeat (a) for simultaneous input and output uncertainty, as modelled by (3.88), and show that this is in fact an instance of structured uncertainty.

3.9   (a)   Using the notation defined in Section 3.11.2, show that

$$DAD^{-1} = \Delta$$

Prove the properties of the structured singular value $\mu(.)$ stated in (3.135)–(3.140) and (3.142)–(3.143).

(b)   Show that $\Delta \in BD_1$ (see Section 3.11.2 definitions), if and only if $\|\Delta\|_\infty \leqslant 1$. (This is needed for the proof of Theorem 3.6.)

3.10   (a)   If Δ has elements $\{\delta_{ij}\}$ $(|\delta_{ij}| \leqslant r_{ij})$ and $\{r_{ij}\}$ are the elements of R, then $\bar{\sigma}(\Delta) \leqslant \bar{\sigma}(\text{abs}(\Delta)) \leqslant \bar{\sigma}(R)$. Show that, if D_1 and D_2 are diagonal, real matrices with positive elements, then $\bar{\sigma}(D_1 \Delta D_2) \leqslant \bar{\sigma}(D_1 R D_2)$. Hence show that

$$\rho(Q\Delta) \leqslant \bar{\sigma}(D_1 R D_2)\bar{\sigma}(D_2^{-1} Q D_1^{-1})$$

(b)   Given a 'full' feedback perturbation Δ about a compensated plant Q, with element-by-element bounds $|\delta_{ij}| \leqslant r_{ij}$, show explicitly how to obtain an equivalent diagonal perturbation $\tilde{\Delta}$, and the corresponding compensated plant transfer function $\tilde{Q}$, with the bounds $|\tilde{\delta}_{ii}| \leqslant 1$.

(c)   Given

$$Q = \begin{bmatrix} 1-j1 & 3+j1 \\ 1+j2 & 0.9-j0.21 \end{bmatrix}$$

and

$$R = \begin{bmatrix} 0.5 & 1 \\ 0.65 & 0.55 \end{bmatrix}$$

(Kouvaritakis and Latchman, 1985), find the corresponding $\tilde{Q}$, as defined in (b), and estimate $\mu(\tilde{Q})$ by using the methods of Safonov and of Kouvaritakis and Latchman.

(*Solution*: Safonov, 5.081; Kouvaritakis and Latchman, 4.732)

3.11 Estimate $\mu(Q)$ for Example 3.5 by using equation (3.152), and compare your estimate with Figure 3.20.

Repeat the assessment of performance robustness for the compensator

$$K = \begin{bmatrix} \frac{3}{4} & -\frac{2}{3} \\ \frac{2}{3} & \frac{3}{4} \end{bmatrix} \begin{bmatrix} \dfrac{10(s+1)}{3s(s+16)} & 0 \\ 0 & \dfrac{9(16s+1)}{32s(s+1)} \end{bmatrix} \begin{bmatrix} \frac{3}{4} & \frac{2}{3} \\ \frac{2}{3} & -\frac{3}{4} \end{bmatrix}$$

This compensator design is taken from Doyle *et al.* (1982).

(*Solution*: Performance robustness achieved for $\alpha > 5.5$)

3.12 (a)   When a sampled-data system is designed using continuous-time techniques, the dynamic effects of a zero-order hold are often approximated by the transfer function

$$\frac{1}{1 + \frac{1}{2}sT}$$

(Franklin *et al.*, 1986). Show that, if the sampling process is considered as a perturbation of the plant model, then the techniques described in this chapter can be used to select a sampling rate, assuming that a continuous-time design has already been performed.

What are the limitations of this approach?

(b)   A plant and compensator are given by

$$g(s) = \frac{1}{s^2}, \qquad k(s) = \frac{0.81(s+0.2)}{s+2}$$

(Franklin *et al.*, 1986). The design specification is that

$$\left| \frac{g(s)k(s)}{1 + g(s)k(s)} \right|_{s=j\omega} \leqslant \left| \frac{1.25}{s^2 + 1.26s + 0.81} \right|_{s=j\omega}$$

If the compensator is to be implemented digitally, estimate the lowest acceptable sampling rate. Obtain digital equivalents of

$k(s)$ and $g(s)$ (see Franklin and Powell, 1980, or Franklin *et al.*, 1986, for methods of doing this) with your chosen sample rate, and check whether the specification is met (with frequency responses evaluated on the unit circle).

References

Åström K.J. and Wittenmark B. (1984). *Computer-Controlled Systems*. Englewood Cliffs NJ: Prentice-Hall.

Bhattacharyya S.P. (1987). *Robust Stabilization Against Structured Perturbations*, Lecture Notes in Control and Information Sciences, Vol. 99. Berlin: Springer-Verlag.

Boyd S., Balakrishnan V. and Kabamba P. (1988). On computing the H_∞ norm of a transfer function. *Proc. 1988 American Control Conference*, Atlanta GA, June 1988.

Daniel R.W., Kouvaritakis B. and Latchman H. (1986). Principal direction alignment: A geometric framework for the complete solution to the μ-problem. *Proceedings of the Institution of Electrical Engineers*, Part D, **133**, 45–56.

Doyle J.C. (1982). Analysis of feedback systems with structured uncertainties. *Proceedings of the Institution of Electrical Engineers*, Part D, **129**, 242–50.

Doyle J.C., Wall J.E. and Stein G. (1982). Performance and robustness analysis for structured uncertainty. In *Proc. IEEE Conf. on Decision and Control*, Orlando FL, pp. 629–36.

Fan M.K.H. and Tits A.L. (1988). m-form numerical range and the computation of the structured singular value. *IEEE Transactions on Automatic Control*, **AC-33**, 284–9.

Francis B.A. (1987). *A Course in H_∞ Control Theory*, Lecture Notes in Control and Information Sciences, Vol. 88. Berlin: Springer-Verlag.

Franklin G.F., Powell J.D. and Emami-Naeini A. (1986). *Feedback Control of Dynamic Systems*. Reading MA: Addison-Wesley.

Franklin G.F. and Powell J.D. (1980). *Digital Control of Dynamic Systems*. Reading MA: Addison-Wesley.

Freudenberg J.S. and Looze D.P. (1986). An analysis of H_∞ optimization design methods. *IEEE Transactions on Automatic Control*, **AC-31**, 194–200.

Glover K. (1986). Robust stabilization of linear multivariable systems: relations to approximation. *International Journal of Control*, **43**, 741–66.

Golub G.H. and van Loan C.F. (1983). *Matrix Computations*. Baltimore MD: Johns Hopkins University Press.

Horowitz I. (1963). *Synthesis of Feedback Systems*. New York: Academic Press.

Horowitz I. (1982). Quantitative feedback theory. *Proceedings of the Institution of Electrical Engineers*, Part D, **129**, 215–26.

Hung Y.S. and MacFarlane A.G.J. (1982). *Multivariable Feedback: A Quasi-classical Approach*, Lecture Notes in Control and Information Sciences, Vol. 40. Berlin: Springer-Verlag.

Kailath T. (1980). *Linear Systems*. Englewood Cliffs NJ: Prentice-Hall.

Kouvaritakis B. and Latchman H. (1985). Necessary and sufficient stability criterion for systems with structured uncertainties: The major principal direction alignment principle. *International Journal of Control*, **42**, 575–98.

MacFarlane A.G.J. and Scott-Jones D.F.A. (1979). Vector gain. *International Journal of Control*, **29**, 65–91.

Osborne E.E. (1960). On preconditioning of matrices. *J. of ACM*, **7**, 338–45.

Pang G.K.H. and MacFarlane A.G.J. (1987). *An Expert Systems Approach to Computer-Aided Design of Multivariable Systems*, Lecture Notes in Control and Information Sciences, Vol. 89. Berlin: Springer-Verlag.

Papoulis A. (1984). *Probability, Random Variables and Stochastic Processes*. New York: McGraw-Hill.

Postlethwaite I., MacFarlane A.G.J. and Edmunds J.M. (1981). Principal gains and principal phases in the analysis of linear multivariable feedback systems. *IEEE Transactions on Automatic Control*, **AC-26**, 32–46.

Priestley M.B. (1981). *Spectral Analysis and Time Series*. London: Academic Press.

Safonov M.G. (1982). Stability margins of diagonally perturbed multivariable feedback systems. *Proc. IEEE, Part D*, **129**, 251–6.

Safonov M. and Doyle J.C. (1984). Minimizing conservativeness of robustness singular values. In *Multivariable Control: New Concepts and Tools* (Tzafestas S.G., ed.), pp. 197–207. Dordrecht: Reidel.

Skogestad S. and Morari M. (1987). Implications of large RGA elements on control performance. *Industrial and Engineering Chemistry Research*, **26**, 2323–330.

Smith M.C. (1982). A generalised Nyquist/root-locus theory for multi-loop feedback systems. *PhD Thesis*, Cambridge University.

Stewart G.W. (1973). *Introduction to Matrix Computations*. New York: Academic Press.

Vidyasagar M. (1985). *Control System Synthesis: A Factorization Approach*. Boston MA: MIT Press.

Wilkinson J.H. (1965). *The Algebraic Eigenvalue Problem*. Oxford: Clarendon Press.

CHAPTER 4

Multivariable Design: Nyquist-like Techniques

4.1 Introduction
4.2 Sequential loop closing
4.3 The characteristic-locus method
4.4 Design example
4.5 Reversed-frame normalization
4.6 Nyquist-array methods

4.7 Achieving diagonal dominance
4.8 Design example
4.9 Quantitative feedback theory
4.10 Control-structure design
Summary
Exercises
References

4.1 Introduction

In this chapter we look at those design techniques that arise as generalizations of the classical SISO techniques reviewed in Chapter 1. When faced with a multivariable feedback design problem, one should certainly begin by trying one or more of the methods described here. They can often be made to work with relatively little effort, and they have the advantage that the designer has more control over the details of the control law than is possible with the methods introduced in Chapters 5 and 6. They can also be understood more easily by those familiar with the classical methods.

In this chapter the design techniques are arranged in an order which is governed by convenience of presentation, rather than the order in which they should be attempted on a design problem. A personal view of the order in which they should be attempted is the following. Try *sequential loop closing* first, since this is the simplest thing to do (but it is also the least powerful, so do not be dismayed if it fails). Then try *Nyquist-array* methods. These allow you to work in the real-coordinate frame of the problem, which in turn allows the understanding of any special features of the plant to be exploited in the design. Finally, turn to the *characteristic-locus* method. This is very powerful, and less conservative than the Nyquist-array approach, but forces you to work in the eigenframe of the plant, which usually makes it much more difficult to relate observed characteristics to the physical behaviour of the plant.

    If none of these approaches has been successful, then it will be necessary to move into deeper water. The *reversed-frame normalization* method may be helpful, but there have not been many successful uses of this to date, and it should probably be regarded as being mainly of theoretical interest. *Quantitative feedback theory* is a much more proven technique, but although it is (briefly) described in this chapter, its complexity is comparable to that of the techniques presented in later chapters.

4.2   Sequential loop closing

The simplest approach to multivariable design is to ignore its multivariable nature. A SISO controller is designed for one pair of input and output variables. When this design has been successfully completed, another SISO controller is designed for a second pair of variables, and so on. Figure 4.1 illustrates the procedure. When designing each controller, the effects of those controllers around which feedback loops have already been closed should of course be taken into account. This is very easy to do, using a computer-aided design facility, since any relevant objects such as Nyquist loci can be readily computed and displayed. In fact, the Nyquist loci of all the remaining open loops can be displayed, and this may help in choosing which loop to tackle next.

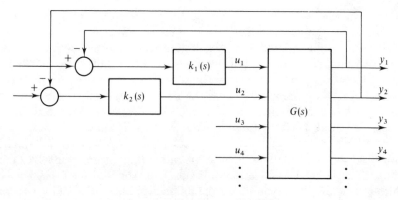

Figure 4.1 Sequential loop closing.

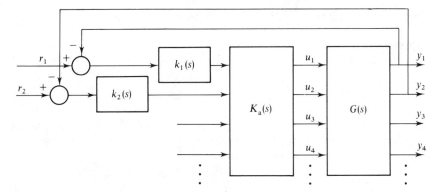

Figure 4.2 Sequential return difference.

Such a sequential design procedure is often adopted in practice. It is appropriate when the controller structure is constrained to be diagonal – that is, to be such that no interactions between loops can take place. (The transfer-function matrix of such a controller is diagonal.) It also has the advantage that it can be implemented by closing one loop at a time, since the design procedure ensures that the system remains stable at each step. This is an important consideration in process control. If a fully cross-coupled controller is used, all feedback loops must be closed simultaneously; if only some of them are closed, there is no guarantee that the system will be stable.

A sequential procedure, though, suffers from a number of drawbacks. It allows only a very limited class of controllers to be designed, and the design must proceed in a very *ad hoc* manner. Design decisions made when closing the first one or two loops may have deleterious effects on the behaviour of the remaining loops. And the only means available for the reduction of inter-action (if this is a requirement) is to use high loop gains.

If we had available powerful methods of deciding how input and output variables should be paired (the **loop-assignment problem**) then sequential loop closing would have a lot to recommend it. But unfortunately this aspect of multivariable design is relatively undeveloped (it is discussed in Section 4.10).

A more sophisticated version of sequential loop closing, proposed by Mayne (1973, 1979), is called the **sequential return-difference** method. Mayne suggests that, before starting to design the individual loop compensators, a cross-coupling stage of compensation should be introduced. This stage should consist of either a constant-gain matrix or a sequence of **elementary operations** (scaled by a common denominator, to obtain a realizable compensator – see Section 2.2), and its purpose is to redistribute the 'diffi-culty of control' among the loops. Figure 4.2 shows the structure of the resulting controller; $K_a(s) = U(s)/d(s)$, where $U(s)$ is unimodular (that is, $\det U(s) = \text{const.}$), and so introduces no zeros.

EXAMPLE 4.1 (Mayne, 1979)

Suppose that

$$G(s) = \frac{1}{(s+1)^2} \begin{bmatrix} 1-s & \frac{1}{3}-s \\ 2-s & 1-s \end{bmatrix} \qquad (4.1)$$

If we try to design a SISO controller for either the first or second loop here, we have difficulties, if the required bandwidth is close to unity, or greater, because the transfer function 'seen' for the design, namely the $(1, 1)$ or $(2, 2)$ element of $G(s)$, has a zero at $+1$. However, $G(s)$ itself has a transmission zero at -1 only, so there should be no inherent difficulty of this kind. If we choose

$$K_a = \begin{bmatrix} 1 & 0 \\ -2 & 1 \end{bmatrix} \qquad (4.2)$$

then

$$Q(s) = G(s)K_a = \frac{1}{(s+1)^2} \begin{bmatrix} \frac{1}{3}+s & \frac{1}{3}-s \\ s & 1-s \end{bmatrix} \qquad (4.3)$$

and we see that no right half-plane zero 'appears' when a SISO compensator $k_1(s)$ is being designed for the first loop. Once the first loop

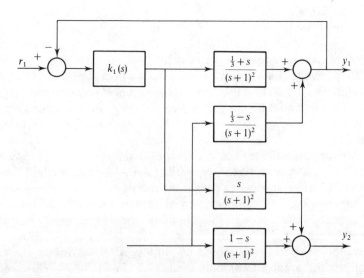

Figure 4.3 Sequential return-difference controller: loop 1 already closed, loop 2 still open.

has been closed, the transfer function 'seen' in the second loop is

$$q_{22}^1(s) = \frac{1-s}{(1+s)^2} + \frac{s}{(s+1)^2} h(s) \frac{\frac{1}{3}-s}{(s+1)^2} \qquad (4.4)$$

(see Figure 4.3), where

$$h(s) = \frac{-k_1(s)}{1 + [(\frac{1}{3}+s)/(s+1)^2] k_1(s)} \qquad (4.5)$$

Now, if we assume that

$$|k_1(s)| \gg \left| \frac{(s+1)^2}{\frac{1}{3}+s} \right|$$

(high gain in the first loop) then

$$h(s) \approx -\frac{(s+1)^2}{\frac{1}{3}+s} \qquad (4.6)$$

and hence

$$q_{22}^1(s) \approx \frac{1}{(3s+1)(s+1)} \qquad (4.7)$$

so that no right half-plane zero is seen when the compensator for the second loop is designed.

This example shows both the main idea and the main weakness of the sequential return-difference approach. The main idea, that of using a first stage of compensation to make subsequent loop compensation easier, is potentially useful. But the main weakness of the method is that little help is available for choosing that first stage of compensation. The available analysis relies on the assumption that there are high gains in the loops which have already been closed (as in the example), and such an assumption can rarely be justified, except at low frequencies.

One rather special case, in which the assumption of high gains is justified, arises when a different bandwidth is required for each loop, and all the bandwidths are well separated from each other. In such a case, if one starts with the fastest (highest-bandwidth) loop, and then designs the second-fastest, and so on, then the assumption of high gain is justified at each step. Suppose, for example, that there are two loops to be closed, with required bandwidths of about 100 and $10 \, \text{rad s}^{-1}$, respectively. Once the faster loop has been designed, the design of the $10 \, \text{rad s}^{-1}$ loop is undertaken, on the assumption that the $100 \, \text{rad s}^{-1}$ loop has been closed. The cross-over frequency for the second loop should be somewhere between 5 and

$10 \, \text{rad} \, \text{s}^{-1}$, and over this frequency range the (open-loop) gain of the first loop is likely to be at least 10. Such a separation of bandwidths arises when there are large differences in the possible rates of change of the output variables; for example, a gas pressure can usually be changed much more quickly than a liquid level.

Mayne (1979) suggests that, if the plant has a state-space realization (A, B, C), the product CB being non-singular (and the matrix D being zero), then the first stage of compensation can be chosen to be

$$K_a = (CB)^{-1} \tag{4.8}$$

The idea behind this is that

$$G(s) \to \frac{CB}{s} \quad \text{as } |s| \to \infty \tag{4.9}$$

(see Section 2.6) and hence

$$G(s)K_a \to \frac{I}{s} \quad \text{as } |s| \to \infty \tag{4.10}$$

so that each loop looks like a first-order SISO system at high frequencies. The problem now is that the approximation (4.9) is likely to hold only at frequencies well beyond those at which loop compensation is required.[1] However, an alternative choice, such as

$$K_a \approx G^{-1}(j\omega_b) \tag{4.11}$$

or

$$K_a \approx j\omega_b \, G^{-1}(j\omega_b) \tag{4.12}$$

may be helpful, where ω_b is the required bandwidth (assuming that a single bandwidth can be specified for the whole system). It will be shown in the next section that a similar first stage of compensation is used in the characteristic-locus design method; a method of performing the approximation required by (4.11) or (4.12) will also be given.

4.3 The characteristic-locus method

4.3.1 Approximate commutative compensators

The basic idea behind approximate commutative compensators (MacFarlane and Kouvaritakis, 1977) is to manipulate the characteristic loci as if they were ordinary Nyquist loci. Let us concentrate first on *how* this can be done, and then on *why* it should be useful to do it.

Suppose we have a square transfer-function matrix $G(s)$, with m inputs and outputs, which has a spectral decomposition

$$G(s) = W(s)\Lambda(s)W^{-1}(s) \qquad (4.13)$$

where $W(s)$ is a matrix whose columns are the eigenvectors, or characteristic directions, of $G(s)$, and

$$\Lambda(s) = \text{diag}\{\lambda_1(s), \lambda_2(s), \ldots, \lambda_m(s)\} \qquad (4.14)$$

where the $\lambda_i(s)$ are the eigenvalues, or characteristic functions, of $G(s)$. If the compensator $K(s)$ is given the structure

$$K(s) = W(s)M(s)W^{-1}(s) \qquad (4.15)$$

where

$$M(s) = \text{diag}\{\mu_1(s), \mu_2(s), \ldots, \mu_m(s)\} \qquad (4.16)$$

then the return ratio is

$$-G(s)K(s) = -W(s)\Lambda(s)M(s)W^{-1}(s) \qquad (4.17)$$

$$= -W(s)N(s)W^{-1}(s) \qquad (4.18)$$

where

$$N(s) = \text{diag}\{v_1(s), v_2(s), \ldots, v_m(s)\} \qquad (4.19)$$

and

$$v_i(s) = \lambda_i(s)\mu_i(s) \qquad (4.20)$$

So we see that if the plant and the compensator share the same eigenvectors, then the system obtained by connecting them in series has eigenvalues which are simply the products of the plant and compensator eigenvalues.

The strategy which suggests itself is to obtain graphical displays of the characteristic loci of the plant, $\{\lambda_i(j\omega): i = 1, \ldots, m\}$, and to design a 'compensator' $\mu_i(j\omega)$ for each $\lambda_i(j\omega)$, using the well-established single-loop techniques. One would then obtain the compensator as the series connection of three systems, corresponding to $W^{-1}(s)$, $M(s)$ and $W(s)$.

It is easily checked that, if $K(s)$ has the structure specified in equation (4.15), then $G(s)$ and $K(s)$ commute, namely

$$G(s)K(s) = K(s)G(s) \qquad (4.21)$$

A compensator having this structure is called a 'commutative' compensator. (Also, this is why we were able to refer earlier to 'the return ratio' rather than 'the return ratio at the output of the plant'.)

Unfortunately, it is quite impractical to attempt to build a commutative compensator. The problem is that the elements of the matrices $W(s)$ and $W^{-1}(s)$ are almost always irrational functions, which have no practical realizations.

EXAMPLE 4.2

The transfer-function matrix

$$G(s) = \begin{bmatrix} \dfrac{1}{s+1} & \dfrac{2}{s+1} \\[2mm] \dfrac{2}{s+2} & \dfrac{1}{s+2} \end{bmatrix}$$

has characteristic functions

$$\lambda_{1,2}(s) = \frac{1}{(s+1)(s+2)}\left[2s+3 \pm \sqrt{(16s^2+48s+33)}\right]$$

The characteristic directions are given by

$$\frac{w_{1i}(s)}{w_{2i}(s)} = -2\left[\frac{-s-1 \pm \sqrt{(16s^2+48s+33)}}{s+2}\right]^{-1}$$

A practical alternative is to give the compensator the structure

$$K(s) = A(s)M(s)B(s) \qquad (4.22)$$

where $A(s)$ and $B(s)$ are chosen to be realizable, and such that

$$A(s) \approx W(s) \qquad (4.23)$$

and

$$B(s) \approx W^{-1}(s) \qquad (4.24)$$

(see Figure 4.4).

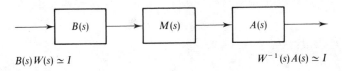

$B(s)W(s) \simeq I$ $W^{-1}(s)A(s) \simeq I$

Figure 4.4 The approximate commutative compensator.

This gives rise to a potentially infinite family of techniques, depending on the algorithm chosen for arriving at the approximations $A(s)$ and $B(s)$. In general, one could envisage using rational approximation theory to arrive at rational matrices $A(s)$ and $B(s)$ which could be realized as dynamic systems. Whichever approximation technique is chosen, we obtain an **approximate commutative compensator**, although in the literature only the simplest case has been investigated, namely that which arises when $A(s)$ and $B(s)$ are constant matrices, and can therefore be realized by networks of amplifiers, without any dynamic elements.

One way of choosing constant matrices A and B is as approximations to $W(s_0)$ and $W^{-1}(s_0)$ at some particular point s_0. Usually one would choose $s_0 = j\omega_0$ for some (real) ω_0, so $W(s_0)$ and $W^{-1}(s_0)$ are usually complex matrices, which need to be approximated by real matrices.

A successful algorithm for computing constant real matrices for use as the matrices A and B is the ALIGN algorithm (Kouvaritakis, 1974). This is based on the observation that a commutative compensator is obtained at a point s_0 even if A is not exactly equal to $W(s_0)$; it is sufficient that the columns of A have the same 'directions' as the columns of $W(s_0)$, namely that they are scalar multiples of the columns of $W(s_0)$. Writing

$$A = (a_1, a_2, \ldots, a_m) \tag{4.25}$$

and

$$W(s_0) = (w_1, w_2, \ldots, w_m) \tag{4.26}$$

we obtain a commutative compensator at s_0 if

$$a_i = w_i z_i, \quad i = 1, 2, \ldots, m \tag{4.27}$$

for some scalar complex numbers z_i, and

$$B = A^{-1} \tag{4.28}$$

If we now define

$$V(s) = W^{-1}(s) \tag{4.29}$$

and write

$$V^{T}(s_0) = (v_1, v_2, \ldots, v_m) \tag{4.30}$$

then

$$v_j^H a_i = 0, \quad i \neq j \tag{4.31}$$

if equation (4.27) holds. Consequently, the ALIGN algorithm chooses the columns of A according to:

$$a_i = \arg\max_{a_i} \frac{|v_i^H a_i|^2}{\sum_{j \neq i} |v_j^H a_i|^2} \tag{4.32}$$

This maximization problem can be solved by solving a generalized eigenvalue problem, as shown by MacFarlane and Kouvaritakis (1977). Let us write the function being maximized in (4.32) as

$$J(a_i) = \frac{a_i^T X a_i}{a_i^T Y a_i} \tag{4.33}$$

where

$$X = v_i v_i^H \tag{4.34}$$

and

$$Y = \sum_{j \neq i} v_j v_j^H \tag{4.35}$$

Since X and Y are Hermitian ($X^H = X$ and $Y^H = Y$), we can replace (4.33) by

$$J(a_i) = \frac{a_i^T C a_i}{a_i^T D a_i} \tag{4.36}$$

where $C = \mathrm{Re}\{X\}$ and $D = \mathrm{Re}\{Y\}$, and C and D are both symmetric and positive-semidefinite.

Differentiating (4.36), we obtain

$$\frac{\partial J(a_i)}{\partial a_i} a_i^T D a_i + 2J(a_i) D a_i = 2C a_i \tag{4.37}$$

which, setting $\partial J/\partial a_i = 0$, gives

$$J(a_i) D a_i = C a_i \tag{4.38}$$

which is a generalized eigenvalue problem (see Section 8.2.8). Thus the maximum value of $J(a_i)$ is obtained as the largest generalized eigenvalue of the pair (C, D), and a_i is obtained as the corresponding generalized eigenvector.

An equivalent formulation of the maximization problem (4.32) is as a least-squares problem (Edmunds and Kouvaritakis, 1979). If we observe that we can impose the constraint

$$|a_i| = |w_i| \tag{4.39}$$

without losing any generality, then we see that in (4.27) we must have

$$z_i = \exp(j\delta_i) \tag{4.40}$$

for some real δ_i. It follows that we can write (if (4.27) holds)

$$A = W(s_0)\operatorname{diag}\{\exp(j\delta_i)\} \tag{4.41}$$

which implies that

$$V(s_0)A = \operatorname{diag}\{\exp(j\delta_i)\} \tag{4.42}$$

The least-squares version of the ALIGN algorithm therefore chooses a_i according to

$$a_i = \arg\min_{a_i, \delta_i} \| V(s_0)a_i - \exp(j\delta_i)e_i \| \tag{4.43}$$

where e_i denotes the ith standard basis vector, and the norm is defined by

$$\|x\|^2 = x^H x \tag{4.44}$$

Note that the solution of (4.43) is not unique, since the same value of the norm is obtained with the pair $(-a_i, \delta_i \pm \pi)$ as with (a_i, δ_i). For the purposes of alignment this does not matter. We shall soon see, however, that the ALIGN algorithm is very useful for obtaining real approximate inverses of complex matrices. When used for this purpose the sign of a_i *does* matter, and it should be chosen so as to give the smaller value of $\| Va_i - e_i \|$.

It can be shown that (4.32) and (4.43) are formally equivalent, although they naturally lead to alternative algorithms.

Use of the approximate commutative compensator, with constant matrices A and B, clearly relies on the eigenvectors of the plant not changing too quickly with frequency, since the compensation would be useless if its effects were predictable at only one frequency. If the eigenvectors do change too quickly, one can try to choose A and B so as to approximate $W(s)$ and $W^{-1}(s)$ at several frequencies rather than one; this is achieved if one chooses

a_i according to

$$a_i = \arg\max_{a_i} \frac{\sum_k p_k |v_i^H(j\omega_k)a_i|^2}{\sum_k p_k \left\{ \sum_{j \neq i} |v_j^H(j\omega_k)a_i|^2 \right\}}$$

(4.45)

where $\{p_1, \ldots, p_N\}$ is a set of (real non-negative) weights which allow the approximation at some frequencies to be emphasized more than at others. If we write the function to be maximized as

$$J(a_i) = \frac{a_i^T \tilde{X} a_i}{a_i^T \tilde{Y} a_i}$$

(4.46)

where

$$\tilde{X} = \sum_k p_k v_i(j\omega_k) v_i^H(j\omega_k)$$

(4.47)

and

$$\tilde{Y} = \sum_k p_k \left\{ \sum_{j \neq i} v_j(j\omega_k) v_j^H(j\omega_k) \right\}$$

(4.48)

then we see that the solution is again obtained by choosing a_i to be the generalized eigenvector which corresponds to the largest generalized eigenvalue of the pair $(\tilde{C}, \tilde{D})$, where $\tilde{C} = \text{Re}\{\tilde{X}\}$ and $\tilde{D} = \text{Re}\{\tilde{Y}\}$. On the other hand, the corresponding formulation of the 'multi-frequency' problem as a least-squares problem does not have a simple solution.

This refinement of the ALIGN algorithm frequently gives no better results than the basic version, which tries to match the eigenvectors at one frequency only. The reason seems to be that the price paid in not obtaining a good approximation to the eigenframe at any frequency often outweighs the advantage gained. If one attempts multi-frequency alignment with a dynamic compensator, however, one can obtain greatly improved frame alignment, as might be expected. We shall pursue this possibility further in Section 4.7.3.

Henceforth we shall use 'ALIGN' to mean the single-frequency alignment algorithm, unless we explicitly state otherwise.

If all the characteristic loci require shaping over approximately the same range of frequencies, then the choice of the frequency at which the ALIGN algorithm should be used does not present a major problem. But it often happens that different loci require compensation at different frequencies. One can then either choose one of these frequencies, and hope that approximate compensation is achieved on all loci satisfactorily, or compensate each locus with a separate approximate commutative compensator:

$$K(s) = K_1(s)K_2(s) \ldots K_m(s) \qquad \text{(4.49)}$$

where

$$K_i(s) = A_i M_i(s) B_i \qquad \text{(4.50)}$$

and $K_i(s)$ is designed to provide compensation at frequency $j\omega_i$. The second alternative leads to a very complicated compensator, and suffers from the difficulty that each factor of the compensator (i.e. each $K_i(s)$) interferes with all the others.

4.3.2 Design procedure

Now that we know how to manipulate the characteristic loci, let us consider why it should be useful to do so. One reason is, obviously, that it may allow us to obtain a stable design by satisfying the generalized Nyquist stability theorem. But merely attaining stability is rarely the objective of feedback design, and is certainly not the sole objective of the classical 'locus-bending' techniques. Do we gain anything if we do not merely achieve the required number of encirclements by the characteristic loci, but also ensure that the characteristic loci stay outside the $M = \sqrt{2}$ circle, for example, or have large magnitudes at low frequencies?

The answer is that we certainly need to give the characteristic loci features such as these, but that even if all the loci look satisfactory – when judged against classical SISO criteria, as if they were Nyquist loci – this may not ensure satisfactory performance of the multivariable feedback system.

Suppose that the plant and compensator together have the spectral decomposition

$$G(s)K(s) = W(s)\operatorname{diag}\{v_i(s)\}W^{-1}(s) \qquad \text{(4.51)}$$

$$= \sum_{i=1}^{m} w_i(s)v_i(s)v_i^{\mathrm{T}}(s) \qquad \text{(4.52)}$$

(using notation as in (4.26) and (4.30)). The sensitivity function is then

$$S(s) = W(s)\operatorname{diag}\left\{\frac{1}{1 + v_i(s)}\right\}W^{-1}(s) \qquad \text{(4.53)}$$

$$= \sum_{i=1}^{m} w_i(s)\frac{1}{1 + v_i(s)}v_i^{\mathrm{T}}(s) \qquad \text{(4.54)}$$

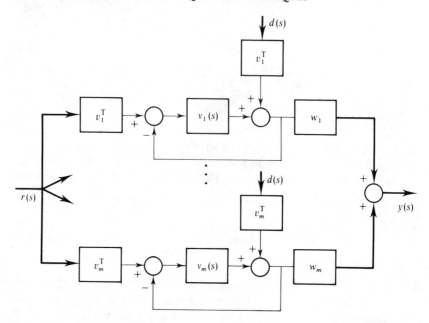

Figure 4.5 Interpretation of the spectral decomposition of a multivariable compensated plant inside a feedback loop as a collection of SISO feedback loops.

and the closed-loop transfer function is

$$T(s) = W(s)\,\text{diag}\left\{\frac{v_i(s)}{1 + v_i(s)}\right\} W^{-1}(s) \tag{4.55}$$

$$= \sum_{i=1}^{m} w_i(s)\,\frac{v_i(s)}{1 + v_i(s)}\,v_i^T(s) \tag{4.56}$$

Expressions (4.54) and (4.56) lead to the following interpretation, which is illustrated by Figure 4.5. The input signal $r(s)$ is 'projected' onto the ith characteristic direction $w_i(s)$. This projection results in the signal $v_i^T(s)r(s)$, which is the input to a (SISO) feedback system having the transfer function $v_i(s)$ in the forward path and unity negative feedback around it. The outputs of these m feedback systems are added together (vectorially) to give the system output $y(s)$. The disturbance signal $d(s)$ is similarly projected onto the characteristic directions, and each component of the disturbance enters the corresponding feedback loop in the usual way.

It seems reasonable to expect that, if each of these m feedback systems is designed according to classical precepts (satisfactory speed of response, adequate rejection of disturbances and so on), then the whole multivariable feedback system will inherit these properties. Unfortunately, this is not always true. As we saw in Chapter 3, the performance of a feedback system depends essentially on the principal gains of functions such as $S(s)$ and $T(s)$. Now, equations (4.53) and (4.55) show that these transfer functions have eigenvalue functions

$$\left\{\frac{1}{1 + v_i(s)}\right\} \quad \text{and} \quad \left\{\frac{v_i(s)}{1 + v_i(s)}\right\}$$

respectively, and we know that, for any matrix X,

$$\underline{\sigma}(X) \leqslant |\lambda(X)| \leqslant \bar{\sigma}(X) \tag{4.57}$$

where $\lambda(X)$ denotes any eigenvalue of X. We can therefore conclude that if the characteristic loci of GK have low magnitudes at low frequencies, then the sensitivity will certainly be large, in some signal directions, at those frequencies. And if some characteristic locus penetrates the $M = \sqrt{2}$ circle, then at least one principal gain of $T(s)$ will exhibit a resonance peak, of magnitude greater than $\sqrt{2}$. But we cannot draw the converse conclusions.

Equations (4.53) and (4.55) also show that GK, S and T all share the same eigenvectors, and in Chapter 3 we saw that the magnitudes of the smallest and largest eigenvalues are close to the smallest and largest singular values if the eigenvectors are nearly orthogonal to each other. We can therefore expect that shaping the $v_i(j\omega)$ will lead to good system performance if the return ratio has low skewness (see Chapter 3).

Note that the discrepancies between characteristic loci and principal gains do not make the shaping of characteristic loci a fruitless activity. On the contrary, manipulating the characteristic loci is often the most straightforward and productive way of designing a multivariable feedback system, or at least of initiating the design. Of course, the properties of the resulting design must be checked by methods more revealing than examination of the characteristic loci.

A frequent requirement for multivariable systems – one which does not arise for SISO systems – is that they should exhibit little interaction, that is, a change in one of the reference signals should cause only the corresponding output to change, without excessive transients occurring on the other outputs. Alternatively, the appearance of a sudden disturbance on one of the outputs should not disturb the other outputs excessively. (For systems with one degree of freedom, these statements are equivalent.) If we had 'high gain' everywhere in the sense that

$$|v_i(s)| \gg 1, \quad \text{for each } i \tag{4.58}$$

then

$$\frac{v_i(s)}{1 + v_i(s)} \approx 1 \tag{4.59}$$

and so, from (4.55), we would have

$$T(s) \approx I \tag{4.60}$$

This generalizes the familiar single-loop argument which shows the benefits of high loop gain, and demonstrates that lack of interaction is a benefit obtained in addition to the usual single-loop benefits. But we know that, beyond some frequency, the condition (4.58) cannot hold, both for physical reasons and because the generalized Nyquist theorem needs to be satisfied. Thus, at these higher frequencies, feedback cannot be relied upon to enforce low interaction: the only thing that can be done is to insert a series compensator whose main purpose is to obtain a 'decoupled' (that is, diagonal) return ratio of $-G(s)K(s)$ at (some of) these frequencies. One way of attempting to decouple the return ratio at one or several frequencies is to attempt to invert $G(s)$ at these frequencies, and we already have an algorithm for doing this – the ALIGN algorithm.

Suppose that we decide to use the ALIGN algorithm to obtain an approximate decoupling of the system at a particular frequency. How should that frequency be chosen? A closed-loop bandwidth ω_b can be achieved (if the characteristic directions span the signal space well) by setting

$$\frac{|v_i(j\omega_b)|}{|1 + v_i(j\omega_b)|} = \frac{1}{\sqrt{2}}, \quad \text{for each } i \tag{4.61}$$

which implies that $|v_i(j\omega_b)|$ lies on the $M = 1/\sqrt{2}$ circle, as shown in Figure 4.6. From the figure it is clear that $|v_i(j\omega_b)|$ cannot exceed 1 by much, and is usually less than 1, so high gain can no longer be relied upon to reduce interaction at the frequency ω_b. However, at frequencies much higher than ω_b the 'loop gains' will usually be very low, so contributions to the output at these frequencies will be unimportant. This leads to the conclusion that 'high-frequency decoupling' should be attempted at or near ω_b.

So the first step of the design procedure is to obtain a 'high-frequency' constant compensator $K_h \approx -G^{-1}(j\omega_b)$. The negative sign is chosen because this shifts the eigenvalues of $G(j\omega_b)K_h$ into the vicinity of -1, rather than $+1$. In the final design the characteristic loci at the frequency ω_b should be closer to -1 than $+1$ (see Figure 4.6), so this choice is likely to make later design steps easier. One effect of the compensator K_h is that it results in all the characteristic loci having similar values near the frequency ω_b, whereas the original plant's loci may be very widely scattered at this frequency. This makes it easier to design an approximate commutative compensator in the next design step, since all the loci require compensation at similar frequencies.

It may be advantageous to adjust the signs of the columns of K_h at this stage. For example, some column of $G(j\omega_b)K_h$ may be very far from a

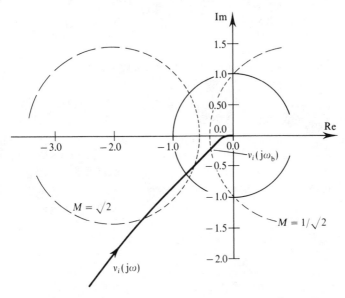

Figure 4.6 A characteristic locus, with the $M=\sqrt{2}$, $M=1/\sqrt{2}$, and $|S|=1$ circles.

column of a real unit matrix, in which case one of the characteristic loci will be very far from -1 at frequency ω_b. In such a case changing the sign of the corresponding column of K_h may bring this characteristic locus into a region in which subsequent compensation becomes easier. In particular, it may change the number of encirclements of -1 by the characteristic loci; it is usually easier to achieve the correct number of encirclements at this stage by means of sign changes, than at a later stage by means of dynamic compensation. Note that a change of sign does not just change the phase of one locus – in general it will alter the loci completely.

Grosdidier *et al.* (1985) have shown that, if the compensator has the structure

$$K(s) = \frac{k}{s} C(s)$$

and $G(s)C(s)$ is proper, then it is always possible to find a $C(s)$ such that the closed-loop system is stable for a range of gains

$$0 < k < k^*$$

if all the eigenvalues of $G(0)C(0)$ have positive real parts. Furthermore, it is impossible to find such a $C(s)$ if any eigenvalue of $G(0)C(0)$ has a negative real part. This strongly suggests that the signs of the columns of K_h should be adjusted to bring all the eigenvalues of $G(0)K_h$ into the right half-plane, if possible.

A suitable overall design procedure is the following, which has been called the **characteristic-locus method**:

(1)   Compute a real $K_h \approx -G^{-1}(j\omega_b)$.

(2)   Design an approximate commutative controller $K_m(s)$ at some frequency $\omega_m < \omega_b$, for the **compensated plant** $G(s)K_h$, such that $K_m(j\omega) \to I$ as $\omega \to \infty$ (the latter requirement is an attempt to ensure that the decoupling effected by K_h is not disturbed too much). Ideally we should like $K_m(j\omega)$ to approach I as $\omega \to \omega_b$, but this is not a realistic goal.

(3)   If the low-frequency behaviour is unsatisfactory (typically because there are excessive steady-state errors), design an approximate commutative controller $K_1(s)$ at some frequency $\omega_1 < \omega_m$, for the compensated plant $G(s)K_hK_m(s)$, such that $K_1(j\omega) \to I$ as $\omega \to \infty$.

(4)   Realize the complete compensator as

$$K(s) = K_hK_m(s)K_1(s) \qquad\qquad (4.62)$$

The subscripts h, m and l denote high, medium and low frequency, respectively. Usually, $K_1(s)$ is used to introduce integral action.

Step 2, the medium-frequency compensation, is concerned with shaping the characteristic loci in the vicinity of -1, and this is where phase lead or lag, or more complicated shaping, is applied to the loci. This means, however, that frequencies ω_m and ω_b are quite close together, and step 2 is likely to interfere with step 1. To reduce the interaction between these two steps it is often necessary to obtain K_h at a frequency rather higher than ω_b. K_m can be designed at a frequency rather lower than the final 0 dB cross-over frequency of the loci, but the dynamic elements within it will inevitably have to exert their influence near this frequency.

The characteristic-locus design technique is best suited to plants which are 'uniform' in the sense that it is feasible to obtain a similar bandwidth for each loop with reasonable control signals. If a plant is not uniform in this sense, for example because it has two outputs which can respond very quickly and two which can respond only rather slowly, then it may be possible to perform two (or more) characteristic-locus designs: the first one would usually be for the fast loops, and the second for the slower loops, the fast-loop design being taken into account. In other words, a kind of sequential loop-closing approach may be appropriate, the characteristic-locus method being used inside the sequential loop-closing approach.

4.4 Design example

(1) *THE PLANT*

Consider the aircraft model AIRC described in the Appendix. This has three inputs, three outputs and five states.

(2) *THE SPECIFICATION*

We shall attempt to achieve a bandwidth of about $10 \, \mathrm{rad \, s^{-1}}$ for each loop, with little interaction between outputs, good damping of step responses and zero steady-state error in the face of step demands or disturbances. We assume a one-degree-of-freedom control structure.

(3) *PROPERTIES OF THE PLANT*

The time responses of the plant to unit step signals on inputs 1 and 2 exhibit very severe interaction between outputs. The poles of the plant (eigenvalues of A) are found to be

$$0, \quad -0.78 \pm 1.03j, \quad -0.0176 \pm 0.1826j$$

so the system is stable (but not asymptotically stable). The number of finite zeros of the plant can be at most $n - m - d$, where n is the McMillan degree (5), m is the number of inputs or outputs (3), and d is the 'rank defect' of the product CB (see Chapter 2). For the present example,

$$CB = \begin{bmatrix} 0 & 0 & 0 \\ -0.12 & 1 & 0 \\ 0 & 0 & 0 \end{bmatrix} \tag{4.63}$$

which has rank $= 1$, and therefore $d = 2$. Thus this plant has no finite zeros, and we do not expect any limitations on performance to be imposed by zeros.

Figure 4.7 shows the three characteristic loci of the plant. Since the range of magnitudes is very large over the frequency range (0.001 to $100 \, \mathrm{rad \, s^{-1}}$), these loci have been shown in the usual polar form, but with logarithmic calibration of the axes. The crescent-shaped dashed line is the $M = \sqrt{2}$ circle, distorted by the logarithmic calibration. The loci have also been displayed for both positive and negative frequencies, since this makes it easier to count encirclements. It can be seen that the loci encircle the point $-1 + 0j$ (which lies inside the $M = \sqrt{2}$ 'circle') twice clockwise and once anticlockwise, giving one net clockwise encirclement. We conclude that if negative feedback were applied around the plant as it stands, the closed loop would be unstable.

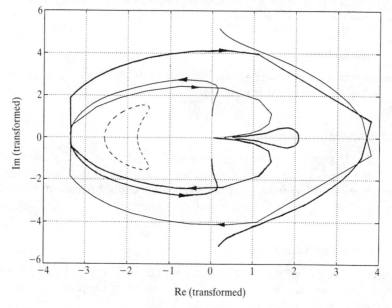

Figure 4.7   Characteristic loci of plant, with logarithmic calibration of the axes (the broken curve is the $M = \sqrt{2}$ 'circle').

(4)   *ALIGNMENT AT* 10 rad s^{-1}

Since we are aiming for a closed-loop bandwidth of 10 rad s^{-1}, we try using the ALIGN algorithm at that frequency in order to reduce interaction there. The resulting (constant) pre-compensator is:

$$K_h = \text{ALIGN}(-G(j10))$$

$$= \begin{bmatrix} -71.535 & -0.0036 & -3.669 \\ -8.5375 & -9.9984 & -0.5376 \\ -189.44 & 0.0065 & -69.378 \end{bmatrix} \qquad (4.64)$$

This gives

$$G(j10)K_h = \begin{bmatrix} 0.983+0.068j & 0.0001-0.003j & -0.0005+0.009j \\ 0.068+0.0004j & -0.005+1.010j & 0.0008+0.0003j \\ -0.009+0.100j & 0.0000-0.0005j & -0.9962-0.086j \end{bmatrix}$$

$$(4.65)$$

which shows that the only significant interactions left at $\omega = 10$ are from input 1 to outputs 2 and 3, and that these interactions have been reduced to

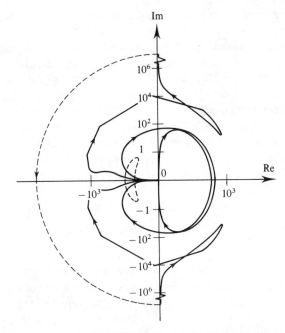

Figure 4.8 Characteristic loci of GK_h, with logarithmic axis calibrations.

10% or less (at that frequency). Note that the (2, 2) element is j rather than -1 (approximately); the power of the ALIGN algorithm is evident here, since it has allowed us to decouple the system at $\omega = 10$, even though there is no good real approximation to $-G(j10)^{-1}$.

In fact, it is more useful to have the second column of $G(j10)K_h$ close to $[0 \ -j \ 0]^T$ rather than to $[0 \ j \ 0]^T$, since this results in one characteristic locus approaching the origin along the negative imaginary axis rather than along the positive imaginary axis, and gives no encirclements of -1 by the set of characteristic loci, as we shall show below. We therefore change the sign of the second column of K_h, to obtain

$$K_h = \begin{bmatrix} -71.535 & 0.0036 & -3.669 \\ -8.5375 & 9.9984 & -0.5376 \\ -189.44 & -0.0065 & -69.378 \end{bmatrix} \qquad \textbf{(4.66)}$$

Figure 4.8 shows the characteristic loci of GK_h, again with logarithmic calibrations. We see that one locus comes up along the imaginary axis from $-j\infty$ (because of the plant pole at 0) and we must therefore be careful about counting encirclements.

If the Nyquist contour is imagined to be indented to the left around the pole at 0, so that one open-loop pole is enclosed within the contour, then for closed-loop stability we need one anticlockwise encirclement of -1 by the characteristic loci. We can see that the image of the indentation does indeed give one such encirclement, either by the usual conformal-mapping argument (since eigenvalues are analytic functions) or by direct calculation: the eigenvalues of $G(-10^{-4}+j0)K_h$ are -10^6, 97 and 160 (approximately). This shows that one characteristic locus becomes large and negative as the indentation crosses the negative real axis, so the image of one locus of $G(s)K_h$, as s goes round the indentation ($s = \varepsilon e^{-j\theta}: \frac{1}{2}\pi \leqslant \theta \leqslant \frac{3}{2}\pi, 0 < \varepsilon \ll 1$) is as shown by the dashed line in Figure 4.8. If we examined the loci in the vicinity of -1 we would see that the other two loci do not encircle it, so the net result is one anticlockwise encirclement. If we closed the feedback loops at this stage the system would be stable.

(5)   *APPROXIMATE COMMUTATIVE COMPENSATOR*

As might be expected from (4.65), however, two of the loci pass extremely close to -1 at $\omega = 10$ and require some compensation. Figure 4.9 shows the loci on a Nichols display, from which we can see that two of the loci require about $40°$ of phase lead at the 0 dB cross-over frequency if they are to remain outside the $M = \sqrt{2}$ circle. We should aim for a cross-over frequency rather lower than the required closed-loop bandwidth, and we choose $5\,\mathrm{rad\,s^{-1}}$ initially. We shall design an approximate commutative compensator, with the compensator K_h still in place. A design was initially obtained at $5\,\mathrm{rad\,s^{-1}}$, but some cut and try showed that the compensator was most effective when

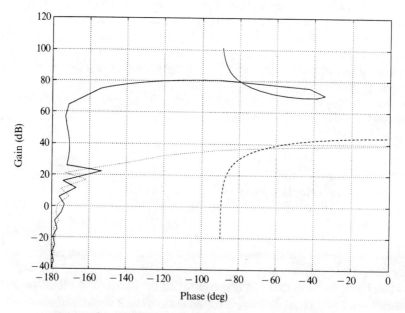

Figure 4.9   Characteristic loci of GK_h, on Nichols chart.

designed at 10 rad s^{-1}: details of this design only are given. We use the ALIGN algorithm again to obtain the constant matrices A and B which appear in equation (4.22). If

$$G(j10)K_h = W\Lambda W^{-1} \tag{4.67}$$

then

$$A = \text{ALIGN}(W^{-1}) = \begin{bmatrix} 0.2426 & -0.2077 & -0.0016 \\ -0.0087 & 0.0079 & 0.9999 \\ 0.6151 & 0.9656 & 0.0010 \end{bmatrix} \tag{4.68}$$

and

$$B = \text{ALIGN}(W) = \begin{bmatrix} 2.1937 & 0.0031 & 0.5587 \\ -1.3900 & -0.0017 & 0.6491 \\ 0.0278 & 1.0000 & 0.0014 \end{bmatrix} \tag{4.69}$$

In order to determine which two of the diagonal elements of Λ require compensation, we have to examine the eigenvalues and eigenvectors of $G(j10)K_h$ when they are evaluated simultaneously by suitable software, to ensure that the ordering of the elements of Λ corresponds to the ordering of the columns of W. In this case it is the $(1,1)$ and $(2,2)$ elements of Λ that require compensation.

We shall try a standard phase-lead compensator of the form

$$\frac{a(1 + sT)}{1 + saT}, \quad a < 1 \tag{4.70}$$

From Chapter 1 we find that to get a maximum phase lead of 40° we need $a = 0.2175$. The maximum phase lead occurs at the frequency $1/T\sqrt{a}$, so to obtain this phase lead at 5 rad s^{-1} we choose $T = 0.429$. We therefore insert, between A and B, the dynamic matrix

$$M(s) = \begin{bmatrix} \dfrac{0.0933s + 0.2175}{0.0933s + 1} & 0 & 0 \\ 0 & \dfrac{0.0933s + 0.2175}{0.0933s + 1} & 0 \\ 0 & 0 & 1 \end{bmatrix} \tag{4.71}$$

In Figure 4.10 the resulting loci are plotted in Nichols form, and the loci are seen to have been given the required phase advance. The intersections with the $M = 1/\sqrt{2}$ circle show that, if they were ordinary Nyquist loci of SISO

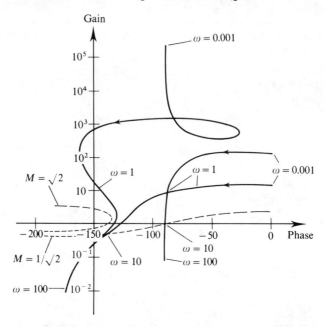

Figure 4.10   Characteristic loci of GK_hK_m, on Nichols chart.

systems, then each of them would give a closed-loop bandwidth of about $10\,\text{rad}\,\text{s}^{-1}$, as required.

Note that our medium-frequency compensator $K_m(s) = AM(s)B$ satisfies $K_m(s) \to I$ as $|s| \to \infty$ (approximately), but $K_m(s)$ clearly has a significant effect at $10\,\text{rad}\,\text{s}^{-1}$, so that the decoupling obtained by K_h may well have been destroyed. To check this we can evaluate $G(j10)K_hK_m(j10)$. To evaluate the interactions at this frequency it is appropriate to examine

$$\text{abs}\,(G(j10)K_hK_m(j10)) = \begin{bmatrix} 0.5664 & 0.0027 & 0.0058 \\ 0.0239 & 0.9999 & 0.0014 \\ 0.0579 & 0.0012 & 0.6797 \end{bmatrix} \qquad \textbf{(4.72)}$$

This shows that decoupling has not been destroyed at $10\,\text{rad}\,\text{s}^{-1}$. We still have an interaction of about 10% from input 1 to output 3, as before.

(6)   LOW-FREQUENCY COMPENSATION

Figure 4.10 shows that one of the characteristic loci has a low-frequency gain of only about 10. Hence we must have $\underline{\sigma}(G(0)K_hK_m(0)) \leqslant 10$, which will certainly not achieve the objective of zero steady-state error in the face of step disturbances. To attain this it is clearly necessary to introduce integral action. We shall apply compensation of the form $(1 + sT)/sT$ to each characteristic locus, since this ensures that the integral action does not affect the loci at

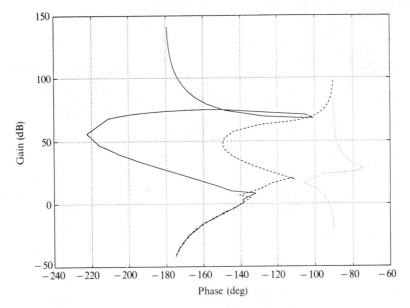

Figure 4.11   Characteristic loci of $GK_hK_mK_1$, on Nichols chart.

frequencies much above $1/T$. Since we do not want frequencies near 5 rad s^{-1} to be affected much, we should choose $T \geqslant 1/0.5 = 2$, say. On the other hand, keeping T small extends the frequency range over which integral action is effective, and so speeds up the elimination of errors. We are therefore driven to choosing $T = 2$. We could choose a different value of T for each locus, but the advantage of having only one value of T is that the commutative controller becomes simply

$$K_1(s) = \frac{1 + sT}{sT} I \qquad \qquad \textbf{(4.73)}$$

and each locus is shaped by $(1 + sT)/sT$ at each frequency. Note that each locus will have an additional phase lag of 45° at $\omega = 0.5$ (compared with Figure 4.10); this is acceptable, but will result in a conditionally stable system, since the resonance lobe exhibited by one of the loci will now cross the negative real axis.

   Figure 4.11 shows the characteristic loci of $G(s)K_hK_m(s)K_1(s)$ in Nichols form, while Figure 4.12 shows its smallest and largest principal gains in Bode form. Figure 4.13 shows the smallest and largest principal gains of the closed-loop transfer function, and shows that the bandwidth requirement has been achieved. Closed-loop time responses to step demands on each of the outputs are shown in Figure 4.14. Interaction is greatly reduced, compared with the open-loop plant, and the responses are well damped.

   Of course it is easy to obtain excellent step responses with a linear model. At this stage one should check typical control-signal levels, to

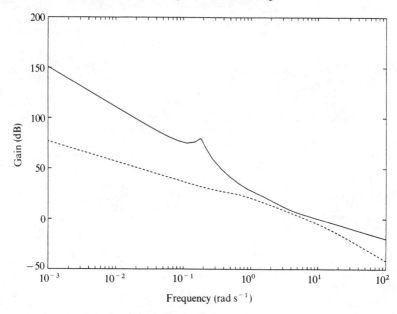

Figure 4.12 Largest and smallest open-loop principal gains.

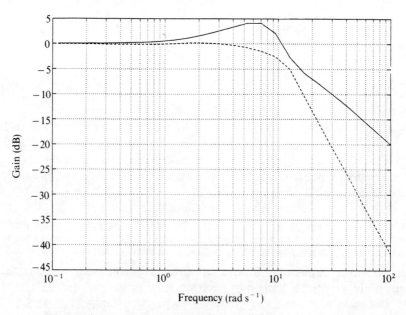

Figure 4.13 Largest and smallest closed-loop principal gains.

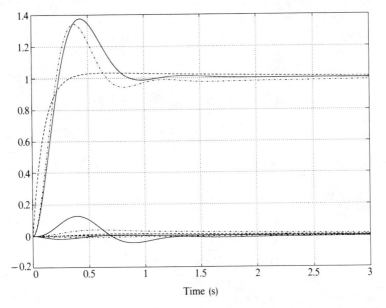

Figure 4.14 Closed-loop step responses to step demand on output 1 (solid curves), output 2 (dashed curves) and output 3 (dash–dotted curves).

determine whether the designed system would behave substantially in accord with the linear model. If it would not, then the original bandwidth specification was probably too ambitious.

(7) *REALIZATION OF THE COMPENSATOR*

We have designed the final compensator as the series connection of five compensators: one for high-frequency alignment (K_h), three for shaping the characteristic loci near the cross-over frequency (K_m) and one for shaping the loci at low frequencies by adding integral action (K_l). The dynamical complexity of the compensator is quite low, however, and it can be realized as a five-state linear system (two states for the phase-lead transfer functions in K_m, and three for the integrators in K_l).

The following is a state-space realization of the complete compensator:

$$KA = \begin{bmatrix} 0 & 0 & 0 & 0 & 0 \\ 0 & 0 & 0 & 0 & 0 \\ 0 & 0 & 0 & 0 & 0 \\ 1.0968 & 0.0016 & 0.2793 & -10.7225 & 0 \\ -0.6950 & -0.0008 & 0.3246 & 0 & -10.7225 \end{bmatrix} \quad (4.74)$$

$$KB = \begin{bmatrix} 1 & 0 & 0 \\ 0 & 1 & 0 \\ 0 & 0 & 1 \\ 2.1937 & 0.0031 & 0.5587 \\ -1.3900 & -0.0017 & 0.6491 \end{bmatrix} \qquad (4.75)$$

$$KC = \begin{bmatrix} -29.3803 & 0.0167 & 1.8067 & 164.6078 & -94.9601 \\ -3.5182 & 5.0006 & -0.2555 & 20.8859 & -11.1922 \\ -78.0162 & -0.0028 & -33.7400 & 743.9029 & 232.0489 \end{bmatrix}$$

$$(4.76)$$

$$KD = \begin{bmatrix} -58.7607 & 0.0333 & -3.6134 \\ -7.0365 & 10.0011 & -0.5110 \\ -156.0324 & -0.0056 & -67.4800 \end{bmatrix} \qquad (4.77)$$

4.5 Reversed-frame normalization

The characteristic-locus design technique is based on the eigenvalue–eigenvector decomposition

$$G(s) = W(s)\Lambda(s)W^{-1}(s)$$

of a square transfer function $G(s)$. Let us now focus instead on the **singular-value decomposition**

$$G(s) = Y(s)\Sigma(s)U^H(s) \qquad (4.78)$$

where

$$\Sigma(s) = \text{diag}\{\sigma_1(s), \ldots, \sigma_m(s)\} \qquad (4.79)$$

the σ_i being the principal gains of $G(s)$, and $U(s)$ and $Y(s)$ being unitary matrices.

The σ_i are real-valued, being in essence measures of 'gain', but we can associate them with 'phase' information in the following way. First we define

$$\theta_i(s) = \arg\min_{\psi_i} \bar{\sigma}[U^H(s)Y(s) - \text{diag}\{\exp(j\psi_i)\}] \qquad (4.80)$$

$$\Theta(s) = \text{diag}\{\exp(j\theta_i(s))\} \qquad (4.81)$$

and

$$m(G(s)) = \bar{\sigma}[U^H(s)Y(s) - \Theta(s)] \qquad (4.82)$$

The point of this is that if $m(G(s)) = 0$, then $U(s)$ and $Y(s)$ are aligned in the sense that, if $u_i(s)$ and $y_i(s)$ are their respective ith columns (namely the ith input and output principal directions, respectively) then

$$y_i(s) = u_i(s) z(s) \tag{4.83}$$

where $z(s)$ is some complex number, and if we normalize $U(s)$ and $Y(s)$ so that $\|y_i(s)\| = \|u_i(s)\|$, then

$$y_i(s) = u_i(s) \exp(j\theta_i(s)) \tag{4.84}$$

(Note that the idea here is the same as that of the least-squares version of the ALIGN algorithm, namely that defined by (4.43).) We shall say that $G(s)$ is **aligned** if $m(G(s)) = 0$.

If we now define

$$\Gamma(s) = \Theta(s)\Sigma(s) \tag{4.85}$$

and

$$Z(s) = Y(s)\Theta^H(s) \tag{4.86}$$

then we can write the decomposition (4.78) as

$$G(s) = Z(s)\Gamma(s)U^H(s) \tag{4.87}$$

This is called the **quasi-Nyquist decomposition** of $G(s)$, and it should be noted that

$$\Gamma(s) = \text{diag}\{\gamma_1(s), \ldots, \gamma_m(s)\} \tag{4.88}$$

where each of the γ_i is now complex-valued, and so carries 'phase' as well as 'gain' information, in some sense.

The sense in which the γ_i carry 'phase' information is made precise by the following result, which implies that the graphs of $\gamma_i(j\omega)$ approximately locate the characteristic loci.

Theorem 4.1:   Let $\lambda(s)$ be an eigenvalue of $G(s)$. Then, for some $\gamma_i(s)$,

$$|\lambda(s) - \gamma_i(s)| \leqslant m(g(s)) \max_j |\gamma_j(s)| \tag{4.89}$$

$$= m(G(s))\bar{\sigma}(G(s)) \tag{4.90}$$

Proof: See Hung and MacFarlane (1982), Chapter 3.4.

that

The loci of $\gamma_i(j\omega)$ are called **quasi-Nyquist loci**, and it should be noted

$$|\gamma_i(j\omega)| = \sigma_i(j\omega) \tag{4.91}$$

Now, suppose that the compensator $K(s)$ has the structure

$$K(s) = U(s)M(s)Z^H(s) \tag{4.92}$$

where

$$M(s) = \text{diag}\,\{\mu_1(s), \ldots, \mu_m(s)\} \tag{4.93}$$

Then

$$G(s)K(s) = Z(s)\Gamma(s)M(s)Z^H(s) \tag{4.94}$$

$$= Z(s)N(s)Z^H(s) \tag{4.95}$$

$$= Y(s)N(s)Y^H(s) \tag{4.96}$$

where

$$N(s) = \text{diag}\,\{v_1(s), \ldots, v_m(s)\} \tag{4.97}$$

and

$$v_i(s) = \gamma_i(s)\mu_i(s) \tag{4.98}$$

Then the return ratios $G(s)K(s)$ and $K(s)G(s)$ are both normal ($[GK]^H GK = GK[GK]^H$, etc.) and aligned ($m(GK) = 0$).

In this case we see that the quasi-Nyquist loci are precisely the characteristic loci, and that their magnitudes are the principal gains of both return ratios. We therefore approach more nearly the classical situation of being able to predict stability, performance and robustness from the same set of loci.

A compensator with the structure defined by equation (4.92) is called a **reversed-frame normalizing** (RFN) compensator. Of course, such a compensator is no more realizable than an exact commutative controller, but, as might be expected, one can obtain useful results by settling for an approximate RFN controller.

Hung and MacFarlane (1982) have developed the above theory and its applications in great detail, and outlined an optimization-based technique for computing approximate RFN controllers. This technique assumes that one knows what a satisfactory return ratio would be, in the form

$$Q(s) = Y(s)N(s)Y^H(s) \tag{4.99}$$

Here $Y(s)$ comes from the singular-value decomposition of the plant, so that only the elements of the diagonal matrix $N(s)$, namely the required quasi-Nyquist loci, need to be specified. Once $Q(s)$ has been specified, the structure of $K(s)$ must be specified in some way (for example, the order of each element) and numerical optimization used to find

$$K(s) = \arg \min_{K(s) \in \kappa} \|Q(s) - G(s)K(s)\| \tag{4.100}$$

where κ denotes the set of compensators over which the search is performed.

Hung and MacFarlane performed least-squares optimization, and specified κ as the set of transfer functions having a matrix-fraction description with a particular canonical form (Hermite form) and McMillan degree. Clearly many alternatives to these particular choices are possible; for example, one could constrain certain elements in $K(s)$ to be zero.

One could also attempt approximate RFN at one specific frequency, and hope that the resulting compensator would be beneficial over a useful range of frequencies, just as we did with the approximate commutative compensator. In particular, if $G(s)$ has no poles at $s = 0$, then $G(0)$ is real, and therefore $U(0)$, $Y(0)$ and $\Sigma(0)$ are all real. In this case an exact (at $s = 0$) RFN compensator can be realized as

$$K(s) = U(0)\,\text{diag}\,\{\mu_1(s), \ldots, \mu_m(s)\}\,Y^H(0) \tag{4.101}$$

(since $Y(0) = Z(0)$). This can be particularly useful for manipulating the principal gains at low frequencies.

EXAMPLE 4.3

A partly compensated return ratio is closely approximated by the transfer function

$$Q(s) = \begin{bmatrix} \dfrac{16.9}{s+5} & \dfrac{36.12}{s+11} \\[2mm] \dfrac{-9.57}{s+5} & \dfrac{-4.17}{s+11} \end{bmatrix} \tag{4.102}$$

It is desired to add integral action by means of a series compensator which will equalize the two principal gains at low frequencies, and give a loop gain of 50 at $\omega = 0.01$.

We find that the singular-value decomposition at $s = 0$ is

$$Q(0) = Y(0)\Sigma(0)U^H(0)$$

$$= \begin{bmatrix} 0.940 & 0.342 \\ -0.342 & 0.940 \end{bmatrix} \begin{bmatrix} 5 & 0 \\ 0 & 1 \end{bmatrix} \begin{bmatrix} 0.766 & 0.643 \\ -0.643 & 0.766 \end{bmatrix} \qquad \textbf{(4.103)}$$

A compensator of the form

$$K(s) = U(0) \begin{bmatrix} \dfrac{s+k}{s} & 0 \\ 0 & \dfrac{s+5k}{s} \end{bmatrix} Y^H(0) \qquad \textbf{(4.104)}$$

would have the desired effect since, at low frequencies, we would have

$$Q(j\omega)K(j\omega) \xrightarrow[\omega \to 0]{} Y(0) \begin{bmatrix} 5 & 0 \\ 0 & 1 \end{bmatrix} \begin{bmatrix} \dfrac{k}{j\omega} & 0 \\ 0 & \dfrac{5k}{j\omega} \end{bmatrix} Y^H(0) \qquad \textbf{(4.105)}$$

so that each of the principal gains would be $5k/\omega$ at low frequencies. To achieve a gain of 50 at $\omega = 0.01$ we choose $k = 0.1$.

Note that this compensator does *not* leave the return ratio un-altered at high frequencies; however, since both principal gains of $K(s)$ become unity at high frequencies ($\omega > 5$, say), the principal gains of $Q(s)$ and $Q(s)K(s)$ *are* the same at high frequencies.

4.6   Nyquist-array methods

4.6.1   Compensator structure

We shall again assume that the plant's transfer function is square. Suppose that a compensator is rational, invertible, and has all its poles and (trans-mission) zeros in the left half-plane (including the origin); one normally uses a compensator of this form if possible. The following theorem shows that the compensator can always be obtained as the series connection of systems of three simple types.

> **Theorem 4.2** (Rosenbrock, 1970):   Let $K(s)$ be square, rational and invertible, and have all its poles and zeros in the open left half-plane.

Then

$$K(s) = K_a K_b(s) K_c(s) \qquad (4.106)$$

and

$$K(s) = K'_c(s) K'_b(s) K'_a \qquad (4.107)$$

where K_a and K'_a are permutation matrices (namely ones which permute rows or columns, or, in other words, reorder the outputs or inputs), $K_b(s)$, $K'_b(s)$ are products of elementary matrices, each elementary matrix having the form

$$K_b^k(s) = I + \alpha_{ij}^k(s) e_i e_j^T, \quad j \neq i \qquad (4.108)$$

where $\alpha_{ij}^k(s)$ is rational and stable, e_i and e_j are the ith and jth standard basis vectors, and $K_c(s)$ and $K'_c(s)$ are diagonal matrices,

$$K_c(s) = \text{diag}\{k_i(s)\} \qquad (4.109)$$

each $k_i(s)$ being rational, non-zero, and with poles and zeros in the open left half-plane only.

Proof: Omitted. It is similar to the proof of the existence of the Smith–McMillan form, given in Chapter 2. See Rosenbrock (1970) for details.

This theorem implies that compensator design can be split into two stages. In the first stage, K_a and $K_b(s)$ are used to make the return ratio diagonally dominant, if possible (but the theorem does not guarantee that this is possible, of course). If dominance is achieved then the system has been decoupled to some extent, and behaves more like a set of independent SISO systems. The second stage begins when dominance has been achieved, and consists of designing a set of separate SISO compensators (the elements of $K_c(s)$, which is diagonal), one for each loop. At this stage no attention is paid to the remaining interactions in the system, except in so far as the Gershgorin bands of the return ratio replace SISO Nyquist loci. Closed-loop stability can be ensured by shaping the Gershgorin bands, using SISO techniques, in such a way that they do not overlap the point $-1 + j0$, and encircle it the appropriate number of times, in accordance with the generalized Nyquist theorem.

Since it is the point $-1 + j0$ rather than the origin that must be left uncovered by the Gershgorin bands of the return ratio $Q_0(s)$, it is, strictly speaking, $I + Q_0(s)$ which must be diagonally dominant rather than $Q_0(s)$ itself. In practice, however, one usually aims to make $Q_0(s)$ diagonally dominant, for several reasons. First, if a high degree of dominance of $Q_0(s)$ is achieved over a considerable range of frequencies (namely, if the Gershgorin bands of $Q_0(s)$ are made very narrow), then there is a good chance that $I + Q_0(s)$ will be diagonally dominant. Secondly, subsequent compensation by diagonal transfer functions can leave the diagonal dominance of $Q_0(s)$ unaltered, as explained below, whereas the dominance of $I + Q_0(s)$ will always be changed. And thirdly, if $Q_0(s)$ is made very dominant, but $I + Q_0(s)$ is not, so that the Gershgorin bands of $Q_0(s)$ are narrow but some of them cover the point $-1 + j0$, then subsequent compensation by a diagonal transfer function can be used to swing the Gershgorin bands away from $-1 + j0$, with little effect on their width.

Design can be pursued using either **direct Nyquist arrays** $G(s)$, K_a, $K_b(s)$ and $K_c(s)$, or **inverse Nyquist arrays** $G^{-1}(s)$, K_a^{-1}, $K_b^{-1}(s)$ and $K_c^{-1}(s)$. In either case stability can be determined from the return ratio evaluated at any point in the loop, provided the Gershgorin bands at that point do not overlap the point $-1 + j0$. But to monitor performance, as well as stability, a particular return ratio must be examined, most commonly that at the output of the plant,

$$Q_0(s) = G(s)K_a K_b(s)K_c(s) \tag{4.110}$$

Since $K_c(s)$ post-multiplies the other transfer functions in equation (4.110), and is diagonal, the effect of each of its elements is to multiply one column of $G(s)K_a K_b(s)$ by the same transfer function. Hence column dominance of $G(s)K_a K_b(s)$ is not destroyed by $K_c(s)$, whereas row dominance may be. When working with direct Nyquist arrays it is therefore usual to attempt to achieve column dominance. When working with inverse arrays, on the other hand, we have

$$Q_0^{-1}(s) = K_c^{-1}(s)K_b^{-1}(s)K_a^{-1}G^{-1}(s) \tag{4.111}$$

Since $K_c^{-1}(s)$ pre-multiplies the other transfer functions, the effect of each of its elements is to multiply one row of $K_b^{-1}(s)K_a^{-1}G^{-1}(s)$ by the same transfer function, so leaving row dominance undisturbed. Therefore, when working with inverse Nyquist arrays it is usual to try to achieve row dominance.

4.6.2 The inverse Nyquist-array (INA) method

The motivation for using inverse rather than direct arrays is that, in principle at least, they allow the closed-loop behaviour to be determined more precisely from open-loop information. Consider $K_c(s)$, the diagonal part of the compensator, to be made up of the series connection of $K_d(s)$, another diagonal dynamic compensator, and F, a diagonal *constant* compensator, so that

$$K_c(s) = K_d(s)F \tag{4.112}$$

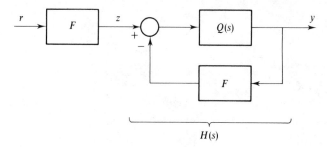

Figure 4.15 Representation of a feedback system which allows the effect of varying feedback gains to be studied.

This allows us to study the effects of changing the gain in each feedback loop while keeping the dynamic characteristics fixed. If we also define

$$Q(s) = G(s)K_a(s)K_b(s)K_d(s) \qquad (4.113)$$

then we have

$$Q_0(s) = Q(s)F \qquad (4.114)$$

and the closed-loop transfer function becomes

$$T(s) = H(s)F \qquad (4.115)$$

where

$$H(s) = [I + Q(s)F]^{-1}Q(s) \qquad (4.116)$$

(See Figure 4.15 for a block-diagram representation of this.)

Rosenbrock (1970) introduced the notation $\hat{H}(s)$ to denote the inverse $H^{-1}(s)$. This is useful because it allows us to refer easily to elements of $H^{-1}(s)$ as $\hat{h}_{ij}(s)$, the point being that $\hat{h}_{ij} \neq (h_{ij})^{-1}$. In this notation we have

$$\hat{H}(s) = \hat{Q}(s) + F \qquad (4.117)$$

and it is the simplicity of this relationship, as compared with (4.116), which underlies the use of inverse Nyquist arrays.

If $F = \text{diag}\{f_1, f_2, \ldots, f_m\}$ then setting $f_i = 0$ represents opening the ith feedback loop. In this case we have

$$\hat{H}(s)|_{f_i=0} = \begin{bmatrix} \hat{q}_{11}+f_1 & \hat{q}_{12} & \cdots & & \cdots & \hat{q}_{1m} \\ \hat{q}_{21} & \hat{q}_{22}+f_2 & & & & \\ \vdots & & \ddots & & & \vdots \\ & & & \hat{q}_{i-1,i-1}+f_{i-1} & & \\ & & & \hat{q}_{i,i} & & \\ & & & \hat{q}_{i+1,i+1}+f_{i+1} & & \\ \vdots & & & & \ddots & \vdots \\ \hat{q}_{m1} & \cdots & & & \cdots & \hat{q}_{mm}+f_m \end{bmatrix}$$

$$\text{(4.118)}$$

Now suppose that z is the vector of signals at the input to $H(s)$ (see Figure 4.15). The transfer function between z_i and y_i (where y is the output of the closed-loop system), with the ith loop still open, is

$$l_i(s) = [(\hat{H}(s)|_{f_i=0})^{-1}]_{ii} \qquad \text{(4.119)}$$

$$= \frac{\det[\hat{H}(s)|_{f_i=0}^{ii}]}{\det[\hat{H}(s)|_{f_i=0}]} \qquad \text{(4.120)}$$

where $\hat{H}(s)|_{f_i=0}^{ii}$ is just $\hat{H}(s)|_{f_i=0}$ with the ith row and ith column removed. Now, if each $f_j \to \infty$ ($j \neq i$) then, expanding by the ith row, we obtain

$$\det[\hat{H}(s)|_{f_i=0}] \to \hat{q}_{ii}(s)\det[\hat{H}(s)|_{f_i=0}^{ii}] \qquad \text{(4.121)}$$

and hence

$$l_i(s) \to \frac{1}{\hat{q}_{ii}(s)} \qquad \text{(4.122)}$$

This result has the following interpretation. If the gains in all the loops except the ith are very high, and the ith loop is open, then $\hat{q}_{ii}(s)$ (which contains open-loop information only) gives a good indication of the *inverse* response between z_i and y_i. The ith loop dynamics (namely the (i, i) element of $K_d(s)$) can therefore be designed independently of the other loops, and inspection of $\hat{q}_{ii}(j\omega)$ for the final design will give a good indication of the behaviour of that loop (provided one is adept at interpreting inverse Nyquist loci – more on that shortly).

In practice the loop gains cannot be made arbitrarily high, and are frequently constrained to be rather low, so it is important to have some measure of the error incurred by using $\hat{q}_{ii}(j\omega)$ as an approximation to $1/l_i(j\omega)$. This measure is provided by drawing Ostrowski bands. These are based on:

***Theorem 4.3* (Ostrowski's Theorem):** Let $A \in C^{m \times m}$ be (a constant matrix which is) diagonally row dominant. Then

$$\left| a_{ii} - \frac{1}{\hat{a}_{ii}} \right| \leqslant \left[\sum_{k \neq i} |a_{ik}| \right] \max_{j \neq i} \left(\sum_{k \neq j} |a_{jk}| \Big/ |a_{jj}| \right) \tag{4.123}$$

and a corresponding inequality holds if A is diagonally column dominant.

Proof: Omitted. See Rosenbrock (1970).

If $\hat{H}(j\omega)$ is row dominant, then we can apply this theorem to obtain

$$\left| \hat{h}_{ii}(j\omega) - \frac{1}{h_{ii}(j\omega)} \right| \leqslant \rho_i(\omega)\phi_i(\omega) \tag{4.124}$$

where

$$\rho_i(\omega) = \sum_{k \neq i} |\hat{h}_{ik}(j\omega)| \tag{4.125}$$

and

$$\phi_i(\omega) = \max_{j \neq i} \left(\sum_{k \neq j} |\hat{h}_{jk}(j\omega)| \Big/ |\hat{h}_{jj}(j\omega)| \right) \tag{4.126}$$

Now, recalling that

$$\hat{h}_{ii}(j\omega) = \hat{q}_{ii}(j\omega) + f_i \tag{4.127}$$

and using the relationship

$$\frac{1}{h_{ii}(j\omega)} = \frac{1}{l_i(j\omega)} + f_i \tag{4.128}$$

(see Exercise 4.3), we can rewrite (4.124) as

$$\left| \hat{q}_{ii}(j\omega) - \frac{1}{l_i(j\omega)} \right| \leqslant \rho_i(\omega)\phi_i(\omega) \tag{4.129}$$

Finally, since $\hat{h}_{ik}(j\omega) = \hat{q}_{ik}(j\omega)$ (if $i \neq k$), we recognize $\rho_i(\omega)$ as the radius of the ith Gershgorin circle of $\hat{Q}(j\omega)$ and, since $\hat{H}(j\omega)$ is assumed to be row dominant, we have $\phi_i(\omega) < 1$.

We see, therefore, that $1/l_i(j\omega)$ is contained within a circle whose centre coincides with that of the ith Gershgorin circle of $\hat{Q}(j\omega)$, and whose

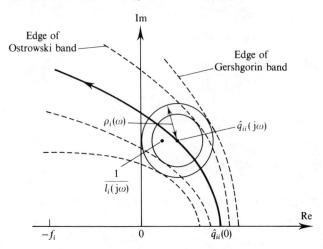

Figure 4.16   An Ostrowski band.

radius is smaller than that of the ith Gershgorin circle by the factor $\phi_i(\omega)$. We call this circle an **Ostrowski circle**, and the union of such circles an **Ostrowski band**. Clearly, the ith Ostrowski band is contained within the ith Gershgorin band, if $\hat{H}(j\omega)$ is diagonally dominant. Diagonal dominance of $\hat{H}(j\omega)$ corresponds to the ith Gershgorin band of $\hat{Q}(j\omega)$ not covering the point $-f_i + j0$. The situation is summarized in Figure 4.16. If $\hat{H}(j\omega)$ is column dominant, then we obtain similar results by interchanging the roles of rows and columns.

   Once diagonal dominance has been obtained, INA design proceeds by attempting to obtain satisfactory Ostrowski bands, treating them as if they were inverse Nyquist loci of SISO systems. The interpretation of inverse Nyquist loci (sometimes called **Whiteley loci**) is less familiar than that of direct loci, but is in fact slightly simpler, and we shall summarize this interpretation for SISO systems. Figure 4.17 shows a typical inverse locus $1/l(j\omega)$, when $l(s)$ is the forward-path (SISO) transfer function, as shown in Figure 4.18, except for a constant gain f, which is considered to be separated out from $l(s)$. The first noticeable feature is that the inverse locus has a small magnitude at low frequencies, and its magnitude increases with frequency. In order to obtain closed-loop stability the locus must satisfy the inverse Nyquist stability criterion, which is just the generalized inverse Nyquist criterion, but with the set of inverse characteristic loci replaced by the single locus. Since we have factored out the gain f from the return ratio, the locus of $1/l(j\omega)$ must encircle the point $-f + j0$ anticlockwise the same number of times as $l(s)$ has right half-plane zeros (which is the same as the locus of $1/(fl(j\omega))$ encircling the point $-1 + j0$ that number of times), and typically the inverse locus will pass 'above' $-f + j0$, as shown in Figure 4.17, if the closed loop is stable.

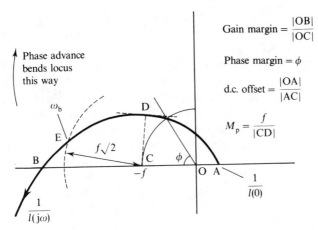

Figure 4.17 Performance prediction from an inverse Nyquist locus.

Applying phase advance to the return ratio 'bends' the inverse locus in a clockwise sense. If we label the point $1/l(0)$ as A, the point at which the locus crosses the negative real axis as B, the point $-f+j0$ as C, and the point on the locus nearest C as D, then SISO performance and robustness measures are obtained as follows. The gain margin is $|\text{OB}/\text{OC}|$ (where $|\text{OC}|$ is the distance from the origin to C, etc.), and the phase margin is the angle ϕ subtended by the circular arc, centred on the origin, drawn from C to the locus. The steady-state error (as a fraction of a constant reference signal) is $|\text{OA}/\text{AC}|$, and the peak dynamic magnification M_p is $f/|\text{CD}|$. The closed-loop ($-3\,\text{dB}$) bandwidth is given by the frequency at point E on the locus, which is the point at which the locus intersects a circle, centred on C, of radius $f\sqrt{2}$. It is easy to show that these relationships hold, and this is left as an exercise for the reader (Exercise 4.3).

The overall design strategy is now clear. First attempt to obtain row diagonal dominance of $\hat{K}_b(s)\hat{K}_a\hat{G}(s)$, using appropriate compensators $\hat{K}_a$ and $\hat{K}_b(s)$. (The next section explains how such compensators may be sought.) Then design a SISO compensator $\hat{k}_i(s)/f_i$ for each of the loops, using SISO techniques, until all the Ostrowski bands are satisfactory (when interpreted as SISO inverse Nyquist loci). Assemble the compensator $\hat{K}_c(s) = \text{diag}\{\hat{k}_i(s)/f_i : i = 1, \ldots, m\}$. As with any design technique, this stage

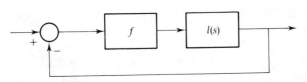

Figure 4.18 Factoring out a gain f from the return ratio.

must be followed by extensive analysis of the proposed design, using both the frequency-domain techniques presented in the previous chapter and time-domain simulation.

The advantage of the inverse Nyquist array method over the direct method is that the use of Ostrowski bands allows the behaviour of individual loops to be predicted, taking into account the effects of interactions with other loops. No such relationship is available for the direct method.[2] A significant disadvantage, on the other hand, is that the final inverse compensator, $\hat{K}_c(s)\hat{K}_b(s)\hat{K}_a$, must correspond to a realizable compensator $K_aK_b(s)K_c(s)$, since it is $K_aK_b(s)K_c(s)$ that is to be implemented. (If it is invertible, the most satisfactory way of performing the inversion is to obtain a state-space realization of $\hat{K}_c(s)\hat{K}_b(s)\hat{K}_a$, and then use the algorithm given in Chapter 8 – unless the complete compensator is diagonal, in which case the individual elements can be inverted.)

Ostrowski bands have one further use, which can be exploited even if the INA technique is not used to design a compensator. Since each band encloses the inverse Nyquist locus corresponding to one loop, when all the other loops are closed with their feedback gains at the design values, it can be used to indicate how the stability of the complete system changes when the gain in that loop changes. In particular, it can be used to indicate whether the system would be stable if one loop failed, while all other loops remained at their design gains. This indication is less conservative than the one obtained by using Gershgorin bands, described in Chapter 3. On the other hand, the Gershgorin bands can be used to predict stability when the gains in all the loops change simultaneously, whereas the Ostrowski bands cannot.

4.6.3   The direct Nyquist-array (DNA) method

Here one works with $G(j\omega)$ rather than $\hat{G}(j\omega)$, and designs K_a, $K_b(s)$ and $K_c(s)$ directly. This has several advantages. First, the designed compensator does not need to be inverted, and one consequence of this is that any structure imposed by the designer, such as setting certain elements to zero, is retained. It also gives the designer more freedom in the choice of compensator, since its inverse does not need to be realizable. Secondly, the plant need not be square, since it need not have an inverse. Thirdly, it may be that the inverse compensator designed by the INA method has right half-plane zeros, particularly if it is designed semi-automatically by one of the methods described in the next section. In this case the compensator itself will be open-loop unstable, which is usually very undesirable. With the DNA method one can keep control over the location of the compensator poles more easily.

However, there is no tool, such as the Ostrowski bands, which allows the designer to take into account the effects of interactions between loops. (Ostrowski bands can still be drawn for direct Nyquist arrays, of course, but their meaning is lost.) Suppose that diagonal dominance of a direct return ratio $GK_aK_b(s)$ has been achieved, and that the diagonal elements of its Nyquist array, together with their Gershgorin bands, are examined in order to choose SISO compensators for the individual loops (that is, to choose the elements of the diagonal compensator $K_c(s)$). Closed-loop stability will be predicted correctly by using the Gershgorin bands. But if the performance of a loop is predicted by treating the locus of its diagonal element as if it were the locus of a SISO system, then the prediction will be correct only if that loop is closed, while *all the other loops remain open*. As soon as the other loops are closed, their behaviour will affect the behaviour of the loop being designed.

This will not matter if, by this stage of the design, interaction between the loops has already been almost eliminated, so that the Gershgorin bands are very narrow, and the system behaves almost like a number of separate SISO systems. Therefore when using the DNA method it is important not to be content with merely achieving diagonal dominance, but to try to obtain as large a degree of dominance as possible before attempting to compensate the individual loops.

This requirement can be relaxed somewhat. We know that closed-loop behaviour will be satisfactory (low interaction, low sensitivity) whenever all the loop gains are large, whatever the details of the design. So we need not strive to maximize diagonal dominance at those (low) frequencies at which we expect the loop gains to be large, namely frequencies significantly lower than the 0 dB cross-over frequency of any loop.

If a high degree of diagonal dominance cannot be obtained, it is possible to go ahead and design a compensator for each loop so as to obtain apparently promising behaviour of the Gershgorin bands. But this leads to a very *ad hoc* design procedure which relies on luck. If the resulting closed-loop behaviour is unsatisfactory, it is not clear what can be done to improve it, at least within the confines of the Nyquist-array method. It may be possible to proceed in a systematic way, however, by switching to some other design method at this point. For example, if the behaviour of only one loop remained unsatisfactory, one could examine the open-loop characteristics of that loop with all the other loops closed, and compensate it by using SISO methods – in effect, switching to sequential loop closing at this point.

4.7 Achieving diagonal dominance

Nyquist-array methods are effective, provided one can obtain the required degree of diagonal dominance fairly easily. In many cases, however, this is very difficult to achieve. Manual cut-and-try methods, which were the ones originally suggested, fail to achieve diagonal dominance (with reasonable effort) more often than not, and this is probably the main reason why the methods have not been employed very widely. In recent years, however, some successful automatic ways of achieving dominance have been proposed, and these promise to give a new lease of life to the Nyquist-array approach.

It should always be remembered, though, that for some plants it would be entirely unreasonable to expect to obtain diagonal dominance, or a high degree of dominance, while using control signals of reasonable amplitude. In a highly manoeuvrable ship, for example, one would not expect to be able to decouple the ship's rate of turning from its roll angle to any appreciable degree. And one should bear in mind that diagonal dominance is not a necessary precondition for feedback design, as the other techniques in this and later chapters show. Thus persistent failure to achieve an adequate degree of diagonal dominance probably indicates that it is time to try a different approach.

4.7.1 Cut and try

It is sometimes possible to examine the display of a Nyquist array and observe that some straightforward transformation will achieve diagonal dominance. A simple but contrived example will show what is involved. Suppose the plant transfer function is

$$G(s) = \begin{bmatrix} \dfrac{-s}{3s+2} & \dfrac{s+1}{3s+2} \\[2ex] \dfrac{s+2}{3s+2} & \dfrac{-s}{3s+2} \end{bmatrix} \tag{4.130}$$

so that the inverse transfer function is

$$\hat{G}(s) = \begin{bmatrix} s & s+1 \\ s+2 & s \end{bmatrix} \tag{4.131}$$

This is clearly neither row dominant nor column dominant anywhere on the Nyquist contour. If we wish to use the INA method we look for operations on the rows of $\hat{G}(s)$ which could produce dominance. (Row operations correspond to pre-multiplications of $\hat{G}(s)$, which correspond to post-multiplications of $G(s)$, and hence to pre-compensation of the plant.) Since $|\hat{g}_{12}(s)| > |\hat{g}_{22}(s)|$ and $|\hat{g}_{21}(s)| > |\hat{g}_{11}(s)|$ for any s, row dominance (and column dominance, in this case) can clearly be achieved by interchanging the two rows. Mathematically, this corresponds to pre-multiplication by

$$\hat{K}_{\mathrm{a}} = \begin{bmatrix} 0 & 1 \\ 1 & 0 \end{bmatrix} \tag{4.132}$$

and hence to post-multiplication by the pre-compensator transfer function

$$K_{\mathrm{a}} = \begin{bmatrix} 0 & 1 \\ 1 & 0 \end{bmatrix} \tag{4.133}$$

so that the compensated plant is represented by

$$G(s)K_a = \begin{bmatrix} \dfrac{s+1}{3s+2} & \dfrac{-s}{3s+2} \\[2ex] \dfrac{-s}{3s+2} & \dfrac{s+2}{3s+2} \end{bmatrix} \qquad (4.134)$$

Physically, this corresponds to nothing more than a re-ordering of the inputs (or a re-assignment of inputs to outputs, if we take the view that the principal role of each input is to control one output).

In this artificial example both the direct and the inverse array have been made equally dominant, and the compensation required in each case is the same. This is not usually true. It is quite possible for a particular compensator to make the direct array dominant, but not the inverse array, and vice versa.

The 'elementary matrices' which figure in Theorem 4.2 are supposed to represent simple transformations devised by the designer. Thus, for example, the transformation 'subtract $2j\omega$ times the first column from the second column' results in the pre-compensator

$$K_b(s) = \begin{bmatrix} 1 & -2s \\ 0 & 1 \end{bmatrix} \qquad (4.135)$$

(for a two-input, two-output system). If column dominance is being sought then each column can be scaled by any scalar function without affecting the degree of dominance, so this compensator can be made realizable by dividing the second column by s:

$$K_b(s) = \begin{bmatrix} 1 & -2 \\ 0 & 1/s \end{bmatrix} \qquad (4.136)$$

In practice, it is rarely possible to make much progress by relying on being able to find such transformations by *ad hoc* means.

An alternative strategy is to try to diagonalize (i.e. completely decouple) a system at one frequency, and hope that the effect will be sufficiently beneficial over a wide range of frequencies to result in diagonal dominance being attained. The easiest way to do this is to invert the system at one frequency. If the system has no poles at the origin, then $K = G^{-1}(0)$ is a realizable (because constant) compensator. At any other frequency $G^{-1}(j\omega)$ is complex, but can be approximated (with luck) by a real constant compensator. Fortunately, we already have an algorithm for performing the required approximation: the ALIGN algorithm introduced in Section 4.3.1 above.

If we extend this idea to diagonalizing the system at several frequencies, and also to using dynamic compensators to do so, then we enter the realm of 'pseudo-diagonalization', to be described in Section 4.7.3.

EXAMPLE 4.4

A plant has the inverse transfer function

$$\hat{G}(s) = \begin{bmatrix} \dfrac{s^2 + s + 2}{s + 1} & 10s + 3 \\ 5s + 1 & \dfrac{s^2 + s + 1}{s + 2} \end{bmatrix} \tag{4.137}$$

It is not diagonally dominant at low frequencies, since

$$\hat{G}(0) = \begin{bmatrix} 2 & 3 \\ 1 & 0.5 \end{bmatrix} \tag{4.138}$$

The compensator

$$\hat{K}_b = G(0) = \begin{bmatrix} -0.25 & 1.5 \\ 0.5 & -1 \end{bmatrix} \tag{4.139}$$

gives $\hat{K}_b \hat{G}(0) = I$, which is obviously diagonally dominant, and it gives column, but not row, dominance at high frequencies, since

$$\hat{K}_b \hat{G}(s) \rightarrow \begin{bmatrix} \dfrac{29}{4} & -1 \\ -\dfrac{18}{4} & 4 \end{bmatrix} s \quad \text{as } |s| \to \infty \tag{4.140}$$

4.7.2 Perron–Frobenius theory

As we saw in Section 2.10, Perron–Frobenius theory allows us to check whether a plant can be made diagonally dominant by input and output scaling. This can clearly be applied to design, but with a proviso: in general, we must be careful about using output scaling as part of the strategy for achieving diagonal dominance. On the face of it, output scaling corresponds to inserting a post-compensator (that is, inserting a compensator between the outputs and the variables being controlled). This is physically impossible, since the meaningful plant outputs (which are variables such as velocity, or thickness of steel strip) cannot be affected by mathematical operations. It is certainly possible to change the scaling of the *measurements* of the output variables, however, and Figure 4.19 shows how this is achieved by placing the post-compensator in the feedback path.

One can use the scalings S^{-1}, at the plant input, and S, in the feedback path, to obtain a diagonally dominant return ratio at the point labelled X in

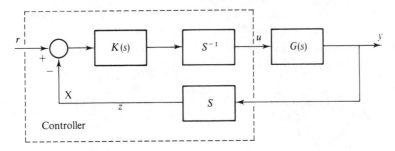

Figure 4.19   Scaling a plant's inputs and outputs.

Figure 4.19. But we must be wary of falling into the trap of believing that this return ratio tells us anything about interaction at the plant output. The output variables may be interacting with each other to a considerable extent, and this interaction may be being hidden by the measurement scaling S. Thus the choice of measurement scaling is really part of the specification of the feedback design problem, and changing this scaling is not usually admissible as a step in a design technique. The output scaling is, in effect, built into the definition of the plant. If the specification is rather loose, however (for example, that closed-loop stability and good long-term following of set points are to be achieved, with no specification on interaction), then output scaling may be considered.

Fortunately, the Perron–Frobenius theory gives useful results, even if only pre-compensation (input scaling) is allowed.

The basic result of Perron–Frobenius theory has already been stated in Section 2.10: that a positive, primitive square matrix has a particular distinguished eigenvalue λ_P (the Perron–Frobenius eigenvalue), which is real, positive and greater than the magnitude of any other eigenvalue; furthermore, the elements of the corresponding eigenvector may all be chosen to be real and positive.

The next result we need is the following:

Theorem 4.4 (Seneta, 1973):   Let M be a positive, primitive square matrix with Perron–Frobenius eigenvalue λ_P. Then, for any vector x with real positive elements

$$\min_i \frac{1}{x_i} \sum_j m_{ij} x_j \leqslant \lambda_P \leqslant \max_i \frac{1}{x_i} \sum_j m_{ij} x_j \qquad (4.141)$$

and both inequalities become equalities if x is chosen to be the Perron–Frobenius right eigenvector of M (that is, if $Mx = \lambda_P x$).

To exploit this, we define the **normalized comparison matrix**

$$M(s) = \mathrm{abs}(G_{\mathrm{diag}}^{-1}(s) G(s)) \qquad (4.142)$$

where $G_{\text{diag}}(s)$ is the matrix containing only the principal diagonal of $G(s)$. (We have already seen a similar matrix in Theorem 2.12.) Now, suppose that we have a diagonal pre-compensator $K(s) = \text{diag}\{k_1(s), \ldots, k_m(s)\}$, such that $G(j\omega)K(j\omega)$ is row dominant. Then

$$\sum_{j \neq i} |g_{ij}(j\omega)k_j(j\omega)| < |g_{ii}(j\omega)k_i(j\omega)| \tag{4.143}$$

Since the (i, j) element of $M(s)$ is

$$m_{ij}(s) = \frac{|g_{ij}(s)|}{|g_{ii}(s)|} \tag{4.144}$$

the inequality (4.143) is equivalent to

$$\sum_{j \neq i} m_{ij}(j\omega) \frac{k_j(j\omega)}{k_i(j\omega)} < 1 \tag{4.145}$$

or, since $m_{ii}(s) = 1$, to

$$\sum_j m_{ij}(j\omega) \frac{k_j(j\omega)}{k_i(j\omega)} < 2 \tag{4.146}$$

Furthermore, the diagonal pre-compensator which gives the greatest degree of row dominance is the one which gives the smallest value of the sum in (4.146).

Now, Theorem 4.4 tells us immediately that the smallest value which can be obtained is $\lambda_P(j\omega)$, and that the best choice for each (diagonal) element of $K(j\omega)$ is the corresponding element of the Perron–Frobenius right eigenvector of $M(j\omega)$. Note that if $\lambda_P(j\omega) \geq 2$, then there is no possibility of obtaining dominance at frequency ω by using a diagonal pre-compensator. Also, the optimal choice of the elements of $K(j\omega)$ gives the same degree of dominance in each row of $G(j\omega)K(j\omega)$.

The exact realization of the compensator $K(j\omega)$ is obviously a problem. It can be overcome quite easily, however, by taking one of two approaches. The approach advocated by Mees (1981) is to define the constant matrix T with elements

$$t_{ij} = \max_{\omega} \{m_{ij}(j\omega)\} \tag{4.147}$$

and apply the previous reasoning to T instead of $M(j\omega)$. If the Perron–Frobenius eigenvalue of T is less than 2, then choose the elements of the constant diagonal compensator K to be the elements of the Perron–

Frobenius right eigenvector of T. This will make TK row dominant:

$$\sum_{j \neq i} t_{ij} k_j < k_i \tag{4.148}$$

(since $t_{ii} = 1$). But $m_{ij}(j\omega) \leqslant t_{ij}$, so (4.145) will certainly hold, and hence $G(j\omega)K$ will be row dominant.

This approach has the advantage of yielding a constant, real pre-compensator. The disadvantage is that the Perron–Frobenius eigenvalue of T is usually greater than that of $M(j\omega)$ at any frequency ω, so it may well be impossible to make TK dominant, while a dynamic diagonal compensator $K(s)$ may exist which makes $G(j\omega)K(j\omega)$ dominant. Munro (1985, 1987) has pointed out that $K(j\omega)$ need not be real, provided the magnitudes of its elements follow the frequency variations of the elements of the Perron–Frobenius eigenvector of $M(j\omega)$. He has therefore suggested that a diagonal $K(s)$ should be chosen whose elements behave approximately like those of the eigenvector. Since the eigenvector is unique up to scaling only, one diagonal element of $K(s)$ can be constant. In practice, it is often effective to display the variation of each element of the eigenvector (suitably normalized, usually by fixing one element) on a Bode plot, and find a simple transfer function whose gain matches this variation approximately. A further advantage of using a dynamic compensator is that the optimal degree of dominance can be approached at each frequency.

When using the INA method, dominance of $\hat{K}(j\omega)\hat{G}(j\omega)$ is sought. In this case the matrix

$$M(s) = \text{abs}(\hat{G}(s)\hat{G}_{\text{diag}}^{-1}(s)) \tag{4.149}$$

is used in place of (4.142), its Perron–Frobenius *left* eigenvector is used to construct $K(j\omega)$, and column dominance is obtained. Everything else remains the same, except that if a dynamic compensator is used then the elements of $\hat{K}(s)$ must be chosen to have realizable inverses.

A drawback of using diagonal compensators to achieve diagonal dominance is that only row dominance can be obtained when using the DNA method, and only column dominance can be obtained when using the INA method. This is exactly the opposite of what we would like, since further diagonal compensation, for the purpose of 'loop shaping', may destroy the dominance which has been achieved. It will not destroy it, however, if the additional compensation is made the same in each loop.

If both pre- and post-compensation is used, as shown in Figure 4.19 (when this is permissible), and the pre-compensator S^{-1} is chosen as described above, then SGS^{-1} will be both row and column dominant (provided $\lambda_p < 2$). (This is the content of Theorem 2.12; the proof is left as an exercise for the reader.) Furthermore, the degree of row and column dominance will be the same. In this case further diagonal pre-compensation will leave the column dominance unaltered.

EXAMPLE 4.5   (Munro, 1985)

Consider the transfer function

$$G(s) = \begin{bmatrix} \dfrac{s+4}{(s+1)(s+5)} & \dfrac{1}{5s+1} \\ \dfrac{s+1}{s^2+10s+100} & \dfrac{2}{2s+1} \end{bmatrix}$$

(4.150)

The inverse Nyquist array $\hat{G}(j\omega)$ is not column dominant. The Perron–Frobenius eigenvalue of $\text{abs}(\hat{G}(j\omega)\hat{G}^{-1}_{\text{diag}}(j\omega))$ is shown on a Bode plot in Figure 4.20. Its value is smaller than 6 dB (i.e. $\lambda_p < 2$) at all frequencies, so it is possible to obtain column dominance by using a diagonal compensator. Figure 4.21 shows the second element of the Perron–Frobenius left eigenvector, when the first element is fixed at 1. It also shows, superimposed, the gain variation of the compensator

$$\hat{k}_2(s) = \frac{0.447s + 3.574}{s + 0.566}$$

(4.151)

This matches the variation of the eigenvector very well, and the inverse compensator $\hat{K}(s) = \text{diag}\{1, \hat{k}_2(s)\}$ gives the Nyquist array for

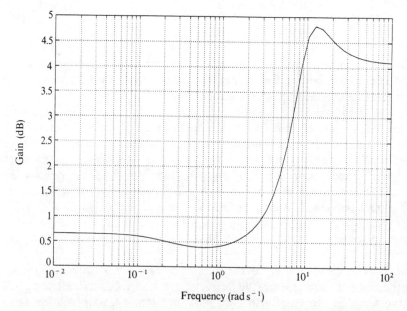

Figure 4.20   Perron–Frobenius eigenvalue of the inverse of system (4.150).

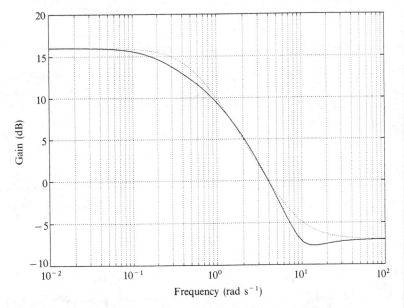

Figure 4.21   The second element of the Perron–Frobenius left eigenvector of the inverse of system (4.150), with the first element fixed at 1 (solid curve), and the gain of the SISO compensator (4.151) (dotted curve).

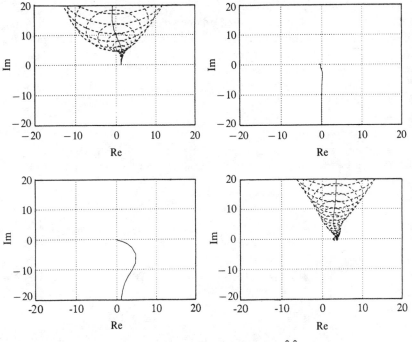

Figure 4.22   Nyquist array of $\hat{K}\hat{G}$.

$\hat{K}(j\omega)\hat{G}(j\omega)$ shown in Figure 4.22. The Gershgorin circles are computed by columns, and the array is seen to be column dominant.

As discussed in Section 2.10, if abs$(G(j\omega))$ is not primitive then $G(j\omega)$ can usually be decomposed into subsystems, and the Perron–Frobenius theory should then be applied independently to those blocks of $G(j\omega)$ which are on its principal diagonal.

Kantor and Andres (1979) have obtained results which are identical to those obtained by Mees (1981), but using the theory of *M-matrices* (Fiedler and Ptak, 1966; Siljak, 1978) instead of Perron–Frobenius theory. They examine the most positive eigenvalue of $I - M$, rather than M. Dominance can be obtained if this eigenvalue is smaller than 1, and again a suitable compensator is obtained from the corresponding eigenvector.

4.7.3 Pseudo-diagonalization

It may not be possible to achieve diagonal dominance by using a diagonal compensator. And it is certainly not possible to change the column dominance of a direct Nyquist array, or the row dominance of an inverse array, by using diagonal pre-compensators. So we need to look for ways of automatically generating compensators with a more general structure. This can be done by choosing some measure of diagonal dominance, some compensator structure, and then optimizing the measure of dominance over this structure.

We shall use the term **pseudo-diagonalization** for any such scheme, although the term is often reserved for the particular scheme proposed by Hawkins (1972). Hawkins assumed that inverse arrays are to be used, but his method can be applied equally well to direct arrays. If we have a plant $G(s)$ and a constant compensator K, with $Q(s) = G(s)K$, then individual elements of Q are given by

$$q_{ij}(j\omega) = g_i^T(j\omega)k_j \tag{4.152}$$

where $g_i^T(s)$ denotes the ith row of $G(s)$, and k_j denotes the jth column of K. Hawkins proposed to minimize

$$J_j = \sum_{k=1}^{N} p_k \left\{ \sum_{i \neq j} |q_{ij}(j\omega_k)|^2 \right\} \tag{4.153}$$

$$= \sum_{k=1}^{N} p_k \left\{ \sum_{i \neq j} |g_i^T(j\omega_k)k_j|^2 \right\} \tag{4.154}$$

subject to the constraint

$$\|k_j\| = 1 \tag{4.155}$$

Here $\{p_k\}$ is a set of weighting coefficients, and (4.153) is formulated so as to achieve column dominance. By using Lagrange multiplier methods, this can be shown to lead to an eigenvalue problem, so the solution can be obtained easily. But a shortcoming of the method is that minimizing J_j may not prevent $|q_{jj}(j\omega_k)|$ from being made small, as well as the off-diagonal elements, so that diagonal dominance may not be obtained.

Suppose, however, that we replace the criterion (4.153) by

$$J_j = \frac{\sum_{k=1}^{N} p_k \left\{ \sum_{i \neq j} |q_{ij}(j\omega_k)|^2 \right\}}{\sum_{k=1}^{N} p_k |q_{jj}(j\omega_k)|^2} \tag{4.156}$$

This is a better measure of column dominance, and has the further advantage that it can be minimized without any constraints on K. Equation (4.156) can be rewritten as

$$J_j = \frac{\sum_{k} p_k \left\{ \sum_{i \neq j} |g_i^T(j\omega_k) k_j|^2 \right\}}{\sum_{k} p_k |g_j^T(j\omega_k) k_j|^2} \tag{4.157}$$

which is just the reciprocal of the cost function which is maximized by the multi-frequency ALIGN algorithm; this has already been described in Section 4.3.1. Note that, since (4.157) is to be minimized, while (4.45) is to be maximized, the multi-frequency ALIGN algorithm can be used without modification to solve the pseudo-diagonalization problem.

Ford and Daly (1979) have extended this approach to dynamic compensators:

$$K(s) = [k_1(s), \ldots, k_m(s)] \tag{4.158}$$

in which each column has the form[3]

$$k_j(s) = k_{0j} + \ldots + k_{\beta j} s^\beta \tag{4.159}$$

and each k_{ij} is a column vector. In this case (4.152) is replaced by

$$q_{ij}(j\omega) = \gamma_i^T(j\omega) \eta_j \tag{4.160}$$

where $\gamma_i^T(j\omega)$ is the row vector

$$\gamma_i^T(j\omega) = [g_i^T(j\omega), j\omega g_i^T(j\omega), \ldots, (j\omega)^\beta g_i^T(j\omega)] \tag{4.161}$$

and η_j is the column vector

$$\eta_j = [k_{0j}^{\mathrm{T}}, \ldots, k_{\beta j}^{\mathrm{T}}]^{\mathrm{T}} \tag{4.162}$$

To maximize column dominance we now minimize the function

$$J_j = \frac{\sum_k p_k \left\{ \sum_{i \neq j} |\gamma_i^{\mathrm{T}} (\mathrm{j}\omega_k) \eta_j|^2 \right\}}{\sum_k p_k |\gamma_j^{\mathrm{T}} (\mathrm{j}\omega_k) \eta_j|^2} \tag{4.163}$$

and the solution is again given by the multi-frequency ALIGN algorithm.

Ford and Daly (1979) also replace the sum $\Sigma_k p_k \{\ldots\}$ by the integration $\int_\Omega p(\omega) \{\ldots\} \, d\omega$. This makes no difference to the development given above, although it clearly affects the way in which the matrices $\tilde{X}$ and $\tilde{Y}$ (in (4.47) and (4.48)) are formed. In either case the weight p_k or $p(\omega)$ is used to emphasize frequencies where dominance must be increased.

If column dominance of $G(s)K(s)$ is achieved, it is not destroyed by any scaling of the columns of $K(s)$. A realizable compensator is therefore obtained by dividing $k_j(s)$ by any polynomial of degree β (or greater). This polynomial should of course be chosen to have stable roots, the locations of which can be chosen to provide desirable loop compensation. The question of realizability can be postponed until the whole compensator, including individual loop shaping, has been designed; less adjustment may be necessary at that stage, since additional poles may have been added by then.

Pseudo-diagonalization can be applied to either direct or inverse Nyquist arrays. (As usual, the roles of rows and columns are interchanged with inverse arrays.) But a practical difficulty arises if a dynamic compensator is found for an inverse array: its inverse needs to be realizable. Each row (of the inverse compensator) can be scaled to make the inverse realizable. One procedure is to ensure that each element of $\hat{K}(\infty)$ is non-zero, obtain a state-space realization, and then obtain the realization of $K(s)$, using the algorithm given in Section 8.3.4.

It should be noted that, in our development of pseudo-diagonalization, we have not assumed that G and K are square. The technique can therefore be used with plants which have unequal numbers of inputs and outputs.

Although we have described pseudo-diagonalization in the context of Nyquist array methods, it can clearly be applied whenever approximate inverses of frequency responses are required. In particular, it can be viewed as an extension of the ALIGN algorithm, and can therefore be applied in the context of the characteristic-locus method.

4.8   Design example

4.8.1   The design

(1)   *PLANT AND SPECIFICATION*

We shall use the same plant model and specification as we used for the characteristic-locus design example of Section 4.4. As the plant we take the three-input, three-output, five-state aircraft model AIRC, described in the Appendix, which is stable, with one pole at the origin, and has no finite transmission zeros. We aim to achieve a bandwidth of about $10\,\mathrm{rad\,s^{-1}}$ for each loop, with little interaction between outputs, good damping of step responses, and zero steady-state error in response to step demands or disturbances, with a one-degree-of-freedom control structure.

(2)   *OBTAINING COLUMN DOMINANCE*

The plant itself is not column dominant. Figure 4.23 shows the column dominance measure

$$\frac{\sum_{i\neq j}|g_{ij}(j\omega)|}{|g_{jj}(j\omega)|} \tag{4.164}$$

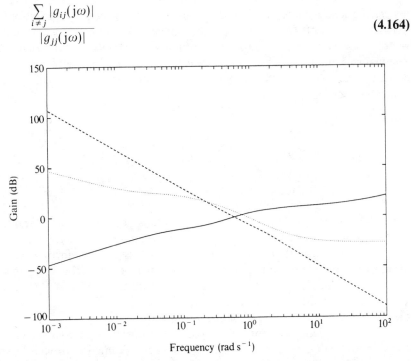

Figure 4.23 Column dominances of the plant: column 1 (solid curve), column 2 (dashed curve) and column 3 (dotted curve).

for each column, in the form of a Bode plot. Values below 0 dB indicate that a column is diagonally dominant; the more negative the value, the greater the degree of dominance. Figure 4.23 shows that column 1 is dominant up to about $1 \, \text{rad s}^{-1}$, but not at higher frequencies, while columns 2 and 3 are dominant beyond about $1 \, \text{rad s}^{-1}$, but not at lower frequencies.

We use pseudo-diagonalization (the algorithm of Ford and Daly (1979)) to obtain column dominance, and apply it to one column at a time. For this purpose, and in the rest of this design example, all frequency responses are evaluated at a set of 50 frequency points, equally spaced on a logarithmic scale between 0.001 and $100 \, \text{rad s}^{-1}$, except for a greater density of points in the region of $0.18 \, \text{rad s}^{-1}$ (chosen for particular emphasis because the plant has a very lightly damped pair of resonant poles at $-0.018 \pm 0.183 \text{j}$).

Applying pseudo-diagonalization to the first column and optimizing over constant compensator elements only, with uniform weighting on all frequencies, was not successful. Diagonal dominance was improved at low frequencies, where it was not needed, but remained almost unchanged at high frequencies. Frequencies above $0.1 \, \text{rad s}^{-1}$ were therefore weighted 10 times as much as lower frequencies, but dominance was still not achieved above $1 \, \text{rad s}^{-1}$. The weighting was increased further to 100 times as much as at lower frequencies, *and* the optimization was performed over first-order dynamic terms of the form $k_{0j} + k_{ij}s$ (where k_{0j} and k_{ij} are column vectors – see equation (4.159)). This produced virtually perfect diagonal dominance, with a dominance measure less than 10^{-6} at all frequencies. The designed first column of the compensator is (except for arbitrary scaling by a scalar transfer function)

$$k_1(s) = \begin{bmatrix} -1.52 \times 10^{-3} s - 9.78 \times 10^{-4} \\ -1.83 \times 10^{-4} s + 3.11 \times 10^{-5} \\ -4.05 \times 10^{-3} s - 1.31 \times 10^{-3} \end{bmatrix}. \tag{4.165}$$

The degree of dominance achieved here is necessarily great, and the practical consequence will certainly be very severe exertion by the control effectors – in particular, by the two control surfaces. Some further experimentation would therefore be worth while in a real design at this stage, to see whether a higher value of the dominance measure (say 10^{-2}) might not be obtained, possibly with a simpler compensator structure. Only one element of the column may need to be dynamic, for example.

For the second column, it was again necessary to optimize over a first-order dynamic structure, and to use frequency weighting: frequencies below $1 \, \text{rad s}^{-1}$ were weighted 100 times as much as higher frequencies. Again, almost perfect dominance was obtained, with a dominance measure better than 2×10^{-4} everywhere. The design obtained for the second column of the

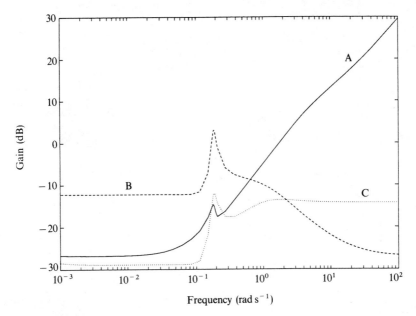

Figure 4.24 Successive improvement of the dominance of column 3 (see text).

compensator is

$$k_2(s) = \begin{bmatrix} 0s - 1.22 \times 10^{-2} \\ -5.74 \times 10^{-2}s - 4.50 \times 10^{-3} \\ 0s - 3.40 \times 10^{-2} \end{bmatrix} \qquad \textbf{(4.166)}$$

Note that first-order structures were not needed for the first and third elements. The computed coefficients of s in these elements were of the order of 10^{-6}.

The third column proved to be the most difficult to compensate. Optimizing over a first-order structure with high weighting on low frequencies ($< 1\,\mathrm{rad\,s^{-1}}$) achieved dominance up to about $1\,\mathrm{rad\,s^{-1}}$, but lost it at higher frequencies (thus reversing the original dominance properties of this column) – see curve A in Figure 4.24. Reducing the weighting on the low frequencies gave no benefit at higher frequencies, so optimization over a second-order structure

$$k_3(s) = k_{03} + k_{13}s + k_{23}s^2 \qquad \textbf{(4.167)}$$

was attempted, with uniform weighting at all frequencies. This achieved dominance everywhere, except in a narrow range of frequencies near $0.2\,\mathrm{rad\,s^{-1}}$ (which is where the plant resonance occurs) and gave poor

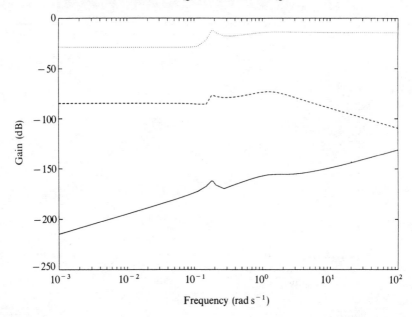

Figure 4.25 Column dominances after compensation: column 1 (solid curve), column 2 (dashed curve) and column 3 (dotted curve).

dominance between 0.1 and $1 \, \text{rad s}^{-1}$ – see curve B in Figure 4.24. The weighting on frequencies between 0.1 and $1 \, \text{rad s}^{-1}$ was therefore increased to 10 times as much as on other frequencies, and this gave diagonal dominance at all frequencies, with a dominance measure less than 0.25 everywhere – see curve C in Figure 4.24. The design obtained for the third column of the compensator is

$$k_3(s) = \begin{bmatrix} 4.21 \times 10^{-2}s^2 - 9.78 \times 10^{-2}s - 1.14 \times 10^{-1} \\ 5.07 \times 10^{-3}s^2 - 2.58 \times 10^{-2}s - 3.34 \times 10^{-2} \\ 2.86 \times 10^{-1}s^2 - 3.08 \times 10^{-2}s - 1.52 \times 10^{-1} \end{bmatrix} \qquad \textbf{(4.168)}$$

Figure 4.25 shows the dominance measures obtained for each of the columns. We shall refer to the compensator designed so far as $K_b(s)$, for consistency with the notation used in Section 4.6.

Note that

$$K_b(s) = [k_1(s), k_2(s), k_3(s)] \qquad \textbf{(4.169)}$$

where $k_1(s)$, $k_2(s)$ and $k_3(s)$ are defined by (4.165), (4.166) and (4.168), respectively.

(3) *LOOP COMPENSATION*

The first two columns of $G(j\omega)K_b(j\omega)$ are so dominant that the Gershgorin circles, superimposed on the (1, 1) and (2, 2) loci, cannot be distinguished from the loci themselves, and the compensation of the first two loops is no different from compensation of a SISO system.

The frequency response of the (1, 1) element looks exactly like that of an integrator, and lies along the negative imaginary axis. Figure 4.26 shows the Bode magnitude response of this element. The only adjustment needed here is an increase in gain, in order to obtain the required bandwidth. An increase of 72 dB (i.e. a gain of 3980) brings the 0 dB cross-over frequency up to 8 rad s^{-1}, and since the phase is $-90°$ everywhere, the closed-loop (-3 dB) bandwidth should be the same, or very slightly higher (see Figure 1.15). Figure 4.26 also shows the gain characteristic after the gain has been increased.

The frequency response of the (2, 2) element, shown as a Nyquist plot in Figure 4.27, is essentially constant at -25 dB ($= 1/17.4$) at all frequencies. This can be compensated easily, by first changing the sign to obtain a positive constant, and then inserting an integrator with enough gain to move the 0 dB cross-over frequency to about 10 rad s^{-1}. Thus the compensation needed for the second loop is $-174/s$. The result looks like a perfect integrator characteristic with the correct cross-over frequency (not shown).

The response of the (3, 3) element is shown as a Nyquist plot in Figure 4.28, together with its Gershgorin band (computed column-wise), and its

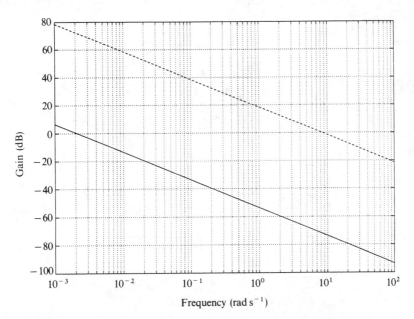

Figure 4.26 Response of (1, 1) element with and without compensation.

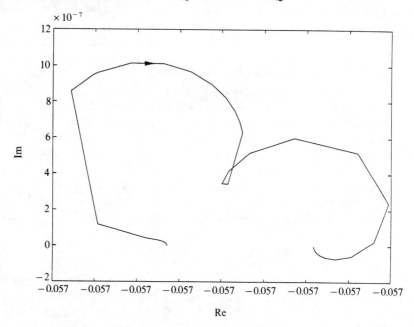

Figure 4.27 Response of (2, 2) element before compensation.

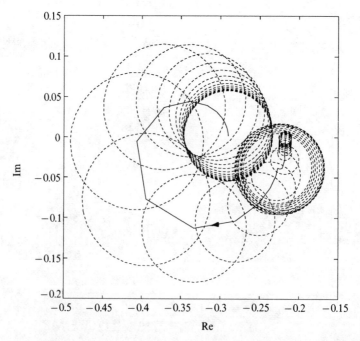

Figure 4.28 Response of (3, 3) element before compensation.

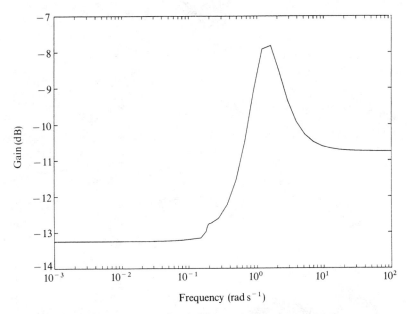

Figure 4.29 Gain of $(3, 3)$ element before compensation.

magnitude is shown in Bode form in Figure 4.29. Suitable compensation of this element is obtained by first changing its sign, so that the locus starts and ends on the positive real axis, and then inserting an integrator with enough gain to add about 10 dB at 10 rad s^{-1}. The compensator $-31.6/s$ achieves this, and the compensated magnitude characteristic is shown in Figure 4.30. For this element the thickness of the Gershgorin band is significant, so we should check that the band does not overlap -1 in order to be sure that our inference of closed-loop stability is correct. Figure 4.31 shows the compensated $(3, 3)$ element, together with its (column) Gershgorin band, in the form of a Nichols chart. This shows that the Gershgorin band is well clear of -1 (i.e. of the point $(0\,\text{dB}, -180°)$) and our design looks satisfactory.

The three loop compensators we have designed at this stage can be assembled into the diagonal matrix

$$K_c(s) = \begin{bmatrix} 3980 & 0 & 0 \\ 0 & -\dfrac{174}{s} & 0 \\ 0 & 0 & -\dfrac{31.6}{s} \end{bmatrix} \tag{4.170}$$

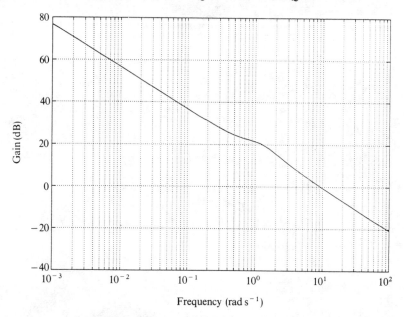

Figure 4.30   Gain of (3, 3) element after compensation.

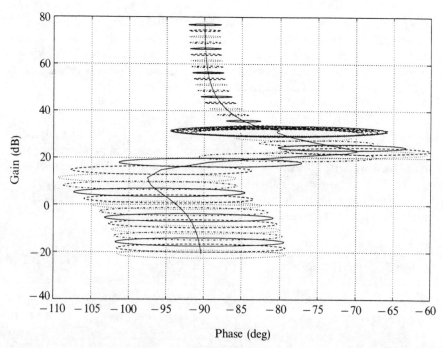

Figure 4.31   Response of (3, 3) element after compensation, with Gershgorin band, on a Nichols chart.

The compensation is not yet finished, because the product $K_b(s)K_c(s)$ is not realizable: each element in the first and third columns has one more zero than poles. We obtain a realizable compensator by inserting some more poles. For simplicity we multiply each element of $K_c(s)$ by $1/(0.05s + 1)$; this gives an additional phase lag of 26° at $10\,\text{rad}\,\text{s}^{-1}$, and so should not reduce stability margins significantly, while reducing the loop gain above $20\,\text{rad}\,\text{s}^{-1}$, thus reducing noise transmission.

(4) REALIZATION OF THE COMPENSATOR

With this modification, the complete compensator has the five-state realization (KA, KB, KC, KD), where

$$KA = \begin{bmatrix} -0.0204 & 0.0898 & -1.1977 & 0.0178 & -0.0834 \\ 0.0299 & -0.2042 & 2.0407 & 0.4695 & -0.8313 \\ -0.3410 & 1.5291 & -19.8216 & 0.0360 & -0.0628 \\ 0.4771 & 1.2183 & 0.0969 & -19.9707 & -0.0531 \\ 0.1436 & -0.4167 & -0.0517 & -0.0098 & -19.9831 \end{bmatrix}$$

(4.171)

$$KB = \begin{bmatrix} 0.0053 & -0.0162 & -0.0250 \\ 0.0458 & 0.0794 & -0.0411 \\ 0.0230 & -0.9959 & -0.0370 \\ -0.4720 & -0.0392 & 0.8783 \\ 0.8801 & 0.0010 & 0.4743 \end{bmatrix}$$

(4.172)

$$KC = \begin{bmatrix} 0 & 0 & 25.7 & -592.1 & 2351.9 \\ 0 & 0 & -197.3 & -77.4 & 297.5 \\ 0 & 0 & -0.2 & 193.8 & 7306.5 \end{bmatrix}$$

(4.173)

$$KD = \begin{bmatrix} -121.3929 & 0 & -26.6305 \\ -14.5671 & 0 & -3.2056 \\ -322.1833 & 0 & -180.7996 \end{bmatrix}$$

(4.174)

4.8.2 Analysis of the design

The characteristic loci of the design are shown in Figure 4.32, together with part of the $\sqrt{2}$ M-circle. These apparently indicate ample stability margins.

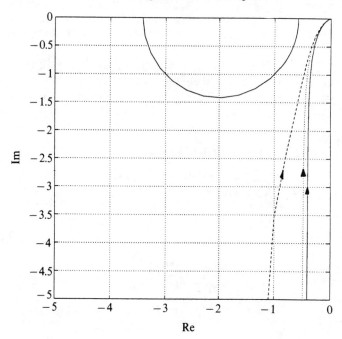

Figure 4.32 Characteristic loci, with 3 dB M-circle.

Figure 4.33 shows the largest and smallest principal gains of the closed-loop transfer function

$$T(s) = G(s)K(s)[I + G(s)K(s)]^{-1} \qquad (4.175)$$

where $G(s)$ is the plant transfer function, and $K(s)$ is the final compensator transfer function. These show that, at the output of the plant, the design specification has been achieved: the low-frequency gain of 0 dB indicates that constant demands will be tracked exactly (in the steady state), the small peak shows that all responses to changes in demanded values, or to disturbances at the output, will be well damped, and the -3 dB bandwidth is seen to be between 10 and 15 rad s^{-1}, depending on the signal direction. The time responses to step demands on each of the outputs are plotted in Figure 4.34 and show excellent behaviour, interaction from the first two outputs being so small as to be invisible.

In spite of the apparently excellent results shown in Figures 4.32–4.34, the design suffers from a serious flaw. The compensator has exactly cancelled the resonant pair of plant poles at -0.018 ± 0.182j, thus making these resonant modes uncontrollable from the compensator input and unobservable from the compensator output, leaving them as modes of the closed-loop system. This is apparent from examination of the eigenvalues of the matrix A of the *closed-loop* system, or of the transmission zeros of the compensator. These modes are not visible as long as we examine the transfer functions $GK(I+GK)^{-1}$ (Figure 4.33) or $KG(I+KG)^{-1}$ (whose largest and smallest

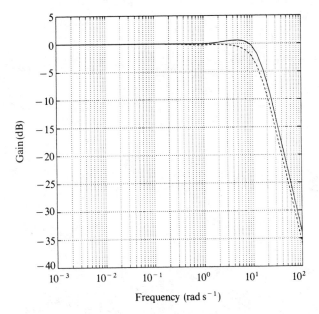

Figure 4.33 Largest and smallest closed-loop principal gains.

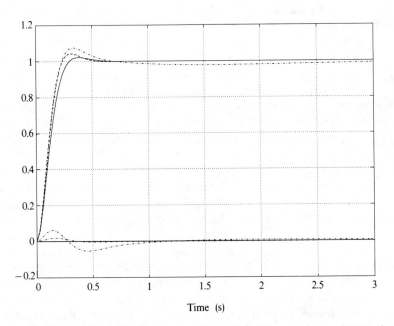

Figure 4.34 Closed-loop step responses to step demand on output 1 (solid curves), output 2 (dashed curves) and output 3 (dash–dotted curves).

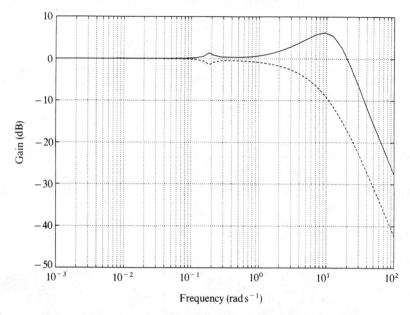

Figure 4.35 Largest and smallest principal gains at plant input.

principal gains are shown in Figure 4.35). But they become apparent once we consider how any disturbances *at the plant inputs* may propagate to the plant outputs, in other words once we examine the transfer function *SG*. Figure 4.36 shows the largest principal gain of this transfer function; a sharp resonant peak at 0.18 rad s^{-1} is very prominent. The effect of this resonance is that all the controlled variables are liable to oscillate with a very slow decay $(e^{-0.018t})$, and low frequency (0.18 rad s^{-1}, or 0.03 Hz), in response to a disturbance which affects the control inputs.[4] Figure 4.37 shows the time responses of the outputs after a unit impulse disturbance on the first input (spoiler angle). Initial large transients are removed quickly, as would be expected from the earlier figures, but the beginning of a slow subsequent oscillation is visible, particularly on output 1 (altitude).

The existence of this resonance in the closed-loop behaviour does not necessarily invalidate the design. Its effect needs to be examined in the light of detailed knowledge of the specific application. Disturbances are likely to appear at the plant inputs, in this case in the form of friction and hysteresis at the control surfaces, and of minor variations in the engine thrust. But the likely magnitudes of such disturbances, and the consequent effects on the aircraft's handling qualities, need to be assessed before the design is rejected.

We can also examine the various principal-gain plots from the point of view of robustness assessment, in the face of (unstructured) perturbations. In terms of the analysis presented in Section 3.11.1, Figure 4.33 tells us that the system remains stable, provided any output multiplicative perturbations Δ_o

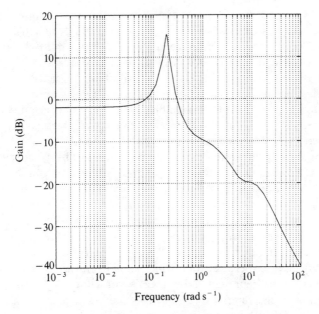

Figure 4.36 Largest (closed-loop) principal gain from plant inputs to plant outputs.

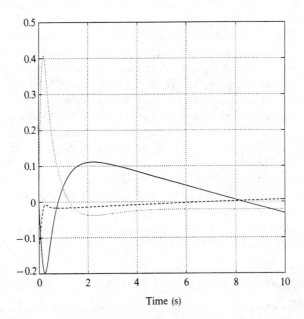

Figure 4.37 Responses to impulse disturbance on plant input 1.

(such that $G = (I + \Delta_o)G_0$, where G_0 is the nominal plant model) satisfy the condition $\|\Delta_o\|_\infty < 1/1.12 \ (= -1\,\text{dB})$, since the peak principal gain value in Figure 4.33 is $+1\,\text{dB}$). Figure 4.35 shows that input multiplicative perturbations Δ_i (where $G = G_0(I + \Delta_i)$) do not cause instability, provided $\|\Delta_i\|_\infty < 1/2.1 \ (= -6.4\,\text{dB})$. (These assessments of stability robustness assume that no unstable poles are introduced by the perturbations.) These stability margins seem quite reasonable (89% relative uncertainty allowed at the output, 48% uncertainty allowed at the input), particularly when one recalls that these margins are conservative.

Figure 4.36, however, shows that the largest stable feedback perturbation Δ_F, such that $G = G_0(I + \Delta_F G_0)^{-1}$, which can be guaranteed not to destabilize the loop has magnitude $\|\Delta_F\|_\infty < 1/5.6 \ (= -15\,\text{dB})$. This allows only 18% relative uncertainty, which looks rather low. But this figure relates only to a narrow band of frequencies around $0.18\,\text{rad}\,\text{s}^{-1}$, and it is legitimate to take the least conservative bound for a given perturbation at each frequency. In other words, the 18% figure need cause concern only if we can identify possible perturbations which can be represented by the use of Δ_F, but could not be represented by the use of either Δ_o or Δ_i. One such perturbation is the migration of the resonant pole pair $-0.018 \pm 0.182j$ across the stability boundary; whether such a perturbation could occur is again a question which cannot be answered here, since it requires detailed knowledge of the aerodynamics of the aircraft being modelled. In short, Figure 4.36 indicates that there is a possible robustness problem, but more detailed analysis is needed in order to evaluate its severity. Such analysis could usefully employ the 'structured uncertainty' approach taken in Section 3.11.2.

The present design can also be criticized on the grounds that it assumes the availability of very large control signals – some of the elements in (4.173) and (4.174) have extremely large magnitudes. This problem is largely induced by the specification, which asks for a bandwidth which is probably unattainably high. The bandwidth specification is rather vague, so some refinement of the design to reduce the achieved bandwidth a little, and perhaps to reduce the high-frequency 'roll-off' rate further by inserting further lags in the controller, could be pursued, and may succeed in reducing short-term control-signal magnitudes by a factor of two or three. But to obtain a reduction of an order of magnitude, which is what is required in practice, would probably require abandoning the goal of a $10\,\text{rad}\,\text{s}^{-1}$ bandwidth and settling for something like $1\,\text{rad}\,\text{s}^{-1}$ instead. Some reduction of control-signal magnitudes may also be obtained by moving to a design with a much lower degree of diagonal dominance.

4.8.3 Comparison with the characteristic-locus design

The design produced by the characteristic-locus method, in Section 4.4, clearly differs in several respects from the DNA design produced in this section. It is most important to emphasize, then, that these differences arise almost entirely from differences in the detailed execution of the design, and *not* from the choice of design technique.

The fact that the DNA design places zeros on top of resonant poles, while the characteristic-locus design shifts all the poles to well-damped locations $(-4.02 \pm 7.79\,\mathrm{j}, \; -3.83 \pm 6.58\,\mathrm{j}, \; -3.16 \pm 0.23\,\mathrm{j}, \; -0.49 \pm 0.01\,\mathrm{j}, \; -9.52, \; -0.53)$, is a consequence of the use of optimization over dynamic structures in the DNA design, which allows the zeros to be introduced, but optimization over only constant structures in the characteristic-locus design, which does not. Had we used pseudo-diagonalization to find the approximate commutative controller, instead of the basic ALIGN algorithm, then such pole–zero cancellation may well have been introduced into the characteristic-locus design too. A major lesson to be learnt from the example, then, is that too free a use of optimization, with *any* of the techniques presented in this chapter, may be unproductive if there are unpleasant modes in the plant (this is particularly true if there are actually unstable modes in the plant, of course)..

The facts that the spread of principal gains is larger in Figure 4.13 than in Figure 4.33, that the peak is larger, and that the 'roll-off' rate is only 20 dB/decade rather than 40, are all results of specific decisions taken during the detailed design, and not of the choice of one design technique or another. Further work could be done on either design to make it appear very similar to the other. Of course, the fact that interaction is lower in the DNA design is again the result of more degrees of freedom having been used in the optimization, and is not an inherent advantage of using Nyquist-array techniques.

4.9 Quantitative feedback theory

The design methods which we have considered so far require design objectives to be stated in terms which are familiar from classical SISO design methods. In the design examples given in Sections 4.4 and 4.8 we have stated these requirements very imprecisely, but we could have been more specific about bandwidths, about allowed values of principal gains of various transfer-function matrices, and about permitted interactions. We could also have specified acceptable time-domain responses of the diagonal elements of the closed-loop transfer function. Some of these specifications may be derived from an accurate knowledge of noise and disturbance statistics, or of possible perturbations to the nominal plant model (by means of the methods described in Chapter 3), but they are more often obtained in a less quantitative manner. Typically they are initially obtained from previous experience with similar plants, and then refined in a **specify–design–analyse** cycle until the closed-loop behaviour is judged to be acceptable.

This approach has been forcefully criticized as inadequate by Horowitz (1982). In Chapter 1 we pointed out that the basic reason for using

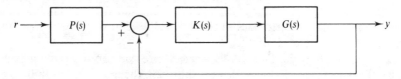

Figure 4.38   A feedback configuration with two degrees of freedom.

feedback is to combat uncertainty, in the form of either unpredictable noises or disturbances, or unpredictable variations in the behaviour of the plant. If we have a quantitative description of the amount of uncertainty which may be present, and a precise specification of the range of behaviours which may be tolerated in the face of such uncertainty, then we should aim to develop a design technique which allows us to proceed systematically to satisfy this specification.

One such technique which has been proposed is known as **quantitative feedback theory** (Horowitz, 1979, 1982; Horowitz and Sidi, 1980) often abbreviated to QFT. (The H_∞ theory described in Chapter 6 is an alternative technique with the same aim.) The QFT approach assumes that the plant uncertainty is represented by a set of templates on the complex plane, each of which encloses within it all the possible frequency responses $g_{ij}(j\omega_k)$ from some input j to some output i at some frequency ω_k, as described in Section 3.10.3. It also assumes that the design specification is in the form of bounds on the magnitudes of the elements of the frequency-response matrices $T(j\omega)$ and/or $S(j\omega)$; for example,

$$a_{ij}(\omega) \leqslant |t_{ij}(j\omega)| \leqslant b_{ij}(\omega) \tag{4.176}$$

where $a_{ij}(\omega)$ and $b_{ij}(\omega)$ are real-valued functions of frequency. The QFT technique leads to a design which satisfies these specifications for all permissible plant variations, while approximately minimizing the transmission of output sensor noise (if such a design is possible).

The real performance specification is more likely to consist of bounds for closed-loop time responses to particular reference or disturbance signals, rather than frequency-domain bounds such as (4.176). Although relatively little is known about mapping time-domain bounds into frequency-domain bounds, Horowitz and Sidi (1972) have obtained sufficient conditions on frequency-domain bounds which imply the satisfaction of time-domain bounds; the frequency-domain bounds obtained in this way do not appear to be unduly conservative.

At the heart of the QFT technique is a method for designing SISO systems. We shall describe this briefly, before going on to the multivariable problem. Figure 4.38 shows a feedback system with two degrees of freedom which is to be designed. For the purposes of this description we shall ignore disturbances and measurement noise, and assume that we are dealing with a

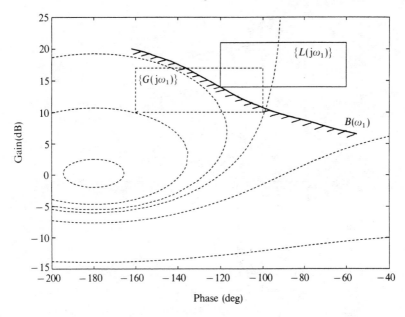

Figure 4.39   Nichols chart with template.

stable, minimum-phase plant. Figure 4.39 shows a Nichols chart on which

$$L(j\omega) = G(j\omega)K(j\omega) \qquad (4.177)$$

will eventually be plotted. The broken curves are contours of

$$|L[1 + L]^{-1}| = M \ (= \text{const.}) \qquad (4.178)$$

namely the familiar M-circles (assuming a SISO system for now). The dashed-line rectangle is the template of all possible values of $G(j\omega_1)$ at some frequency ω_1. Now, suppose that the design specification, at this frequency, is

$$a(\omega_1) \leqslant |T(j\omega_1)| \leqslant b(\omega_1) \qquad (4.179)$$

where $T(s)$ is the transfer function

$$T(s) = L(s)[1 + L(s)]^{-1}P(s) \qquad (4.180)$$

If we assume that we can implement the pre-filter $P(s)$ with negligible uncertainty, which we generally can, then the variation in $|T(j\omega_1)|$, as $G(j\omega_1)$ ranges over its possible values, will be the same as the variation in $|L(j\omega_1)[1 + L(j\omega_1)]^{-1}|$. We therefore determine all possible positions on the Nichols chart to which the uncertainty template of $G(j\omega_1)$ could be

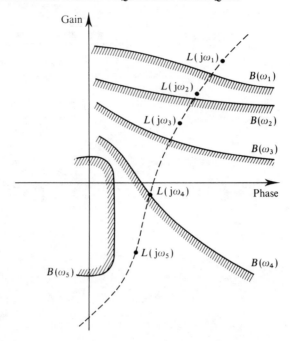

Figure 4.40 A set of template boundaries dictated by the specification, together with a satisfactory $L(j\omega)$ locus.

translated (without any distortion or change of size or orientation), such that it would not intersect any pair of M-circles whose values differed by more than $|b(\omega_1)|/|a(\omega_1)|$. The solid-line rectangle shows the template translated to one such position. This template is labelled $\{L(j\omega_1)\}$ since it shows the set of all possible values of $L(j\omega_1)$ which could occur with *one* particular compensator $K(s)$ (assuming that $K(s)$ can be implemented with negligible uncertainty). The use of this compensator would reduce the variation in the closed-loop response to the required amount. (Note that at this stage we do not worry about the actual values attained by $|L(1 + L)^{-1}|$, since later we can design $P(s)$ so as to get the correct range of $|T|$.)

   We have shown just one possible template, $\{L(j\omega_1)\}$, but an infinite set of such templates is possible, each of which reduces the variation of $|L(1 + L)^{-1}|$ to the required amount. This set has a boundary which corresponds to the curve labelled $B(\omega_1)$ in Figure 4.39. The hatching shows the side on which values of $L(j\omega_1)$ may *not* lie.

   Repeating this construction for a set of frequencies $\{\omega_1, \ldots, \omega_k\}$ generates a set of boundaries $\{B(\omega_1), \ldots, B(\omega_k)\}$, as shown in Figure 4.40. Design now proceeds by finding a loop-gain function $L(s)$ whose frequency response satisfies all these bounds. It can be shown that a solution exists (for minimum-phase plant) which is optimal in the sense that $L(j\omega)$ lies on the boundary $B(\omega)$ for each ω, but this optimal solution may require a compensator $K(s)$ of great complexity. Design trade-offs are generally made to obtain reasonably simple compensators, at the expense of having a larger loop gain (at some frequencies) than is required. A typical $L(j\omega)$ locus is shown by the broken line in Figure 4.40. The choice of $L(s)$ is restricted by the requirement that the compensator should be realizable, so $|L(j\omega)|$ must fall at least as quickly as $|G(j\omega)|$ at high frequencies, and of course the Nyquist stability criterion must be satisfied. The loop-gain function $L(s)$ must also satisfy Bode's gain–phase relations, so arbitrary loci (on the Nichols chart) cannot be attained.

Once $K(s)$ has been designed, the inequalities

$$a_i(\omega_i)c_i(\omega_i) \leqslant |L(j\omega_i)[1 + L(j\omega_i)]^{-1}| \leqslant b_i(\omega_i)c_i(\omega_i) \qquad (4.181)$$

will be satisfied for $\{\omega_1, \ldots, \omega_k\}$ for some $c_i(\omega_i)$. In order to meet the design specifications

$$a_i(\omega_i) \leqslant |T(j\omega_i)| \leqslant b_i(\omega_i) \qquad (4.182)$$

the pre-filter is chosen to have the gain behaviour

$$|P(j\omega_i)| \approx [c_i(\omega_i)]^{-1} \qquad (4.183)$$

If specifications also exist in the form of bounds on the sensitivity function

$$S(s) = [1 + L(s)]^{-1} \qquad (4.184)$$

then a similar approach is used to meet these, and a combination of specifications on both $T(s)$ and $S(s)$ can be handled. (Note that T is not the complementary sensitivity in this subsection, and that T and S are not completely determined by each other.)

This design method is extended to multivariable problems as follows. Suppose that the design specifications are of the form given in (4.176), where $t_{ij}(s)$ is the (i, j) element of the closed-loop transfer function $T(s)$. From Figure 4.38, we have

$$(I + GK)y = GKPr \qquad (4.185)$$

or, if G is square

$$(\hat{G} + K)y = KPr \qquad (4.186)$$

where we have used $\hat{G}$ to denote G^{-1}, as before. We shall consider t_{uv}, namely the transfer function from the vth input to the uth output. If $r_j = 0$ for $j \neq v$,

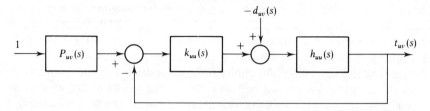

Figure 4.41 Interpretation of t_{uv} (equation (4.191)) as the output of a feedback system.

then the uth element of the vector KPr is given by

$$(KPr)_u = (KP)_{uv}r_v \tag{4.187}$$

$$= \sum_l k_{ul}p_{lv}r_v \tag{4.188}$$

The uth element of the vector $(\hat{G} + K)y$ is given by

$$([\hat{G} + K]y)_u = \sum_l (\hat{g}_{ul} + k_{ul})t_{lv}r_v \tag{4.189}$$

since $y_l = t_{lv}r_v$.

Now we impose the constraint that $k_{ij} = 0$ for $i \neq j$, namely that the compensator K is diagonal. Then, equating (4.188) with (4.189) (because of (4.186)), we obtain

$$r_v \sum_{l \neq u} \hat{g}_{ul}t_{lv} + r_v(\hat{g}_{uu} + k_{uu})t_{uv} = k_{uu}p_{uv}r_v \tag{4.190}$$

If we now define $h_{ij} = 1/\hat{q}_{ij}$, then (4.190) can be rewritten as

$$t_{uv} = \frac{h_{uu}k_{uu}p_{uv}}{1 + h_{uu}k_{uu}} - \frac{h_{uu}d_{uv}}{1 + h_{uu}k_{uu}} \tag{4.191}$$

where

$$d_{uv} = \sum_{l \neq u} \frac{t_{lv}}{h_{ul}} \tag{4.192}$$

The expression for t_{uv} has been written in this way because it now shows t_{uv} as the output of the SISO system shown in Figure 4.41, when the reference input is an impulse, and the disturbance (at the 'plant' input) is $-d_{uv}$.

The idea now is to design k_{uu} and p_{uv}, using the technique already developed for SISO systems. The term d_{uv} is not initially known, since it depends on the elements $\{t_{lv}: l \neq u\}$, which themselves depend on the details of the design. If it is assumed that the specification (4.176) is eventually attained, however, then bounds can be obtained on the possible variation of d_{uv}, which allows the design of k_{uu} and p_{uv} to proceed. This seems like a dangerously circular argument, but it can in fact be shown to be sound. Let us write (4.191) as

$$t_{uv} = \tau_{uv} - \tau_{duv} d_{uv} \tag{4.193}$$

and suppose that we can find k_{uu} and p_{uv} such that

$$|\tau_{uv}(j\omega)| - |\tau_{duv}(j\omega)| d_{uve}(\omega) > a_{uv}(\omega) \geqslant 0 \tag{4.194}$$

and

$$|\tau_{uv}(j\omega)| + |\tau_{duv}(j\omega)| d_{uve}(\omega) < b_{uv}(\omega) \tag{4.195}$$

where d_{uve} is an upper bound on all possible values of $|d_{uv}|$ which may occur for all possible plants

$$d_{uve}(\omega) = \sup_{G(j\omega)} \sum_{l \neq u} \frac{b_{lv}(\omega)}{|h_{ul}(j\omega)|} \tag{4.196}$$

Suppose also that (4.194) and (4.195) hold for all values of u and v – that is, suppose that m compensator elements $k_{uu}(s)$ and m^2 pre-filter elements $p_{uv}(s)$ have been found for which these inequalities hold. Then a result from functional analysis (Schauder's fixed-point theorem) can be used to show that (4.176) is also satisfied, and so the design specification will have been met.

A noteworthy feature of the QFT approach to multivariable design is that the compensator $K(s)$ has only m elements – only the diagonal elements are designed, and all the off-diagonal elements are fixed at zero. Each diagonal element $k_{ii}(s)$ appears in the expressions for the m elements $t_{ij}(s)$ of the transfer function $T(j = 1, \ldots, m)$ – that is, it appears in m separate SISO design problems. Each of these problems generates a set of boundaries $\{B_j(\omega_1), \ldots, B_j(\omega_k)\}$ $(j = 1, \ldots, m)$ which must be achieved by the loop-gain function $h_{ii}(j\omega)k_{ii}(j\omega)$, and the compensator $k_{ii}(s)$ must be chosen to satisfy the most stringent of these.

This strategy recognizes that the use of cross-couplings inside the feedback loop, for the purpose of controlling interaction, is inherently fragile if there is significant uncertainty about the plant model. Loop gains are used to reduce the amount of uncertainty. Once it has been reduced to an acceptable amount, then cross-coupling is used *outside* the loop, in the pre-filter $P(s)$, to obtain the required pattern of interactions.

As with the sequential loop-closing approach (see Section 4.2), success in designing a suitable $K(s)$ may depend on finding the best pairing of plant inputs with outputs.

We have given a very brief description of the QFT approach to feedback design, and have outlined only the simplest version of it. The theory has been elaborated to allow the use of off-diagonal elements in $K(s)$, the specification of responses to disturbances, trade-off between the bandwidths of various loops, and has even been extended to non-linear problems. It is rather complex, but much of the complexity can be encapsulated in computer software.

4.10   Control-structure design

For the design techniques presented in this chapter, it has been assumed that the manipulated and controlled variables (plant inputs and outputs) have already been chosen. In fact, an important part of multivariable design (particularly in the process industries) is the choice of those variables which it is most appropriate to control, and the choice of those variables which can be manipulated most effectively. Once such a choice has been made, it is frequently necessary to choose appropriate pairs of inputs and outputs – that is, to assign each input, or set of inputs, to the control of one particular output or set of outputs. This is known as the **input/output assignment** or **pairing** problem.

If a fully cross-coupled multivariable controller is to be used, then the pairing of inputs and outputs is unimportant since the cross-couplings can distribute control actions between inputs and outputs in the most effective way. But it is frequently necessary to use **decentralized control**, in which either no cross-couplings are permitted (the compensator transfer function being constrained to be diagonal) or the cross-coupling between loops is to be minimized (generally resulting in a block-diagonal compensator transfer function). In this case finding a suitable pairing is of great importance; if an input were to be paired with an output over which it had little or no influence, then satisfactory control would clearly be unattainable. Finding a good pairing of input and output variables is also important when decentralized control is dictated by the design technique, rather than by implementation requirements. For example, both the sequential loop-closing approach to multivariable design, and the version of the QFT approach presented in the previous section, can be made easier by a good initial pairing of inputs with outputs.

Problems such as choosing the input and output variables, and deciding how to pair them, are problems of **control-structure design**. In spite of their supreme importance, very little progress has been made towards solving such problems systematically. Some promising advances have recently been made, however, and we shall summarize them in this section.

An analytical tool for attacking the control-structure problem, which has been available for over twenty years, is the **relative-gain array**, or RGA (Bristol, 1966). Suppose that the transfer function from input j to output i of a plant is g_{ij} when all loops are open, and h_{ij} when all outputs except the ith output are tightly controlled. Then the (i, j) element of the RGA Γ is

$$\gamma_{ij} = \frac{g_{ij}}{h_{ij}} \tag{4.197}$$

This element can be evaluated from the open-loop transfer function of the plant, as follows. Suppose that a change Δu_j in the (transform of the) jth input causes a change Δy_k in the kth output. Then

$$\Delta u_j = \sum_k \hat{g}_{jk} \Delta y_k \tag{4.198}$$

where $\hat{G} = G^{-1}$, as before. (We assume that the plant is square.) But if all the loops except the ith are tightly controlled, then $\Delta y_k = 0$ for all $k \neq i$. Thus (4.198) simplifies to

$$\Delta u_j = \hat{g}_{ji} \Delta y_i \tag{4.199}$$

from which we obtain

$$h_{ij} = \frac{1}{\hat{g}_{ji}} \tag{4.200}$$

and (4.197) becomes

$$\gamma_{ij} = g_{ij} \hat{g}_{ji} \tag{4.201}$$

In terms of matrix operations, the relative gain array is defined by

$$\Gamma(s) = G(s) .* \hat{G}^{T}(s) \tag{4.202}$$

where '$.*$' denotes element-by-element multiplication (the Schur or Hadamard product).

The assumption that all the other loops are perfectly controlled is of course highly artificial unless it is restricted to a specified range of frequencies. In fact the RGA is almost invariably defined in terms of steady-state gain information only – namely, for $s = 0$. From now on we shall use the term 'relative gain array' (or RGA) to mean $\Gamma(0)$ only, except where specified otherwise. There are two reasons for this restriction. First, if integral control is used then the assumption of perfect control at zero frequency is entirely realistic, and integral control is used very commonly. Secondly, the RGA has been used most widely in the process industries; in these industries it is often difficult to obtain reliable dynamic models, whereas steady-state gain information is readily available.

The following properties of the RGA can be derived easily:

Theorem 4.5:   The $m \times m$ RGA $\Gamma(s) = G(s) . * \hat{G}^{\mathrm{T}}(s)$, with elements $\{\gamma_{ij}\}$, has the following properties:

(1) $$\sum_{j=1}^{m} \gamma_{ij} = \sum_{i=1}^{m} \gamma_{ij} = 1$$

(2)   Permutations of the rows or columns of $G(s)$ result in the same permutations of $\Gamma(s)$.

(3)   Input or output scaling leaves $\Gamma(s)$ unchanged. That is, if S and T are diagonal matrices, $G_2(s) = SG(s)T$, and $\Gamma_2(s) = G_2(s) . * \hat{G}_2^{\mathrm{T}}(s)$, then $\Gamma_2(s) = \Gamma(s)$.

(4)   If $G(s)$ is triangular (and hence also if it is diagonal), then $\Gamma(s) = I$.

Proof: The proof is left as an exercise for the reader.

The RGA was introduced as a measure of interaction which could be used both to select those input and output variables to be used for control, and to decide on how to pair them, if a decentralized control structure were required. It has been used widely and successfully in the process industries (Shinskey, 1979, 1981; McAvoy, 1983), but until recently virtually no theoretical explanation of its utility has been available. It has been known, for example, that choosing variables which result in large or negative elements in the RGA leads to difficulties in controlling the plant, and that input and output variables should be paired so that the diagonal elements of the RGA are as close as possible to unity.

Significant progress towards providing such an explanation has been made by Grosdidier *et al.* (1985), Nett and Manousiouthakis (1987), and Skogestad and Morari (1987). Of their results, we shall summarize those most relevant to the control-structure design problem, but we shall omit proofs and derivations. Not all of these results involve the RGA; some are formulated in terms of the steady-state gain matrix $G(0)$. Most of the results relate to the feedback loop shown in Figure 4.42, in which the compensator has the structure

$$K(s) = \frac{k}{s} C(s) \tag{4.203}$$

so that integral action in each loop is assumed, and $G(s)C(s)$ is proper (that is, $\| G(\infty)C(\infty) \| < \infty$). We shall assume that $G(s)$ has m inputs and m outputs, and that $\Gamma(0)$ denotes its RGA.

The first result is the following. Suppose that $C(s)$ is diagonal (decentralized control), $G(s)$ is stable, and the closed-loop system remains stable

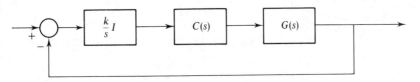

Figure 4.42 Feedback loop with compensator of structure $K(s)=\dfrac{k}{s}C(s)$.

if any $m-1$ feedback loops are opened. Then the system is unstable with all loops closed, for any $k > 0$, if (and only if when $m \leqslant 2$)

$$\det[G(0)] \Big/ \prod_{i=1}^{m} g_{ii}(0) < 0 \tag{4.204}$$

This condition may be met by some input/output pairings but violated by others, since interchanging any pair of rows or columns in $G(s)$ changes the sign of $\det[G(0)]$ and puts two new elements onto the principal diagonal.

The next result again assumes $C(s)$ to be diagonal: if $\gamma_{ii}(0) < 0$ for any i, then for any $k > 0$ the closed loop has at least one of the following properties:

- it is unstable
- loop i is unstable when all the other loops are opened
- it is unstable when loop i is opened

So if the RGA has any negative entries on its principal diagonal, and integral action is employed, then it is impossible to obtain complete integrity (stability in the presence of loop failures) as long as a completely decentralized control structure is used. It is also impossible to design each loop independently in this case.

If $m = 2$ then it is always possible to obtain $\gamma_{ii}(0) > 0$ by a suitable pairing of inputs and outputs, but if $m \geqslant 3$ this may not be so. In such a case the use of a cross-coupled controller is strongly indicated.

EXAMPLE 4.6

If

$$G(0) = \begin{bmatrix} 1 & 1 & -0.1 \\ 0.1 & 2 & -1 \\ -2 & -3 & 1 \end{bmatrix}$$

then

$$\Gamma(0) = \begin{bmatrix} -1.89 & 3.59 & -0.7 \\ -0.13 & 3.02 & -1.89 \\ 3.02 & -5.61 & 3.59 \end{bmatrix}$$

and it is impossible to make $\gamma_{ii}(0) > 0$ for $i = 1, 2, 3$ by any permutation of rows or columns.

We shall say that the system $Q(s) = G(s)C(s)$ is **integral-controllable**[5] if the system shown in Figure 4.42 is stable for any k such that $0 < k \leqslant k^*$ (for some $k^* > 0$), and has zero steady-state error (the latter condition rules out cancellation of the integrators by zeros). Systems which are integral-controllable can be tuned on-line, turning the gain up gradually from zero.

Two more results are known on integrity in the face of loop failures. If $m = 2$ and $\gamma_{ii}(0) > 0$ ($i = 1, 2$), then there exists a diagonal $C(s)$ such that $Q(s) = G(s)C(s)$ is integral-controllable, and each loop remains integral-controllable if the other loop fails. If $m = 3$, $\gamma_{ii}(0) > 0$ ($i = 1, 2, 3$) and a diagonal $C(s)$ is such that $Q(s)$ is integral-controllable and $q_{ii}(0) > 0$, then the failure of any one loop leaves the other loops still integral-controllable.

The next result shifts the emphasis to attainable performance rather than stability. If $g_{ii}(s)$ has no zeros in the right half-plane, $Q(s)$ is strictly proper ($|q_{ij}(\infty)| = 0$), $|g_{kl}(s)|$ approaches zero at least as fast as $|g_{ii}(s)|$ for large $|s|$ and all k, l, then $h_{ii}(s)$ has either right half-plane zeros or poles if $\gamma_{ii}(0) < 0$, where $h_{ii}(s)$ is the transfer function from u_i to y_i when all the other loops are closed with integral control. This result does not give the location of any right half-plane zeros; in particular cases these may turn out to be quite harmless for the design of the ith loop, but it certainly reinforces the belief that negative diagonal elements of the RGA are to be avoided if possible.

The last batch of results about the RGA concerns the robustness of a design to modelling errors or changes in the plant's behaviour. The idea is that large entries in the RGA indicate a plant which is poorly conditioned with respect to inversion. If the controller for such a plant is intended to invert the plant's transfer function at some frequencies, in order to reduce interactions, then its performance (and even its stability) is likely to be very sensitive to changes in the behaviour of the plant. Examples of such controllers are those designed by the ALIGN algorithm or by pseudo-diagonalization.

The basic result is:

Theorem 4.6:

$$\frac{\partial \hat{g}_{ji}}{\hat{g}_{ji}} = -\gamma_{ij} \frac{\partial g_{ij}}{g_{ij}} \tag{4.205}$$

Proof:

$$\hat{g}_{ji} = \frac{G^{ij}}{\det(G)}$$

where G^{ij} is the cofactor of g_{ij}. Hence

$$\ln \hat{g}_{ji} = \ln G^{ij} - \ln \det(G)$$

$$= \ln G^{ij} - \ln \left\{ \sum_k g_{ik} G^{ik} \right\}$$

Differentiating,

$$\frac{1}{\hat{g}_{ji}} \frac{\partial \hat{g}_{ji}}{\partial g_{ij}} = 0 - \frac{G^{ij}}{\det(G)}$$

$$= -\hat{g}_{ji}$$

and therefore

$$\frac{\partial \hat{g}_{ji}}{\hat{g}_{ji}} = - g_{ij} \hat{g}_{ji} \frac{\partial g_{ij}}{g_{ij}}$$

$$= - \gamma_{ij} \frac{\partial g_{ij}}{g_{ij}} \qquad \blacksquare$$

This theorem shows that a change in g_{ij} results in a much larger relative change in $\hat{g}_{ji}$, if $|\gamma_{ij}| \gg 1$. It is not surprising, then, that large elements in the RGA can be related to large values of the condition number

$$\text{cond}(G) = \frac{\bar{\sigma}(G)}{\underline{\sigma}(G)} \qquad (4.206)$$

(which is discussed in some detail in Chapter 8). The condition number depends on the scaling employed, whereas the RGA does not, so we define the **optimal condition number** $\text{cond}^*(G)$ to be

$$\text{cond}^*(G) = \min_{S, T} \text{cond}(SGT) \qquad (4.207)$$

where S and T are diagonal matrices. If we also define the 1-norm

$$\| \Gamma \|_1 = \max_j \sum_{i=1}^{m} |\gamma_{ij}| \qquad (4.208)$$

then the following result is available. If $m = 2$, then

$$\text{cond}^*(G) = \| \Gamma \|_1 + \sqrt{(\| \Gamma \|_1^2 - 1)} \qquad \text{(4.209)}$$

It is *conjectured* that, for any m,

$$\sum_{i,j} |\gamma_{ij}| - \frac{1}{\text{cond}^*(G)} \leqslant \text{cond}^*(G) \leqslant \sum_{i,j} |\gamma_{ij}| \qquad \text{(4.210)}$$

This result is known to hold for $m = 2$.

These and similar results can be used to show that, if the RGAs of both the plant and the controller have large elements, then the closed-loop system has little stability robustness in the face of uncorrelated perturbations at the plant input (that is, if the plant transfer function is $G = G_0(I + \Delta)$, where G_0 is the nominal model, and Δ is diagonal). Such perturbations are almost always present, because of uncertainty about the exact behaviour of control actuators. If $K(s)$ approximately inverts $G(s)$, then the RGA of $K(s)$ will be approximately the same as that of $G(s)$. So, if the plant's RGA contains large elements, an inverting (decoupling) compensator should not be used. A diagonal compensator has an RGA with no element larger than 1, and may therefore be able to provide reasonable stability robustness even if the plant's RGA has large elements. It is unlikely to produce good performance in the face of disturbances, however, since it cannot remove interaction. (Good performance in response to reference or set-point changes may still be obtained if a control structure with two degrees of freedom is used, as in the QFT design method.) One can conclude, then, that large elements in the plant's RGA certainly indicate a plant which is difficult to control.

It should be noted that all the results presented from Theorem 4.6 onwards hold at any frequency, and there is no need to restrict them to $s = 0$. In fact, the RGA of the plant should be inspected at frequencies up to the largest 0 dB cross-over frequency required, and a little higher, to check for inherent sensitivity problems.

The RGA is the major analytical tool developed specifically for the control-structure design problem, but other information is available which can be used to make decisions about the structure.[6]

A Nyquist array, for example, may sometimes provide a quick visual indication of how a plant may be subdivided into several, almost independent sub-processes. Principal-gain plots for various blocks of the plant's transfer-function matrix may indicate which inputs are the most effective for controlling particular outputs, and may also indicate whether there is an obvious decomposition of the plant into 'fast' and 'slow' sub-plants.

If $m > 3$, then the number of possible control structures becomes extremely large if one considers all possible decentralized schemes – that is, all possible ways of dividing an m-input, m-output plant into p subsystems, each of which has m_i inputs and outputs, with $\sum_{i=1}^{p} m_i = m$. Even if one takes each

$m_i = 1$ (i.e. completely decentralized) there are still $m!$ possible input/output pairings to try, and the number of all possible schemes is much larger than this. For example, with $m = 4$ there are 131 possibilities, and with $m = 5$ there are 1496. In practice it is impossible to try more than a small fraction of all the possible combinations 'at random', and there is as yet no systematic way of searching through all the possibilities. The RGA helps a little in this respect, but the main guidance towards choosing an appropriate structure must come from a physical understanding of the plant.

SUMMARY

In this chapter we have presented several approaches to multivariable design, all of which generalize, in one way or another, the classical techniques of feedback design associated with the names of Nyquist and Bode. Each approach ultimately replaces a multivariable problem by a set of SISO design problems. In the sequential loop-closing approach, the Nyquist-array approach and the characteristic-locus approach there is a preliminary stage in which the aim is to reduce high-frequency interaction and make subsequent design steps easier.

In each of these methods, this preliminary decoupling stage can be aided by use of the ALIGN algorithm, or one of its generalizations, which allows the system to be decoupled across a range of frequencies by either a constant or a dynamic compensator. If a dynamic compensator is obtained automatically in this way, however, there is a danger that it may have zeros which cancel unstable (or very lightly damped) poles of the plant.

If a precise specification of both the model uncertainty and permissible closed-loop behaviour is available, then the use of the quantitative feedback theory should be considered. This is more complicated than the other approaches, but it is in many ways more systematic and more powerful.

Before embarking on a multivariable feedback design, one should always consider carefully the design of the controller structure. Is it possible to divide the problem into smaller multivariable problems, or into a mixture of SISO and multivariable problems, or, best of all, into a set of SISO problems only? The answer depends not only on the plant, of course, but also on the design specifications. Is it possible to choose a better set of input and output variables, or a better pairing of the existing set? The relative-gain array gives some help in answering such questions, but this important problem area of multivariable design remains largely unsolved.

Almost all of the material in this chapter can be applied to plant described by irrational transfer functions – in particular, to plant containing time delays. The 'input' to most of the techniques described is the frequency response of the plant, together with knowledge of any unstable poles.

Notes

1.   Much of the research literature contains similar suggestions, based on system characteristics at arbitrarily high frequencies. Such suggestions ignore the facts that real control systems are subject to power limitations, which manifest themselves as bandwidth limitations, and that models are valid over limited frequency ranges only.

2.   Ford (1985) has given some bounds for the extent of interaction, in terms of direct Nyquist arrays; these do not help to predict the *diagonal* elements of the closed-loop response, however.

3.   Ford and Daly (1979) also include negative powers of s in $k_j(s)$. But since each column can be scaled by an arbitrary scalar function, equation (4.159) is sufficiently general.

4.   This example illustrates nicely the point of Definition 2.7: that stability of a feedback system requires *each* of the transfer functions $GK(I + GK)^{-1}$, $KG(I + KG)^{-1}$, $G(I + KG)^{-1}$ and $K(I + GK)^{-1}$ to be stable. In this case they are all stable; however, if we defined a generalized notion of stability (see Section 2.11) to mean that 'all poles must have a damping factor of at least 0.2', then our design would be unstable in this sense, but only one of the four transfer functions would exhibit the instability.

5.   The concept of integral controllability was introduced by Grosdidier *et al.* (1985).

6.   Other tools introduced recently to help attack the control-structure design problem are the block relative-gain array (Manousiouthakis *et al.*, 1986) and the structured singular-value interaction measure (Grosdidier and Morari, 1986).

EXERCISES

4.1   Repeat the characteristic locus design exercise performed on the AIRC model (see the Appendix) in Section 4.4, but aim for a closed-loop bandwidth of about 3 rad s^{-1}.

4.2   Design a compensator for the TGEN model (see the Appendix), assuming a control structure with one degree of freedom, and aiming for a closed-loop bandwidth of about 0.5 rad s^{-1}, using one or more of the following methods:

(a)   sequential loop closing

(b)   characteristic locus

(c)   direct Nyquist array

The sensitivity should not exceed 0.1 for $\omega < 0.05$ (for any signal direction), and the output-sensor noise amplification should not exceed 0.002 for $\omega > 50$ (for any signal direction).

4.3 (a) A SISO negative-feedback system has a forward-path transfer function $l(s)$ and a feedback transfer function $f(s)$. Its closed-loop transfer function is $h(s)$. Show that

$$\frac{1}{h(s)} = \frac{1}{l(s)} + f(s)$$

(b) Show that the relationships indicated in Figure 4.17 hold for the (SISO) feedback loop shown in Figure 4.18.

4.4 Prove Theorem 2.12, and prove also that the degrees of row dominance and column dominance of XGX^{-1}, when X is chosen optimally, are the same.

4.5 Refer to Example 4.5, with the system defined by equation (4.150).

(a) Using suitable software, display a Nyquist array of $\hat{G}(j\omega)$, and check that it is not column dominant. Compute the appropriate Perron–Frobenius eigenvalues and eigenvectors and check the correctness of Figures 4.20, 4.21 and 4.22.

(b) Using the INA technique, design further compensation for each loop which will give a closed-loop bandwidth of about $10\,\text{rad s}^{-1}$ and well-damped responses. Include integral action in your compensator.

(c) Investigate whether column dominance of $\hat{K}\hat{G}(j\omega)$ can be obtained using a constant, diagonal $\hat{K}$.

(d) Investigate whether (a) and (b) can be repeated using the DNA method instead of the INA method.

4.6 Given

$$G = \begin{bmatrix} 2 & -4 \\ -1 & 3 \end{bmatrix}$$

find a diagonal matrix K, with positive elements, such that GK is row-dominant.

4.7 Prove the properties of the RGA listed in Theorem 4.5.

4.8 Show that the plant

$$G(s) = \begin{bmatrix} 1 & 0 \\ \dfrac{1}{s-1} & 1 \end{bmatrix}$$

cannot be stabilized by a fully decentralized compensator of the form

$$K(s) = \begin{bmatrix} k_1(s) & 0 \\ 0 & k_2(s) \end{bmatrix}$$

Does the relative gain array indicate the presence of this difficulty? (The pole at $+1$ is known as a 'fixed mode' in the literature on decentralized control.)

4.9   Examine the RGA for the TGEN model (see the Appendix). Comment on its implications, if a compensator is to be designed to give a closed-loop bandwidth of about 0.5 rad s^{-1}.

4.10   Examine the possibility of using decentralized control for the AIRC model (see the Appendix). Do not assume that the same speed of response is required for each controlled output.

References

Bristol E.H. (1966). On a new measure of interaction for multivariable process control. *IEEE Transactions on Automatic Control*, **AC-11**, 133–4.

Edmunds J.M. and Kouvaritakis B. (1979). Extensions of the frame alignment technique and their use in the characteristic locus design method. *International Journal of Control*, **29**, 787–96.

Fiedler M. and Ptak V. (1966). On matrices with non-positive off-diagonal elements and positive principal minors. *Czechoslovak Mathematical Journal*, **12**, 382–400.

Ford M.P. (1985). Bounds on interaction in the Nyquist array method. *GEC Journal of Research*, **3**, 200–3.

Ford M.P. and Daly K.C. (1979). Dominance improvement by pseudodecoupling. *Proceedings of the Institution of Electrical Engineers*, **126**, 1316–20.

Grosdidier P., Morari M. and Holt B.R. (1985). Closed-loop properties from steady-state gain information. *Industrial and Engineering Chemistry Fundamentals*, **24**, 221–35.

Grosdidier P. and Morari M. (1986). Interaction measures for systems under decentralized control. *Automatica*, **22**, 309–19.

Hawkins D.J. (1972). Pseudodiagonalisation and the inverse Nyquist array method. *Proceedings of the Institution of Electrical Engineers*, **119**, 337–42.

Horowitz I. (1979). Quantitative synthesis of uncertain multiple input-output feedback systems. *International Journal of Control*, **30**, 81–106.

Horowitz I. (1982). Quantitative feedback theory. *Proceedings of the Institution of Electrical Engineers*, Part D, **129**, 215–26.

Horowitz I. and Sidi M. (1972). Synthesis of feedback systems with large plant ignorance for prescribed time domain tolerances. *International Journal of Control*, **16**, 287–309.

Horowitz I. and Sidi M. (1980). Practical design of multivariable feedback systems with large plant uncertainty. *International Journal of Systems Science*, **11**, 851–75.

Hung Y.S. and MacFarlane A.G.J. (1982). *Multivariable Feedback: A Quasi-classical Approach*, Lecture Notes in Control and Information Sciences, Vol. 40. Berlin: Springer-Verlag.

Kantor J.C. and Andres R.P. (1979). A note on the extension of Rosenbrock's Nyquist array techniques to a larger class of transfer function matrices. *International Journal of Control*, **30**, 387–93.

Kouvaritakis B. (1979). Theory and practice of the characteristic-locus design method. *Proceedings of the Institution of Electrical Engineers*, **126**, 542–8.

McAvoy T.J. (1983). *Interaction Analysis*. Research Triangle Park NC: Instrument Society of America.

MacFarlane A.G.J. and Kouvaritakis B. (1977). A design technique for linear multi-variable feedback systems. *International Journal of Control*, **25**, 837–74.

Manousiouthakis V., Savage R. and Arkun Y. (1986). Synthesis of decentralized control structures using the concept of block relative gain. *American Institute of Chemical Engineers Journal*, **32**, 991–1003.

Mayne D.Q. (1973). The design of linear multivariable systems. *Automatica*, **9**, 201–7.

Mayne D.Q. (1979). Sequential design of linear multivariable systems. *Proceedings of the Institution of Electrical Engineers*, **126**, 568–72.

Mees A.I. (1981). Achieving diagonal dominance. *Systems and Control Letters*, **1**, 155–8.

Munro N. (1985). Recent extensions to the inverse Nyquist array method. In *Proc. 24th IEEE Conf. on Decision and Control*, Miami FL, pp. 1852–7.

Munro N. (1987). Computer-aided design I: The inverse Nyquist array design method. In *Multivariable Control for Industrial Applications* (O'Reilly J., ed.), pp. 211–28. Stevenage: Peter Peregrinus.

Nett C.N. and Manousiouthakis V. (1987). Euclidean condition and block relative gain: connections, conjectures, and clarifications. *IEEE Transactions on Automatic Control*, **AC-32**, 405–7.

Rosenbrock H.H. (1970). *State-Space and Multivariable Theory*. London: Nelson.

Rosenbrock H.H. (1974). *Computer-Aided Control System Design*. New York: Academic Press.

Seneta E. (1973). *Non-negative Matrices*. London: Allen & Unwin.

Shinskey F.G. (1979). *Process Control Systems*. New York: McGraw-Hill.

Shinskey F.G. (1981). *Controlling Multivariable Processes*. Research Triangle Park NC: Instrument Society of America.

Siljak D.D. (1978). *Large-Scale Dynamic Systems*. New York: North-Holland.

Skogestad S. and Morari M. (1987). Implications of large RGA elements on control performance. *Industrial and Engineering Chemistry Research*, **26**, 2323–30.

CHAPTER 5

Multivariable Design: LQG Methods

5.1 Introduction
5.2 The solution of the LQG problem
5.3 Performance and robustness of optimal state feedback
5.4 Loop transfer recovery (LTR)
5.5 Design procedure for square plant

5.6 Shaping the principal gains
5.7 Some practical considerations
5.8 Design example
5.9 Non-minimum-phase plant
Summary
Exercises
References

5.1 Introduction

In this chapter we shall begin examining design methods based on optimal control theory. The particular theory we shall be concerned with here is that of the so-called **linear quadratic Gaussian** or LQG problem. We shall state the main results of this theory, but for fuller details we refer the reader to Kwakernaak and Sivan (1972), Anderson and Moore (1972), Davis and Vinter (1985), IEEE (1971), Åström and Wittenmark (1984), Franklin and Powell (1980), Banks (1986); many others could have been mentioned. These sources establish the underlying theory, whereas here we shall be concerned with developing a design method based on the theory. This method is one which allows the designer to shape the principal gains of the return ratio, at either the input or the output of the plant, to achieve required performance or robustness specifications. Stability is obtained automatically, so it is not necessary to worry about the phase characteristics – as we shall see, the characteristic loci do not need to be examined.

The problem addressed by the theory is the following. Suppose that we have a plant model in state-space form,

$$\dot{x} = Ax + Bu + \Gamma w \tag{5.1}$$

$$y = Cx + v \tag{5.2}$$

where w and v are 'white noises', namely zero-mean Gaussian stochastic processes which are uncorrelated in time (but may be correlated with each other), having covariances

$$E\{ww^T\} = W \geqslant 0, \qquad E\{vv^T\} = V > 0 \tag{5.3}$$

We shall assume in this chapter that w and v are in fact uncorrelated with each other, namely that

$$E\{wv^T\} = 0 \tag{5.4}$$

In (5.1) u represents the vector of control signals; in (5.2) y is the vector of measured (but corrupted by v) outputs. The problem is then to devise a feedback-control law which minimizes the 'cost'

$$J = \lim_{T \to \infty} E\left\{ \int_0^T (z^T Q z + u^T R u) dt \right\} \tag{5.5}$$

where

$$z = Mx \tag{5.6}$$

is some linear combination of the states, and

$$Q = Q^T \geqslant 0, \qquad R = R^T > 0 \tag{5.7}$$

are weighting matrices.

The solution to the LQG problem is prescribed by the **separation principle** (or 'certainty equivalence', in the literature on econometrics), which states that the optimal result is achieved by adopting the following procedure. First, obtain an optimal estimate $\hat{x}$ of the state x (optimal in the sense that $E\{(x - \hat{x})^T(x - \hat{x})\}$ is minimized), and then use this estimate as if it were an exact measurement of the state to solve the deterministic linear quadratic control problem. The point of this procedure is that it reduces the problem to two sub-problems, the solutions to which are known.[1]

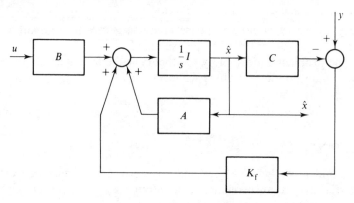

Figure 5.1   The Kalman filter.

The solution to the first sub-problem, that of estimating the state, is given by Kalman-filter theory. Figure 5.1 shows the block diagram of a Kalman filter, which is seen to have the structure of a state observer; it is distinguished from other observers by the choice of the gain matrix K_f, a formula for which will be given in the next section. Note that the *inputs* to the Kalman filter are the plant input and output vectors, u and y, and that its *output* is the state estimate vector $\hat{x}$.

The second sub-problem is to find the control signal which will minimize the (deterministic) cost

$$\int_0^\infty (z^T Q z + u^T R u)\, dt \tag{5.8}$$

on the assumption that

$$\dot{x} = Ax + Bu \tag{5.9}$$

The solution to this is to let the control signal u be a linear function of the state:

$$u = -K_c x \tag{5.10}$$

where the **state-feedback matrix** K_c is given according to the formula which appears in the next section.

Now, if the control problem to be solved can really be represented by equations (5.1)–(5.7) – that is, if only well-defined stochastic processes are acting on the plant and measurements, the concern is to minimize the cost function (5.5), and the available linear model is reliable, then there is little more to be said. Usually, though, there are the familiar aspects of the control problem that are not captured by this mathematical formulation. There

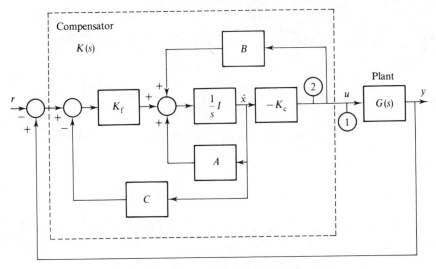

Figure 5.2 The LQG compensator structure.

are model uncertainties, non-linearities, various kinds of disturbance and possibly many constraints on realistic solutions, none of which can easily be given a mathematical representation. In this case one may still be interested in using the LQG theory as a method for synthesizing controllers, but with the matrices W, V, Q and R which appear in the problem formulation considered as 'tuning parameters' which are to be adjusted until a satisfactory design is obtained, rather than as representations of aspects of the real problem.

This is the view we shall take in this chapter. From the perspective of the design problem as we have formulated it in earlier chapters, we are simply designing a compensator with a particular internal structure. This structure is the series connection of a Kalman filter with a state-feedback matrix; a block diagram of it is shown in Figure 5.2.

To judge the suitability of a candidate design with this structure, we shall use methods introduced in earlier chapters. In particular, we shall make much use of principal gains.

5.2 The solution of the LQG problem

The optimal state-feedback matrix K_c is given by

$$K_c = R^{-1}B^{\mathrm{T}}P_c \tag{5.11}$$

where P_c satisfies the **algebraic Riccati equation**

$$A^{\mathrm{T}}P_c + P_c A - P_c BR^{-1}B^{\mathrm{T}}P_c + M^{\mathrm{T}}QM = 0 \tag{5.12}$$

and $P_c = P_c^T \geqslant 0$. (In general there are many solutions to (5.12), but only one of them is positive-semidefinite.)

The Kalman-filter gain matrix K_f is given by

$$K_f = P_f C^T V^{-1} \tag{5.13}$$

where P_f satisfies another algebraic Riccati equation (dual to (5.12)):

$$P_f A^T + A P_f - P_f C^T V^{-1} C P_f + \Gamma W \Gamma^T = 0 \tag{5.14}$$

and

$$P_f = P_f^T \geqslant 0$$

The matrices K_c and K_f exist, and the closed-loop system is internally stable, provided the systems with state-space realizations $(A, B, Q^{1/2} M)$ and $(A, \Gamma W^{1/2}, C)$ are **stabilizable** and **detectable** – namely, provided any uncontrollable or unobservable modes are asymptotically stable.

Several algorithms are available for solving (5.12) and (5.14). Here we give one put forward by MacFarlane (1963) and Potter (1966), which has the advantage of requiring only linear algebra. More sophisticated versions, with improved numerical properties, are described in Section 8.3.10.

Form the **Hamiltonian matrix** (of dimensions $2n \times 2n$, if n is the number of states)

$$H = \begin{bmatrix} A & -BR^{-1}B^T \\ -M^T Q M & -A^T \end{bmatrix} \tag{5.15}$$

which has the property that if λ is an eigenvalue, then so is $-\lambda$ (with the same multiplicity). Let U be the matrix of eigenvectors of H, ordered so that the left-most n columns correspond to eigenvalues with negative real parts, and the right-most n columns therefore correspond to eigenvalues with positive real parts. (If $(A, B, Q^{1/2} M)$ is minimal, H has no eigenvalues on the imaginary axis.) Now partition U into $n \times n$ blocks:

$$U = \begin{bmatrix} U_{11} & U_{12} \\ U_{21} & U_{22} \end{bmatrix} \tag{5.16}$$

The solution to (5.12) is then given by

$$P_c = U_{21} U_{11}^{-1} \tag{5.17}$$

To solve the dual equation (5.14), replace A by A^T, B by C^T, R by V and $M^T Q M$ by $\Gamma W \Gamma^T$, and apply the same algorithm. There will be no eigenvalues on the imaginary axis if $(A, \Gamma W^{1/2}, C)$ is minimal.

If we substitute (5.10) into (5.9) we obtain

$$\dot{x} = (A - BK_c)x \qquad (5.18)$$

as the equation of the closed-loop system which would result if full-state optimal feedback were implemented. The eigenvalues of $A - BK_c$ can be shown to be the (open) left half-plane eigenvalues of H, which shows the hypothetical state-feedback scheme to be asymptotically stable.

From Figure 5.1, the state equation of the Kalman filter is seen to be

$$\frac{d}{dt}\hat{x} = (A - K_f C)\hat{x} + Bu + K_f y \qquad (5.19)$$

The eigenvalues of $A - K_f C$ can be shown to be the left half-plane eigenvalues of the Hamiltonian matrix for the filter problem, which shows the Kalman filter to be an asymptotically stable observer.

Now, the equation of the combined Kalman-filter/optimal state-feedback scheme is

$$\frac{d}{dt}\begin{bmatrix} x \\ \hat{x} \end{bmatrix} = \begin{bmatrix} A & -BK_c \\ K_f C & A - K_f C - BK_c \end{bmatrix}\begin{bmatrix} x \\ \hat{x} \end{bmatrix} + \begin{bmatrix} \Gamma w \\ K_f v \end{bmatrix} \qquad (5.20)$$

If we define ε as the state estimation error,

$$\varepsilon = x - \hat{x} \qquad (5.21)$$

then (5.20) can be rewritten as

$$\frac{d}{dt}\begin{bmatrix} x \\ \varepsilon \end{bmatrix} = \begin{bmatrix} A - BK_c & BK_c \\ 0 & A - K_f C \end{bmatrix}\begin{bmatrix} x \\ \varepsilon \end{bmatrix} + \begin{bmatrix} \Gamma w \\ \Gamma w - K_f v \end{bmatrix} \qquad (5.22)$$

which shows that the closed-loop eigenvalues of the LQG-compensated plant are just the union of the eigenvalues of the optimal state-feedback scheme with those of the Kalman filter. The overall scheme is therefore internally stable (under the stated assumptions).

5.3 Performance and robustness of optimal state feedback

Suppose that all the states of the plant are available for feeding back through K_c, so that we can dispense with the Kalman filter, as shown in Figure 5.3. The return ratio at the input to the plant is then

$$-H_c(s) = -K_c(sI - A)^{-1}B \qquad (5.23)$$

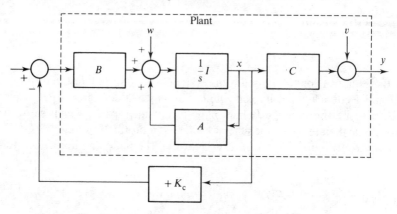

Figure 5.3 Hypothetical optimal full state-feedback controller.

so that the return difference at the input is

$$F_c(s) = I + K_c(sI - A)^{-1} B \qquad (5.24)$$

By adding and subtracting sP_c to (5.12), we obtain

$$-(-sI - A^T)P_c - P_c(sI - A) - P_c BR^{-1} B^T P_c + M^T QM = 0 \quad (5.25)$$

Now, multiplying on the left by $B^T(-sI - A^T)^{-1}$ and on the right by $(sI - A)^{-1} B$ gives

$$
\begin{aligned}
& -B^T P_c(sI - A)^{-1} B - B^T(-sI - A^T)^{-1} P_c B - B^T(-sI - A^T)^{-1} \\
& \quad \times P_c BR^{-1} B^T P_c(sI - A)^{-1} B + B^T(-sI - A^T)^{-1} \\
& \quad \times M^T QM(sI - A)^{-1} B = 0 \qquad (5.26)
\end{aligned}
$$

or

$$-RH_c(s) - H_c^T(-s)R - H_c^T(-s)RH_c(s) + G_c^T(-s)QG_c(s) = 0$$
$$(5.27)$$

where

$$G_c(s) = M(sI - A)^{-1} B \qquad (5.28)$$

The first three terms in (5.27) can be rewritten as $R - [I + H_c^T(-s)] R [I + H_c(s)]$, so, remembering that $F_c(s) = I + H_c(s)$, we can rewrite (5.27) as

$$F_c^T(-s)RF_c(s) = R + G_c^T(-s)QG_c(s) \qquad (5.29)$$

(Note that $G_c(s)$ is not the plant's transfer function unless $M = C$. Equation (5.29) was first derived by MacFarlane (1970).)

From (5.29) it is clear that

$$F_c^H(j\omega) R F_c(j\omega) \geqslant R \tag{5.30}$$

Safonov and Athans (1977) have shown that, if

$$R = \rho I \tag{5.31}$$

so that

$$F_c^H(j\omega) F_c(j\omega) \geqslant I \tag{5.32}$$

then there is at least 60° of phase margin in each input channel (that is, pure phase changes of 60° can be tolerated in each input channel simultaneously without losing stability), and infinite gain margin (in the classical sense: the gain in each channel can be increased indefinitely without losing stability). The loop may be conditionally stable, but it has a margin of at least 6 dB against gain reductions. Furthermore, it follows immediately from (5.32) that

$$\underline{\sigma}(F_c(j\omega)) \geqslant 1 \tag{5.33}$$

which in a SISO system would imply that the sensitivity never exceeds 1 (since $F_c(s)$ is also the return difference at the plant output in a SISO system). We also have that

$$\bar{\sigma}\{H_c(j\omega)[I + H_c(j\omega)]^{-1}\} = \bar{\sigma}\{I - F_c^{-1}(j\omega)\} \tag{5.34}$$

$$\leqslant 1 + \bar{\sigma}\{F_c^{-1}(j\omega)\} \tag{5.35}$$

$$\leqslant 2 \tag{5.36}$$

since (5.33) implies that $\bar{\sigma}\{F_c^{-1}(j\omega)\} \leqslant 1$. Applying the small gain theorem (Chapter 3), we see from (5.36) that the full-state optimal feedback system is robust against any unstructured multiplicative perturbation Δ, at the input of the plant, for which

$$\bar{\sigma}(\Delta(j\omega)) \leqslant \tfrac{1}{2} \tag{5.37}$$

For single-input plants these results can be given striking graphical interpretations in terms of the Nyquist locus, as first shown by Kalman (1964). The inequality (5.33) becomes $|1 + H_c(j\omega)| \geqslant 1$, which means that the Nyquist locus stays outside the circle with centre -1, radius 1, as shown in

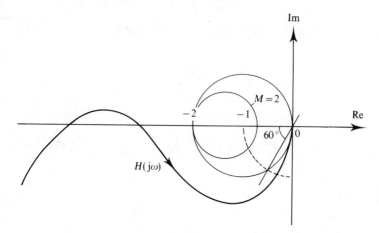

Figure 5.4 Nyquist locus at input of optimally regulated single-input plant (with full state feedback).

Figure 5.4. The inequality (5.36) becomes

$$\left| \frac{H_c(j\omega)}{1 + H_c(j\omega)} \right| \leqslant 2 \tag{5.38}$$

which states that the Nyquist locus remains outside the $M = 2$ circle, namely the circle with centre $-4/3$ and radius $2/3$. It is seen from Figure 5.4 that, for a single-input plant, the results obtained by Safonov and Athans (1977) are a simple consequence of (5.33).

Figure 5.4 also shows that the full state-feedback controller can achieve its impressive properties only because it behaves like a first-order system at high frequencies. Above the 0 dB cross-over frequency the phase is close to $-90°$, and hence the gain 'roll-off' rate is at most -20 dB/decade. The same is true when there are several inputs: for some ω_0 and ε,

$$\underline{\sigma}(H_c(j\omega)) \geqslant \frac{\varepsilon}{\omega}, \quad \omega > \omega_0 \tag{5.39}$$

A remark is in order about the assumption (5.31). This is not as restrictive as it appears, because we can apply (possibly non-diagonal) scaling to our plant, replacing B by $BR^{1/2}$, and equivalently replacing R (in the cost function (5.5)) by I. This procedure is particularly appropriate if it is known that the model uncertainties in some input directions are significantly different from those in other directions.

Since equations (5.13) and (5.14) are dual to (5.11) and (5.12), respectively, it is apparent that we can obtain a similar result to (5.29) for the return difference of the Kalman filter. The return ratio of the filter, evaluated at the input to the gain matrix K_f (see Figure 5.1), is

$$-H_f(s) = -C(sI - A)^{-1}K_f \tag{5.40}$$

and the corresponding return difference is

$$F_f(s) = I + C(sI - A)^{-1}K_f \tag{5.41}$$

Proceeding as before, we obtain

$$F_f(s)VF_f^T(-s) = V + G_f(s)WG_f^T(-s) \tag{5.42}$$

where $G_f(s) = C(sI - A)^{-1}\Gamma$, and we conclude that the Kalman filter, viewed as a feedback system, has good robustness and performance properties.

5.4 Loop transfer recovery (LTR)

Since both the optimal state-feedback regulator and the Kalman filter have such good properties, it might be expected that LQG compensators would generally yield good robustness and performance. Unfortunately this is not the case: Doyle (1978) has shown that LQG designs can exhibit arbitrarily poor stability margins. The usual advice given on state-feedback/observer design is to make the observer dynamics much faster than the desired state-feedback dynamics. One might therefore expect that, if the eigenvalues of $A - K_f C$ were pushed sufficiently far into the left half-plane, then one would obtain nice state-feedback properties at the plant input. This also turns out to be false, as shown by Doyle and Stein (1979). They demonstrate that in some cases stability margins are actually reduced as the observer (filter) dynamics are made faster.

Fortunately there is a way of designing the Kalman filter so that the full state-feedback properties are 'recovered' at the input of the plant. What is needed is for some of the filter's eigenvalues to be placed at the *zeros* of the plant, the remainder being allowed to become arbitrarily fast; we shall outline a procedure for achieving this, taken from Kwakernaak (1969) and Doyle and Stein (1979, 1981). Since the procedure relies on the 'cancellation' of some of the plant dynamics (in particular the zeros) by the filter dynamics, it is guaranteed to work only with minimum-phase plants. If right half-plane zeros exist in the plant the procedure may still work, particularly if these zeros lie beyond the operating bandwidth of the system as finally designed.

The robustness and performance properties at the input of the plant are determined by the return ratio at the point marked 1 in Figure 5.2, whereas the return ratio (5.23), which is the one we would like to have, is the return ratio at the point marked 2. (It may seem surprising that the return ratio at point 2 is as simple as $-K_c(sI-A)^{-1}B$. The reason is that the error dynamics of the filter are not excited by the control input u – in fact they are uncontrollable from u. If the only signal acting on the system is u – the loop being broken at point 1 – and if the matrices A, B and C in the filter match those of the plant exactly, then the signal at the input to the gain matrix K_f is zero.) In general, the return ratios at points 1 and 2 are quite different. The **loop transfer recovery** or LTR procedure which we shall now describe forces the return ratio at point 1 to approach that at point 2.

Let the transfer function of the LQG compensator shown in Figure 5.2 be $K(s)$. The return ratio at point 1 is then

$$K(s)G(s) = -K_c(sI-A+BK_c+K_fC)^{-1}K_fC(sI-A)^{-1}B \quad \text{(5.43)}$$

If we let

$$\phi(s) = (sI-A)^{-1} \quad \text{(5.44)}$$

and

$$\Psi(s) = (sI-A+BK_c)^{-1}$$

then (5.43) becomes (writing ϕ instead of $\phi(s)$, etc.)

$$KG = -K_c[\Psi^{-1}+K_fC]^{-1}K_fC\phi B \quad \text{(5.45)}$$

$$= -K_c[\Psi-\Psi K_f(C\Psi K_f+I)^{-1}C\Psi]K_fC\phi B \quad \text{(5.46)}$$

$$= -K_c\Psi K_f[I-(C\Psi K_f+I)^{-1}C\Psi K_f]C\phi B \quad \text{(5.47)}$$

$$= -K_c\Psi K_f[I+C\Psi K_f]^{-1}C\phi B \quad \text{(5.48)}$$

In (5.46) and (5.48) we have used the matrix-inversion lemma (see Chapter 3, Note 3):

$$(A+BCD)^{-1} = A^{-1}-A^{-1}B(DA^{-1}B+C^{-1})^{-1}DA^{-1}$$

Now suppose that we obtain K_f by choosing the covariance matrix W (in equations (5.3)) as

$$W = W_0 + q\Sigma \quad \text{(5.49)}$$

where $\Sigma = \Sigma^T \geq 0$, and q is a real, positive parameter. Here W_0 could be an estimate of the true process-noise (w) covariance, for example; in order to obtain LTR we shall need to increase q, in theory to arbitrarily large values.

Substituting for W in (5.14), we obtain

$$\frac{P_f A^T}{q} + \frac{A P_f}{q} - \frac{P_f C^T V^{-1} C P_f}{q} + \frac{\Gamma W_0 \Gamma^T}{q} + \Gamma \Sigma \Gamma^T = 0 \tag{5.50}$$

Now, it can be shown (Kwakernaak and Sivan, 1972, p. 307) that, if $C(sI - A)^{-1} \Gamma W^{1/2}$ has no transmission zeros in the right half-plane, and if it has at least as many outputs as rank (Σ), then

$$\lim_{q \to \infty} \frac{P_f}{q} = 0 \tag{5.51}$$

This result may be thought of as meaning that, as q is increased, so the Kalman filter is being 'told' that an increasing proportion of the variance in the plant output is due to state variations, and a decreasing proportion to measurement errors. In these circumstances the filter is able to track the state trajectory relatively well: P_f is (the filter's estimate of) the covariance of the state estimation error, and (5.51) states that this is increasing more slowly than the process-noise covariance (if the stated assumptions hold).

From (5.50) and (5.51) we have

$$\lim_{q \to \infty} (q \Gamma \Sigma \Gamma^T)^{\frac{1}{2}} = P_f C^T V^{-1/2} \tag{5.52}$$

But

$$K_f = P_f C^T V^{-1} \tag{5.53}$$

so

$$K_f \to q^{1/2} \Gamma \Sigma^{1/2} V^{-1/2} \quad \text{as } q \to \infty \tag{5.54}$$

In particular, if we choose

$$\Gamma = B, \qquad \Sigma = I \tag{5.55}$$

and *provided* $C(sI - A)^{-1} B$ *has no zeros in the right half-plane*, then

$$K_f \to q^{1/2} B V^{-1/2} \quad \text{as } q \to \infty \tag{5.56}$$

Substituting this back in (5.48) gives

$$KG \to -q^{1/2} K_c \Psi B V^{-1/2} [I + q^{1/2} C \Psi B V^{-1/2}]^{-1} C \phi B \quad \text{as } q \to \infty \tag{5.57}$$

$$\xrightarrow[q \to \infty]{} -K_c \Psi B V^{-1/2} [C \Psi B V^{-1/2}]^{-1} C \phi B \tag{5.58}$$

provided $C\Psi B$ is square – that is, if the plant has as many inputs as outputs. Maintaining this assumption, and noting that

$$\Psi = \phi[I + BK_c\phi]^{-1} \tag{5.59}$$

we obtain

$$\lim_{q \to \infty} KG = -K_c\Psi B(C\Psi B)^{-1}C\phi B \tag{5.60}$$

$$= -K_c\phi[I + BK_c\phi]^{-1}B(C\phi[I + BK_c\phi]^{-1}B)^{-1}C\phi B \tag{5.61}$$

$$= -K_c\phi B[I + K_c\phi B]^{-1}(C\phi B[I + K_c\phi B]^{-1})^{-1}C\phi B \tag{5.62}$$

$$= -K_c\phi B \tag{5.63}$$

which is the return ratio at the point 2 in Figure 5.2, as required.

The eigenvalues of the resulting Kalman filter are the zeros of

$$\det(sI - A + K_f C) = \det(sI - A + q^{1/2}BV^{-1/2}C) \tag{5.64}$$

$$= \det[(sI - A)(I_n + q^{1/2}(sI - A)^{-1}BV^{-1/2}C)] \tag{5.65}$$

$$= \det(sI - A)\det(I_m + q^{1/2}C(sI - A)^{-1}BV^{-1/2}) \tag{5.66}$$

where n is the state dimension of the plant, and m is the number of inputs or outputs.

If $|s|$ remains bounded, then

$$\det(sI - A + K_f C) \xrightarrow[q \to \infty]{} q^{m/2}\det(sI - A)\det[G(s)V^{-1/2}] \tag{5.67}$$

$$= q^{m/2}p(s)\det[G(s)]\det(V^{-1/2}) \tag{5.68}$$

where $p(s)$ is the pole polynomial of the plant. But $p(s)\det[G(s)]$ is just the zero polynomial of the plant, so it is seen that some of the filter's eigenvalues remain bounded, and are in fact 'attracted' to the plant's zeros as q is increased. The remaining eigenvalues become unbounded as q increases indefinitely.

5.5 Design procedure for square plant

The result derived above leads to the following two-step approach to design.

Step 1. Do the design on the deterministic Linear Quadratic problem alone, manipulating the weighting matrices Q and R until a return ratio $-K_c(sI - A)^{-1}B$ is obtained which would be satisfactory at the plant input. The emphasis of the design here is on aspects such as obtaining appropriate 0 dB cross-over frequencies for the principal gains, possibly 'balancing' the principal gains, and adjusting the low-frequency behaviour. Robustness properties are obtained automatically. Of course, when analysing a trial design one may prefer to examine the principal gains of closed-loop quantities such as the return difference $I + K_c(sI - A)^{-1}B$, or its inverse.

Step 2. Synthesize ('design' seems inappropriate here, since the procedure is virtually automatic) a Kalman filter by setting $\Gamma = B$, $W = W_0 + qI$ and $V = I$; alternatively, choose $W = I$ and $V = \rho I$. Increase q (or reduce ρ) until the return ratio at the input of the compensated plant has converged sufficiently closely to $-K_c(j\omega I - A)^{-1}B$ over a sufficiently large range of frequencies. Do not make q larger than is necessary, since this would force the gain 'roll-off' rate to be only 20 dB/decade at unnecessarily high frequencies, which would cause unnecessary noise amplification at these frequencies, and actually reduce the robustness of the design in the face of high-frequency unstructured perturbations.

It is often desired to design the return ratio at the output of the plant rather than the input; in fact for design of the sensitivity function and of performance this is more appropriate. In this case one can follow a procedure which is dual to that described above:

Step 1. Design a Kalman filter by manipulating the covariance matrices W and V until a return ratio $-C(sI - A)^{-1}K_f$ is obtained which would be satisfactory at the plant output.

Step 2. Synthesize an optimal state-feedback regulator by setting $M = C$, $Q = Q_0 + qI$ and $R = I$ (or $Q = I$ and $R = \rho I$), and increase q (reduce ρ) until the return ratio at the output of the compensated plant has converged sufficiently closely to $-C(j\omega I - A)^{-1}K_f$ over a sufficiently large range of frequencies.

5.6 Shaping the principal gains

In order to exploit the LTR technique, we must know how to modify Q and R in order to bring about desirable changes in $K_c(sI - A)^{-1}B$ or, if we are using the dual procedure, how to modify W and V in order to bring about

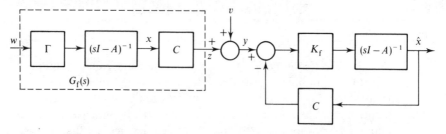

Figure 5.5   Generation and optimal filtering of y (for $u \equiv 0$).

desirable changes in $C(sI - A)^{-1} K_f$. To obtain an intuitive grasp of this, consider the Kalman filter.

Let us assume that $u \equiv 0$, for simplicity. Since u acts equally on x and $\hat{x}$, we do not lose anything essential by this simplification. Figure 5.5 shows the stochastic process y being generated, and being processed by the Kalman filter to produce the state estimate $\hat{x}$. The dashed-line rectangle shown in Figure 5.5 has the transfer function $G_f(s)$ which appears in equation (5.42). Now, since the process noise v is uncorrelated with either Cx or (past values of) $C\hat{x}$, the statistical properties of the state estimate $\hat{x}$ are unchanged if we move v into the feedback path of the Kalman filter, as shown in Figure 5.6. This now looks like a feedback system which is to track (in a sense) the 'reference input' z, while rejecting the measurement errors v, and we are already familiar with the trade-off between these two conflicting objectives.

Roughly speaking, if the power spectral density of z increases relative to that of v at any frequency, then the loop gain should also increase at that frequency, because more importance can be attached to tracking z than to suppressing v. So we expect to increase the bandwidth of the Kalman filter by increasing $\Gamma W \Gamma^T$ relative to V, and to decrease the bandwidth by increasing V. Indeed, from equation (5.50) we see that if W is made very large (q large in (5.50)) then an increase in W by a factor α leads to an increase in P_f, and hence in K_f, by a factor $\sqrt{\alpha}$.

Such adjustments are useful for 'speeding up' or 'slowing down' the Kalman filter uniformly in all signal directions. But they do not help if the shapes of the principal-gain plots need to be changed. To do this we can do one of two things: modify the matrices $\Gamma W \Gamma^T$ and V in a more sophisticated way, or modify the plant model by augmenting it with additional dynamics. We shall now examine these possibilities.

One common requirement is to make the principal gains of the return ratio similar to each other in the vicinity of the required 0 dB cross-over frequency. This is appropriate if the plant has potentially uniform dynamics (with realistic control signals), or if the unstructured uncertainty is 'uniform' (that is, similar in all output directions).

We can use $\Gamma W \Gamma^T$ to increase the smallest principal gain of the sensitivity matrix, or decrease the largest one near some particular frequency.

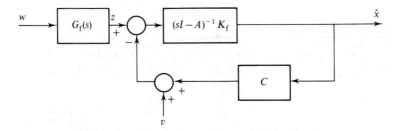

Figure 5.6 Statistically equivalent model to that shown in Figure 5.5.

Suppose we choose $V = I$. Then, from (5.42), we have

$$F_f(j\omega) F_f^H(j\omega) = I + G_f(j\omega) W G_f^H(j\omega) \tag{5.69}$$

from which it follows that

$$\sigma_i[F_f(j\omega)] = \{1 + \sigma_i^2[G_f(j\omega) W^{1/2}]\}^{1/2} \tag{5.70}$$

Hence

$$\bar{\sigma}[F_f^{-1}(j\omega)] = \{1 + \underline{\sigma}^2[G_f(j\omega) W^{1/2}]\}^{-1/2} \tag{5.71}$$

and

$$\underline{\sigma}[F_f^{-1}(j\omega)] = \{1 + \bar{\sigma}^2[G_f(j\omega) W^{1/2}]\}^{-1/2} \tag{5.72}$$

so we can reduce $\bar{\sigma}[F_f^{-1}(j\omega)]$ by increasing $\underline{\sigma}[G_f(j\omega) W^{1/2}]$, etc. But the point is not merely to reduce all the principal gains of $F_f^{-1}(j\omega)$, but to reduce the largest one, relative to the smallest. One way of doing this is first to obtain the singular-value decomposition of $G_f(j\omega) W^{1/2}$ at the frequency at which the adjustment is to be made (ω_1, say)

$$G_f(j\omega_1) W^{1/2} = U \Sigma V^H \tag{5.73}$$

$$= \sum_{i=1}^{m} u_i \sigma_i v_i^H \tag{5.74}$$

where $\{u_i\}$ and $\{v_i\}$ are the input and output principal directions, respectively.

Now replace $W^{1/2}$ by $W^{1/2}(I + \alpha v_j v_j^H)$. Then we have

$$G_f(j\omega_1) W^{1/2}(I + \alpha v_j v_j^H) = \sum_{i \neq j} u_i \sigma_i v_i^H + (1 + \alpha)\sigma_j u_j v_j^H \tag{5.75}$$

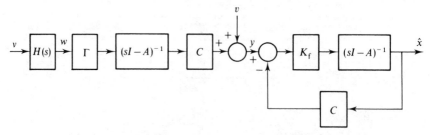

Figure 5.7   Generation and filtering of y when w is not white.

so that the jth principal gain has been changed by the factor $1 + \alpha$ (at frequency ω_1), while all the other principal gains have been left unchanged. So to reduce $\underline{\sigma}$ choose $j = 1$, and $\alpha < 0$; to increase σ choose $j = m$, and $\alpha > 0$.

The problem with this approach is that v_j is usually a complex vector, whereas we wish to keep $W^{1/2}$ real. Once again we are faced with the problem of approximating a complex matrix by a real matrix, and as before we can employ the ALIGN algorithm to solve it. In this case other algorithms may be more appropriate, however, since we really want to approximate v_j rather than 'align' with it. In particular, $\mathrm{Re}\,\{v_j\}$ is sometimes an adequate approximation. A further possibility is to approximate v_j by the left singular vector of the matrix $[\mathrm{Re}\,\{v_j\}\ \mathrm{Im}\,\{v_j\}]$ which corresponds to its largest singular value.[2]

It is not usually sufficient to modify the matrices W and V. The gains in particular frequency regions can be manipulated by altering the power spectral density matrices of z and v (see Figures 5.5 and 5.6) in those regions, and this can be achieved in practice by augmenting the plant model with additional dynamics.

Let us recall equation (5.42),

$$F_f(s)\,VF_f^T(-s) = V + G_f(s)\,WG_f^T(-s)$$

and note that the expression on the right-hand side of this equation is just the power spectral density matrix of y, the measured plant output. If the process noise w is not white, but has a power spectral density $W(s)$, then (5.42) generalizes to

$$F_f(s)\,VF_f^T(-s) = V + G_f(s)\,W(s), G_f^T(-s) \tag{5.76}$$

that is, the expression on the right remains the power spectral density of the measured plant output.[3] In this case we can replace Figure 5.5 by Figure 5.7, which shows w being generated as the output of a transfer function $H(s)$ driven by a white process v, where

$$H(s)\,H^T(-s) = W(s) \tag{5.77}$$

and $E\{vv^T\} = I$. This is no different, in essence, from the situation depicted in Figure 5.5, with $C(sI - A)^{-1}\Gamma$ replaced by $C(sI - A)^{-1}\Gamma H(s)$. In order to perform this replacement in practice, we need a state-space model of $H(s)$:

$$\dot{\zeta} = A_w\zeta + B_w v, \qquad w = C_w\zeta + D_w v \tag{5.78}$$

If we now use (A_p, B_p, C_p) to denote the state-space model of the plant, then, recalling that

$$\dot{x} = A_px + B_pu + \Gamma w, \qquad y = C_px + v \tag{5.79}$$

we obtain

$$\begin{bmatrix} \dot{x} \\ \dot{\zeta} \end{bmatrix} = \begin{bmatrix} A_p & \Gamma C_w \\ 0 & A_w \end{bmatrix}\begin{bmatrix} x \\ \zeta \end{bmatrix} + \begin{bmatrix} B_p \\ 0 \end{bmatrix}u + \begin{bmatrix} \Gamma D_w \\ B_w \end{bmatrix}v \tag{5.80}$$

$$y = [C_p \quad 0]\begin{bmatrix} x \\ \zeta \end{bmatrix} + v$$

These equations have the form of (5.1) and (5.2), and so fit into the LQG framework.

One common application of this augmentation is to force 'integral action' into the LQG compensator. The purpose of integral action is to obtain zero sensitivity at $\omega = 0$, by obtaining infinite open-loop gains at $\omega = 0$. Our previous considerations lead us to expect that this will be achieved if we give the w process an unbounded power spectral density at $0 -$ that is, if we place eigenvalues of A_w at the origin. More formally, we require that

$$\underline{\sigma}[F_f(0)] = \infty \tag{5.81}$$

which implies (because of (5.76) and (5.77)) that

$$\underline{\sigma}[G_f(0)H(0)] = \infty \tag{5.82}$$

which is achieved by placing poles of $H(s)$ at the origin, unless $G_f(s)$ already has them there.

In practice, one would usually start by introducing the augmented model (5.80) at an early stage of the design, at a later stage manipulating the covariance matrices to balance the gains and adjust the cross-over frequency, etc. The augmented model is carried through the various stages of the computations.

There is a minor complication which arises during the recovery step, however. It is seen from (5.80) that the dynamics of ζ are not controllable from the plant input. So if we place eigenvalues of A_w at the origin, we violate

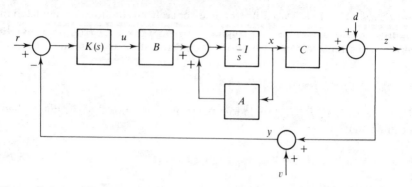

Figure 5.8   Feedback loop with plant output disturbances and measurement noise.

the assumption of asymptotic stabilizability which is required when solving the Riccati equation (5.12) at the recovery stage. Problems with the solution algorithm can be avoided by placing the eigenvalues of A_w not exactly at the origin, but just to the left of it, the distance being much smaller than the required bandwidth.[4]

To boost the gain over a small range of frequencies we can introduce some lightly damped eigenvalues of A_w (resonant poles). To produce a 'notch', namely a drop in the gain over a small frequency range, we can either introduce zeros into the spectral density of w (by suitable choice of B_w and C_w) or, more easily, augment the model in such a way that the spectral density of the measurement noise v is increased over that frequency range. The reader should be able to work out the details of this, on the basis of the material we have already presented.

In this section we have been assuming that the dual of the LTR procedure described in Section 5.4 is followed, namely that the Kalman filter's return ratio is manipulated until a satisfactory design has been obtained, and that it is then 'recovered' at the plant output by synthesizing a state-feedback controller automatically. Of course, everything we have said in this section is directly applicable to LTR at the plant input, if the roles of W and Q, and of V and R, are interchanged. In particular, since we have used augmentation of the plant model to replace the constant covariances W and V by power spectral densities $W(s)$ and $V(s)$, we can use similar augmentation to replace the constant weighting matrices Q and R by frequency-dependent weights $Q(s)$ and $R(s)$.

Now let us return to the general LQG problem, without assuming that the LTR procedure is to be used. Suppose that we have plant disturbances d instead of state disturbances w, as in Figure 5.8, and that these disturbances have power spectral density $D(s)$. Define z to be the plant output before corruption by measurement noise, as in Figure 5.8. The measurement noise v need not be white, and we can assume d and v to be generated by the processes

$$\dot{\zeta} = A_d \zeta + B_d v, \qquad d = C_d \zeta \tag{5.83}$$

$$\dot{\eta} = A_v \eta + B_v \mu, \qquad v = C_v \eta + \theta \tag{5.84}$$

where $\{\theta\}$, $\{\mu\}$ and $\{v\}$ are white-noise processes.

Combining this with equations (5.1) and (5.2) we obtain

$$\begin{bmatrix} \dot{x} \\ \dot{\zeta} \\ \dot{\eta} \end{bmatrix} = \begin{bmatrix} A_p & 0 & 0 \\ 0 & A_d & 0 \\ 0 & 0 & A_v \end{bmatrix} \begin{bmatrix} x \\ \zeta \\ \eta \end{bmatrix} + \begin{bmatrix} B \\ 0 \\ 0 \end{bmatrix} u + \begin{bmatrix} 0 & 0 \\ B_d & 0 \\ 0 & B_v \end{bmatrix} \begin{bmatrix} v \\ \mu \end{bmatrix}$$

$$\tag{5.85}$$

$$y = \begin{bmatrix} C_p & C_d & C_v \end{bmatrix} \begin{bmatrix} x \\ \zeta \\ \eta \end{bmatrix} + \theta \tag{5.86}$$

and

$$z = \begin{bmatrix} C_p & C_d & 0 \end{bmatrix} \begin{bmatrix} x \\ \zeta \\ \eta \end{bmatrix} \tag{5.87}$$

We have given equations (5.85)–(5.87) explicitly in order to confirm that we still have equations in the form of (5.1), (5.2) and (5.6), so that we are still within the LQG framework. Note that we require $E\{\theta\theta^T\} > 0$, $E\{\theta v^T\} = 0$ and $E\{\theta\mu^T\} = 0$ in order to satisfy our earlier assumptions. We also require A_d and A_v to have all their eigenvalues in the open left-half plane, since ζ and η are uncontrollable from u.

From (3.61) we obtain

$$J = E\{z^T Q z + u^T R u\}$$

$$= \frac{1}{2\pi} \int_{-\infty}^{\infty} \{\mathrm{tr}\,[QZ(\omega)] + \mathrm{tr}\,[RU(\omega)]\}\,d\omega \tag{5.88}$$

where $Z(\omega)$ and $U(\omega)$ denote the power spectral densities of z and u, respectively. Furthermore, using (3.66), we can express this as

$$J = \frac{1}{2\pi} \int_{-\infty}^{\infty} \left\{ \sum_i \sigma_i^2 [Q^{1/2} T_{zd}(j\omega) D^{1/2}(j\omega)] \right.$$

$$+ \sum_i \sigma_i^2 [Q^{1/2} T_{zv}(j\omega) V^{1/2}(j\omega)]$$

$$+ \sum_i \sigma_i^2 [R^{1/2} T_{ud}(j\omega) D^{1/2}(j\omega)]$$

$$+ \left. \sum_i \sigma_i^2 [R^{1/2} T_{uv}(j\omega) V^{1/2}(j\omega)] \right\} d\omega \tag{5.89}$$

where

$$z(s) = T_{zd}(s)d(s) + T_{zv}(s)v(s) \tag{5.90}$$

and

$$u(s) = T_{ud}(s)d(s) + T_{uv}(s)v(s) \tag{5.91}$$

From Figure 5.8 (or from Chapter 1) we have that

$$T_{zd}(s) = S(s) \tag{5.92}$$

$$T_{zv}(s) = I - S(s) \tag{5.93}$$

$$T_{ud}(s) = -K(s)F_i^{-1}(s) \tag{5.94}$$

$$T_{uv}(s) = T_{ud}(s) \tag{5.95}$$

where

$$S(s) = [I + G(s)K(s)]^{-1} \tag{5.96}$$

and

$$F_i(s) = I + K(s)G(s) \tag{5.97}$$

This shows that the solution to the LQG problem which we have set up minimizes the principal gains of the sensitivity $S(j\omega)$, of the closed-loop response $I - S(j\omega)$ and of $K(j\omega)F_i^{-1}(j\omega)$, each of these being weighted by $D(j\omega)$, $V(j\omega)$, Q and R, as shown by (5.89). By further augmenting the plant dynamics (in addition to the augmentation shown in (5.85)) we can replace Q and R by frequency-sensitive weights $Q(j\omega)$ and $R(j\omega)$. Safonov *et al.* (1981) advocate pursuing LQG design from this point of view, adjusting each of the four 'parameter' matrices of the problem to shape the closed-loop principal gains as required. If the greatest singular value of $I - S(j\omega)$ is too large, for instance, indicating poor robustness properties, then $V^{1/2}(j\omega)Q^{1/2}(j\omega)$ should be used to increase the weighting in the direction of the singular vector associated with the largest singular value (at the appropriate frequency).

An explicit connection between this approach to LQG and the LTR approach has been pointed out by Stein and Athans (1987). We saw in Section 5.5 that, in order to recover a Kalman filter's return ratio at the plant output, we set $Q = I$ and $R = \rho I$, and reduce ρ towards zero. In this case we get, from (5.89)

$$\lim_{\rho \to 0} J_{\text{LTR}} = \frac{1}{2\pi} \int_{-\infty}^{\infty} \sum_i \{\sigma_i^2 [T_{zd}(j\omega) D^{1/2}(j\omega)]$$

$$+ \sigma_i^2 [T_{zv}(j\omega) V^{1/2}(j\omega)]\} d\omega \qquad (5.98)$$

and we see that the LTR procedure, applied at the plant output, trades off the sensitivity $S(j\omega)$ against the closed-loop response $I - S(j\omega)$, with a relative weighting $W(j\omega) = D^{1/2}(j\omega) V^{-1/2}(j\omega)$. (See Exercise 5.2 for the dual result.)

Stein and Athans (1987) also develop the theory of 'loop shaping' – namely, how to implement the weighting $W(s)$ – in a more formal way than we have done here.

5.7 Some practical considerations

It is evident from equation (5.62) that a compensator obtained by the LTR procedure inverts the stable plant dynamics. (It cannot cancel right half-plane poles or zeros, since this would lead to an internally unstable system, and LQG compensators are known to guarantee the internal stability of the closed loop.) One should therefore be careful to specify a return ratio to be recovered which is as compatible as possible with the plant dynamics. If the plant has lightly damped resonant poles, while the recovered return ratio shows no trace of the resonance, then the compensator will inevitably contain zeros which cancel those poles. As discussed in Chapter 4, this may or may not be acceptable, depending on how accurately the resonant poles can be located.

Trying to suppress the effects of lightly damped poles or zeros also leads to difficulties with the recovery procedure itself. The step from equation (5.62) to equation (5.63) is not valid at poles or zeros of the plant, or indeed of the return difference $I + K_c \phi B$, since some of the inverses will not exist at such points. Although the limiting process (5.60)–(5.63) is valid in any neighbourhood of a pole or a zero, the rate of convergence (with q) may be very slow. The practical consequence of this is that if one tries to suppress (or insert) dynamics which correspond to poles or zeros located very near the imaginary axis, then one may need to increase the 'recovery parameter' q to extremely large values in order to get a good degree of approximation at frequencies close to those pole or zero locations. And this, in turn, leads to recovery of the 20 dB/decade 'roll-off' rate well beyond the loop bandwidth, and thus to unnecessarily low attenuation of measurement noise. This situation can be ameliorated, however, either by further dynamic augmentation of the plant model, or by some judicious low-pass filtering, applied in an *ad hoc* manner.

If the plant to be controlled has unequal numbers of inputs and outputs, then the LTR procedure cannot be applied directly. It remains possible to adopt the viewpoint of Safonov *et al.* (1981), and to adjust the various 'parameter' matrices appearing in (5.89). It is also sometimes possible to apply the LTR procedure, after augmenting the plant with dummy inputs or outputs. If the plant has more outputs than inputs, it may be augmented with dummy inputs (taking care not to introduce any right half-plane zeros), and the return ratio $K(s)G(s)$ at the plant input can be shaped by the LTR procedure. In this case (5.51) holds, so the ideal state-feedback return ratio can be recovered at the plant input. The dual of (5.51) does not hold, however, so recovery cannot be obtained at the plant output. If the plant has more inputs than outputs then it may be augmented with dummy outputs, and recovery can be obtained at the output, but not at the input. Further details of suitable augmentation procedures are given by Doyle and Stein (1981).

5.8   Design example

We shall now use the LQG/LTR approach to design a compensator for the AIRC aircraft model (see the Appendix). Initially, we shall adopt similar specifications to those used in Chapter 4: we aim for a bandwidth of about $10 \, \text{rad} \, \text{s}^{-1}$, and integral action in each loop, with well-damped responses.

We shall design the return ratio of a Kalman filter, and recover this at the output of the compensated plant. We shall judge trial designs entirely on the basis of principal-gain plots, since the stability margins of the Kalman filter are guaranteed to be good.

5.8.1   Kalman-filter design

We need to choose the matrices Γ, W and V which appear in (5.1) and (5.3), solve the corresponding Riccati equation (5.14), and obtain the Kalman-filter gain K_f from (5.13). We shall write

$$K_f = \text{LQE}(A, \Gamma, C, W, V) \tag{5.99}$$

to denote that K_f is obtained from Γ, W and V in this way,[5] with A and C taken from the plant model. It is generally advisable to start with simple choices of Γ, W and V, inspect the resulting Kalman filter return ratio, adjust Γ, W and V accordingly, and so gradually improve the return ratio. One of the simplest possible choices is $\Gamma = B$, $W = I_3$ and $V = I_3$. The choice $\Gamma = B$ corresponds to disturbances on the plant acting through its inputs, rather than directly on the state, and leads to the covariance matrix W having the same dimension as the number of loops (if the plant is square), rather than the number of states.

We have

$$K_{f1} = \mathrm{LQE}(A, B, C, I_3, I_3) \tag{5.100}$$

$$= \begin{bmatrix} 0.9897 & 0.0732 & -0.2507 \\ 0.0732 & 0.9436 & -0.0642 \\ -0.2507 & -0.0642 & 1.7807 \\ -0.4934 & -0.0485 & 1.6190 \\ -0.8076 & -0.2483 & 2.2215 \end{bmatrix} \tag{5.101}$$

The return ratio $-C(sI - A)^{-1}K_{f1}$ has principal gains as shown in Figure 5.9. One of these shows a slope of $-20\,\mathrm{dB/decade}$ at very low frequencies, because of the pole at the origin which this plant contains, but the other two show constant values of gain at low frequencies, of about 16 dB and 2 dB, respectively. The open-loop bandwidth is seen to be about $1\,\mathrm{rad\,s^{-1}}$, or rather less, in some signal directions.

The first thing to do is to insert integral action, by augmenting the plant model. As we mentioned before, placing poles of the augmented model at the origin leads to problems in the recovery step later, so in this case we place them at -0.001, which is virtually at the origin, when compared with the required bandwidth of $10\,\mathrm{rad\,s^{-1}}$. Choosing the simplest possible augmentation of the plant gives

$$A_w = -0.001 I_3, \qquad B_w = I_3$$

$$\tag{5.102}$$

$$C_w = I_3, \qquad D_w = 0_{3,3}$$

As was the case with the characteristic-locus and the Nyquist-array design methods examined in Chapter 4, we have inserted three integrators rather than two, even though the plant itself contains an integrator. We could have introduced just two integrators, and chosen B_w and C_w carefully, to ensure that integration was present for every signal direction, but this would have made the procedure vulnerable to errors in plant modelling, which could result in integration being absent from some loop.

Also, it is often advisable to have integral action at each compensator output, in order to overcome any friction which may be present in the actuators.

We could also have chosen C_w more carefully, with the aim of adjusting the low-frequency gains.

The augmented model is, using equation (5.80),

$$A_a = \begin{bmatrix} A & \Gamma C_w \\ 0 & A_w \end{bmatrix}, \qquad B_a = \begin{bmatrix} B \\ 0 \end{bmatrix}$$

$$C_a = [C \quad 0], \qquad D_a = 0_{3,3}, \qquad \Gamma_a = \begin{bmatrix} \Gamma D_w \\ B_w \end{bmatrix} \qquad (5.103)$$

Now we have

$$K_{f2} = LQE(A_a, \Gamma_a, C_a, I_3, I_3) \qquad (5.104)$$

$$= \begin{bmatrix} 1.4129 & 0.0909 & -0.1648 \\ 0.0909 & 1.3501 & -0.0539 \\ -0.1648 & -0.0539 & 2.0429 \\ -0.4556 & 0.0062 & 2.1018 \\ -1.2024 & -0.2823 & 2.4314 \\ -0.6484 & -0.0329 & 0.7587 \\ 0.0225 & 0.9964 & 0.0624 \\ -0.7573 & 0.0576 & -0.6463 \end{bmatrix} \qquad (5.105)$$

The principal gains of the return ratio $-C_a(sI - A_a)^{-1}K_{f2}$ are shown in Figure 5.9. Integral action is clearly visible in each signal direction, and the gains at 0.001 rad s^{-1} have all been increased by 60 dB, as would be expected for integrators of the form $1/s$. Note that at frequencies above 1 rad s^{-1}, where the gain of the augmenting system (A_w, B_w, C_w, D_w) is comparable with that of the plant, the gains are almost unchanged.

Next we shall attempt to increase the smallest principal gain at low frequencies ($\omega < 0.1$ rad s^{-1}, say) in order to speed up the elimination of steady-state errors. Increasing it by a factor of 10 (20 dB) would make the two smaller principal gains similar at low frequencies, and we shall aim to do this. We use (5.71) but note that, since $\bar{\sigma}[F_f^{-1}(j0.001)] \approx -55$ dB $\ll 0$ dB, we can replace (5.71) by

$$\underline{\sigma}[C_a(j0.001 I - A_a)^{-1}K_f] \approx \underline{\sigma}[C_a(j0.001 I - A_a)^{-1}\Gamma_a] \qquad (5.106)$$

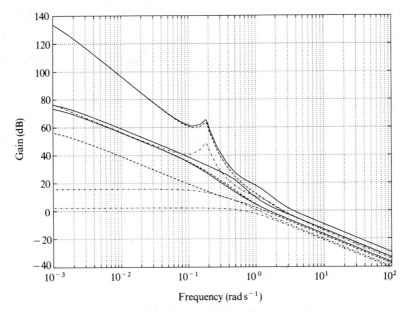

Figure 5.9 Open-loop principal gains of initial Kalman filter, with filter gains K_{f1} (dash–dotted curves), K_{f2} (dashed curves) and K_{f3} (solid curves).

In other words we can, as usual, work with open-loop gains instead of closed-loop gains, if the loop gains are large (or small) enough. The singular-value decomposition (5.73) is, in this case,

$$C_a(j0.001I - A_a)^{-1}\Gamma_a = U\Sigma V^H \qquad (5.107)$$

which gives

$$v_3 = \begin{bmatrix} 0.509 \\ 0.158 + 0.0005j \\ 0.846 - 0.0005j \end{bmatrix} \qquad (5.108)$$

as the third column of V.

In this case $\text{Re}\{v_3\}$ is a very good approximation to v_3, so, setting $\alpha = 9$ (since $1 + \alpha = 10$, see (5.75)), we have

$$W_3^{1/2} = I_3 + 9\,\text{Re}\{v_3\}\,\text{Re}\{v_3^T\} \qquad (5.109)$$

Now we set $W_3 = W_3^{1/2}(W_3^{1/2})^T$ and obtain the corresponding Kalman-filter gain as

$$K_{f3} = \text{LQE}(A_a, \Gamma_a, C_a, W_3, I_3) \tag{5.110}$$

$$= \begin{bmatrix}
2.7724 & -0.3719 & -0.5012 \\
-0.3719 & 1.5375 & 0.1460 \\
-0.5012 & 0.1460 & 2.5049 \\
-1.1123 & 0.5385 & 3.2734 \\
-4.6051 & 1.2121 & 4.4222 \\
-3.9426 & 1.6448 & 2.8887 \\
-0.8835 & 1.4814 & 0.7096 \\
-7.2415 & 3.1669 & 3.0434
\end{bmatrix} \tag{5.111}$$

Figure 5.9 shows the principal gains of $-C_a(sI - A_a)^{-1}K_{f3}$. The previous smallest principal gain is seen to have been increased by almost exactly 20 dB at low frequencies, while the other two have remained unchanged. The smallest principal gain has been increased significantly at all frequencies below 1 rad s^{-1}, the increase at 0.1 rad s^{-1} being about 17 dB, and has remained almost unchanged at higher frequencies. At the same time, the largest principal gain has been increased by about 4 dB at high frequencies.

Figure 5.9 indicates that the 0 dB cross-over frequency of the compensated system is between 1.3 and 4 rad s^{-1}, whereas we would like it to be about 7 rad s^{-1} ($= 10/\sqrt{2}$, approximately) or higher in order to obtain a bandwidth of about 10 rad s^{-1}. If all the gains were increased by 10 dB (that is, by a factor of $\sqrt{10}$) at all frequencies, then the largest principal gain would have a 0 dB cross-over frequency very close to 10 rad s^{-1}. Increasing W_3 by a factor of 10 gives a 10 dB increase at low frequencies, but only about 5 dB near 10 rad s^{-1}, and it turns out that an increase by a factor of 100 is needed to obtain a 10 dB increase near 10 rad s^{-1}. Figure 5.10 shows the largest principal gains obtained using W_3, $10\,W_3$ and $100\,W_3$.

Now let

$$K_{f4} = \text{LQE}(A_a, \Gamma_a, C_a, 100\,W_3, I_3) \tag{5.112}$$

and let us examine the closed-loop properties which would result if we recovered the return ratio $-C_a(sI - A_a)^{-1}K_{f4}$ at the plant output. Figure 5.11 shows the principal gains of both the sensitivity

$$S_f(s) = [I + C_a(sI - A_a)^{-1}K_{f4}]^{-1} \tag{5.113}$$

and the closed-loop transfer function

$$T_f(s) = I - S_f(s) \tag{5.114}$$

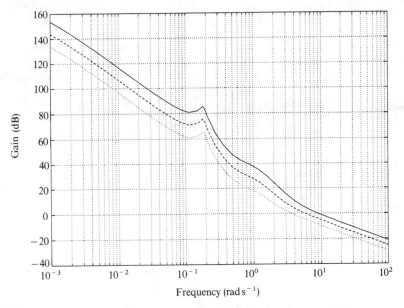

Figure 5.10 Open-loop principal gains with W_3, $10W_3$ and $100W_3$.

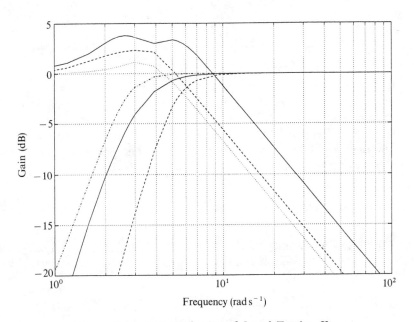

Figure 5.11 Principal gains of S_f and T_f using K_{f4}.

This shows that the bandwidth specification has been approximately satisfied, the -3 dB frequency of $T_f(s)$ being between about 6.5 and 12 rad s^{-1}, and that of $S_f(s)$ being between about 2.5 and 5.5 rad s^{-1}. Reasonably good stability margins are of course guaranteed, and we can see from Figure 5.11 that $\|T_f\|_\infty = 1.6$ (4 dB), approximately. We could therefore terminate the Kalman-filter design at this point, and move on to the recovery step. However, in order to demonstrate that quite fine control can be exercised over the details of closed-loop performance, we shall suppose that we wish to improve the sensitivity further.

Specifically, we shall try to make all the principal gains of the sensitivity S_f behave similarly as they approach 0 dB, by giving them all a value of $1/\sqrt{2}$ (-3 dB) at 5.5 rad s^{-1}. To do this, we need to obtain the singular-value decomposition (see (5.73))

$$C_a(j5.5I - A_a)^{-1}\Gamma_a(100\,W_3)^{1/2} = U\Sigma V^H \tag{5.115}$$

which gives

$$\Sigma = \text{diag}\{0.935, 0.379, 0.119\} \tag{5.116}$$

and

$$V = \begin{bmatrix} 0.723 & -0.481 & 0.496 \\ 0.300 - 0.0305j & 0.859 - 0.110j & 0.395 - 0.0621j \\ 0.621 - 0.0368j & 0.140 + 0.0026j & -0.769 + 0.0562j \end{bmatrix}$$

$$\tag{5.117}$$

($W_3^{1/2}$ is available from (5.109). To obtain the square root of W in general requires an algorithm such as Cholesky decomposition to be available – see Golub and Van Loan (1983).) We proceed by using equations (5.70) and (5.75), as we did earlier, but now we cannot work with open-loop approximations since the loop gains are close to unity. The principal gains of the Kalman filter's return difference, with the current design, are

$$\{\sigma_i[F_f(j5.5)]\} = \{1.369, 1.069, 1.007\} \tag{5.118}$$

and we wish to force these to have the values $\{\sqrt{2}, \sqrt{2}, \sqrt{2}\}$. From equations (5.70) and (5.75) we find that this requires $\alpha_1 = 0.069$, $\alpha_2 = 1.639$ and $\alpha_3 = 7.409$. Again, V is given to a good approximation by its real part, so we set

$$W_5^{1/2} = 10\,W_3^{1/2}(I + \alpha_1\,\text{Re}\,\{v_1 v_1^H\})(I + \alpha_2\,\text{Re}\,\{v_2 v_2^H\})$$

$$\times\,(I + \alpha_3\,\text{Re}\,\{v_3 v_3^H\}) \tag{5.119}$$

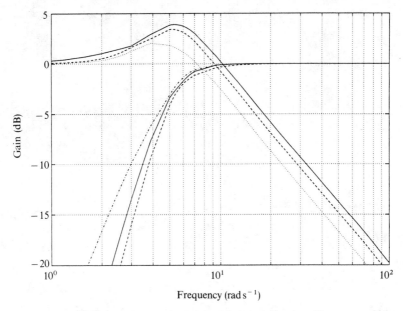

Figure 5.12 Principal gains of S_f and T_f using K_{f5}.

Then $W_5 = W_5^{1/2}(W_5^{1/2})^T$, and

$$K_{f5} = \text{LQE}(A_a, \Gamma_a, C_a, W_5, I_3) \tag{5.120}$$

$$= \begin{bmatrix} 9.7235 & -0.2568 & 0.5814 \\ -0.2568 & 7.2970 & -0.1420 \\ 0.5814 & -0.1420 & 9.5072 \\ 6.5138 & -1.2568 & 45.3725 \\ -46.8171 & 1.9765 & 6.0590 \\ -101.7203 & 7.3047 & -11.6414 \\ -11.1370 & 27.7714 & -1.4828 \\ -265.8522 & 22.2337 & -115.6172 \end{bmatrix} \tag{5.121}$$

Figure 5.12 shows the principal gains of the sensitivity S_f and the closed-loop transfer function T_f of the new design, and it can be seen that the principal gains have indeed been brought very close to each other in the neighbourhood of $5.5\,\text{rad}\,\text{s}^{-1}$. This has been achieved at the expense of a larger bandwidth of T_f, but with very little increase in $\|T_f\|_\infty$, and hence very

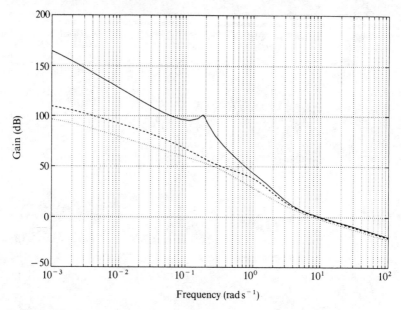

Figure 5.13   Open-loop principal gains after balancing at 5.5 rad s^{-1}.

little deterioration of stability margins. In fact, some benefit has been obtained from the fact that measurement noise is now being amplified $(\sigma_i(T_f) > 1)$ over a smaller range of frequencies than before. As expected, the system looks more uniform in all signal directions; for example, the range of -3 dB frequencies of T_f has been reduced from $6.4 \leqslant \omega \leqslant 12.2$ to $10.7 \leqslant \omega \leqslant 15.0$.

Figure 5.13 shows the principal gains of the open-loop return ratio $-C_a(sI - A_a)^{-1}K_{f5}$; all the gains have clearly been 'squeezed together' near 10 rad s^{-1}, and at higher frequencies. Figure 5.14 shows the characteristic loci of this return ratio, together with the unit circle centred on -1. All the loci remain outside or on the boundary of this circle, as predicted by Kalman-filter theory, from which (and from Figure 5.12) we know that $\sigma_i(S_f) \leqslant 1$, and a necessary condition for this to be true is that the characteristic loci do not penetrate into the circle.

5.8.2   Recovery at the plant output

Having obtained a satisfactory return ratio for the Kalman filter, we now recover this return ratio at the output of the plant. The plant has no

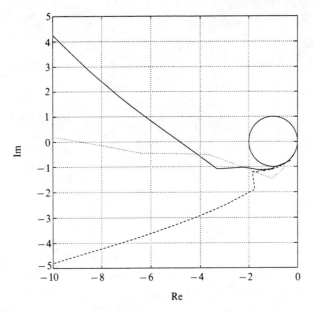

Figure 5.14 Characteristic loci of Kalman filter's return ratio.

transmission zeros in the right half-plane (in fact it has no finite zeros at all), so arbitrarily good recovery should be possible.

To obtain LTR, we solve the Riccati equation (5.12) with $M = C_a$, $Q = I$ and $R = \rho I$, and obtain the corresponding state-feedback matrix K_c from (5.11), using B_a in place of B. We shall write

$$K_c = \text{LQR}(A_a, B_a, C_a^T C_a, \rho I_3) \tag{5.122}$$

to denote that K_c has been obtained in this way.[5] Perfect recovery should be obtained in the limit, as ρ is reduced to zero.

Once K_c has been found, a state-space realization of the compensator $K(s)$ is given by (KA, KB, KC, KD), where

$$KA = A_a - B_a K_c - K_{f5} C_a \tag{5.123}$$

$$KB = K_{f5}, \qquad KC = -K_c, \qquad KD = 0_{3,3}$$

(see Figure 5.2). Figure 5.15 shows the principal gains of $G(s)K(s)$, together with those of $C_a(sI - A_a)^{-1}K_{f5}$, for $\rho = 10^{-2}$, 10^{-4}, 10^{-6} and 10^{-8}. The

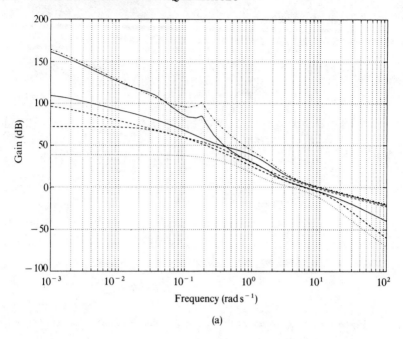

(a)

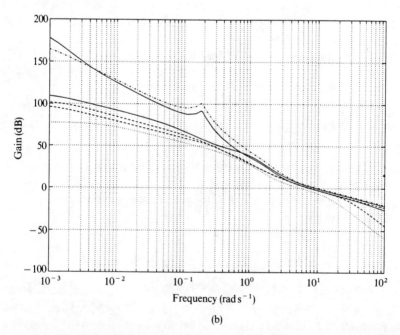

(b)

Figure 5.15 Loop transfer recovery for (a) $\rho = 10^{-2}$, (b) $\rho = 10^{-4}$, (c) $\rho = 10^{-6}$ and (d) $\rho = 10^{-8}$.

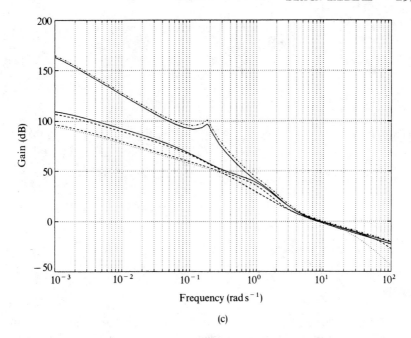

(c)

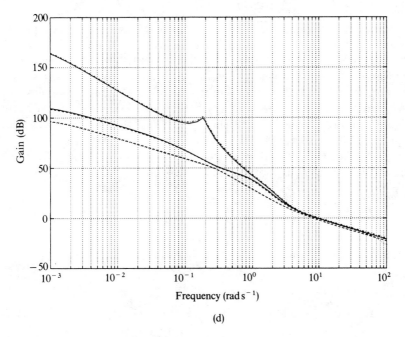

(d)

Figure 5.15 (cont.)

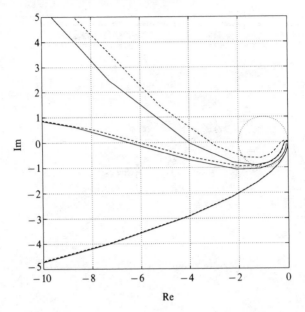

Figure 5.16 Characteristic loci at output of compensated plant, for $\rho = 10^{-6}$ (dashed curves) and $\rho = 10^{-8}$ (solid curves).

value $\rho = 10^{-6}$ appears to give very good recovery, and one might think that there is no need to reduce ρ further. However, Figure 5.16 shows the characteristic loci of $G(s)K(s)$ for $\rho = 10^{-6}$ and 10^{-8}, and it is seen that the increased 'roll-off' rate which is visible in Figure 5.15(c) reduces stability margins considerably. Figure 5.17 shows the principal gains of the sensitivity $S(s)$ and of the closed-loop transfer function $T(s)$ for the two cases. Figure 5.17(a), for $\rho = 10^{-6}$, shows lower high-frequency gain than Figure 5.17(b), but significantly reduced stability margins and increased sensitivity. We shall therefore adopt the compensator with $\rho = 10^{-8}$. The corresponding optimal state-feedback matrix is

$$K_c = [K_{c1}, I_3]$$

where

$$K_{c1} = \begin{bmatrix} -5960 & -1158 & 7853 & 36.3 & 82.0 \\ -700 & 9933 & 910 & 4.3 & 9.8 \\ -8000 & -7.8 & -6205 & -79.4 & 180 \end{bmatrix} \qquad \textbf{(5.124)}$$

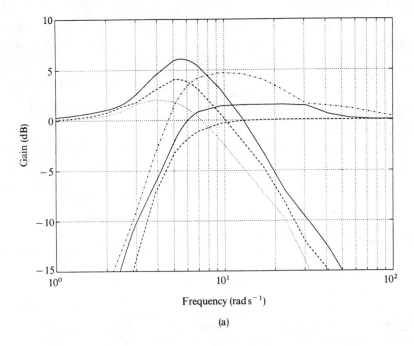

(a)

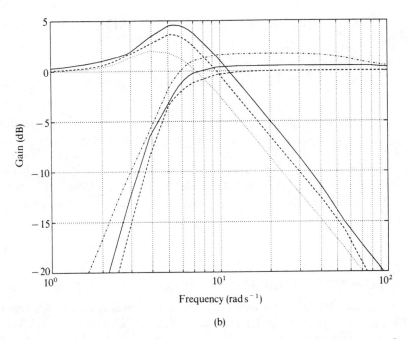

(b)

Figure 5.17   Principal gains of S and T with (a) $\rho = 10^{-6}$ and (b) $\rho = 10^{-8}$.

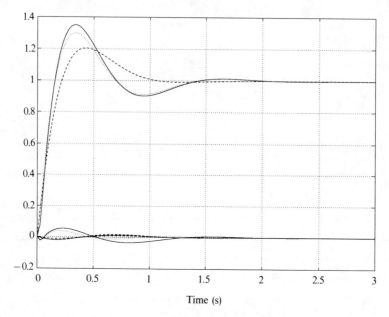

Figure 5.18 Closed-loop step responses to step demand on output 1 (solid curves), output 2 (dashed curves) and output 3 (dotted curves).

When implementing the compensator, A_a in (5.123) should be replaced by

$$\tilde{A}_a = \begin{bmatrix} A & \Gamma C_w \\ 0 & 0 \end{bmatrix} \tag{5.125}$$

(and not by A_a as given by (5.103)) in order to obtain true integration. With this change, the closed-loop responses which result from step demands being applied to each of the three outputs are as shown in Figure 5.18. The closed-loop poles are located at $-3.661 \pm 3.668j$, $-2.600 \pm 4.369j$, $-2.606 \pm 4.676j$, $-5.195 \pm 0.154j$, $-48.14 \pm 48.14j$, $-156.9 \pm 156.9j$ and -10072, which shows that the compensator does not attempt to cancel the plant's resonant poles (which lie at $-0.018 \pm 0.183j$).

5.8.3 Comparison with previous designs

When using the LQG/LTR technique, the design process is quite different to that required with the characteristic-locus or Nyquist-array methods. There is no reduction of the multivariable problem to a set of SISO problems, and the fact that stability and stability margins are guaranteed leads to an absence of explicit concern with phase compensation.

We have not attempted to reduce interaction, as we did in the designs examined in Chapter 4. In the present case, Figure 5.18 shows that interaction has been removed, even without having taken explicit steps to do so.

The state dimension of the LQG compensator is the same as that of the augmented plant model, and hence is 8 for our design, which is larger than that required for either of the two design examples presented in Chapter 4. It may be possible to find a lower-order approximation to this compensator, which has a very similar frequency response, using the model-reduction algorithm discussed in Section 8.3.9.

5.9   Non-minimum-phase plant

The LTR procedure can be applied to unstable plant without difficulty, since the unstable poles are shifted into the left half-plane by the feedback, and no unstable pole–zero cancellations are implied. But recovery cannot be achieved for plants with right half-plane zeros, as already mentioned. There are two ways of viewing the reason for this. The first is that recovery of an arbitrary return ratio would in general require unstable pole–zero cancellations to be introduced by the compensator, and no LQG compensator will do this. The second is to note that the state-feedback return ratio $-K_c(sI - A)^{-1}B$ and the Kalman-filter return ratio $-C(sI - A)^{-1}K_f$ both have all their zeros in the left half-plane. If this were not so, these return ratios would not have infinite gain margins. But any return ratio at the plant input or output which does not involve unstable pole–zero cancellations must have the right half-plane zeros of the plant, and therefore cannot be forced to approach either the state-feedback or the Kalman-filter return ratio.

This second point of view suggests that recovery may be achievable at those frequencies at which the plant's response is very close to that of a minimum-phase plant – that is, at frequencies which are small compared with the distance from the origin to any right half-plane zero. This is indeed the case, so the simplest strategy to use with non-minimum-phase plants is to follow the usual LTR procedure and hope for the best. This will certainly give closed-loop stability, it will still give an optimal trade-off between $S(j\omega)$ and $I - S(j\omega)$ if LTR is applied at the plant output, and it will give some degree of recovery of the ideal return ratio. If the right half-plane zeros lie well outside the required bandwidth, then adequate recovery of the ideal characteristics should be achieved at all significant frequencies. If any of them lie within the required bandwidth, then of course we can expect problems, whatever design technique we choose.

The explanation of the success of the approach described above lies in the theory of **optimal root loci** (Kwakernaak and Sivan, 1972). This shows that, as the LTR 'recovery parameter' q (or ρ) is made large (or small, respectively), each left half-plane zero of the plant 'attracts' one of the compensator poles. (The mirror image, about the imaginary axis, of each right half-plane zero, also 'attracts' one compensator pole.) All the 'minimum-phase dynamics' of the plant are therefore cancelled by the LTR compensator.

An alternative and more complicated strategy is to approximate the plant by one having only left half-plane zeros. If the plant is factored as

$$G(s) = A(s)G_0(s) \tag{5.128}$$

where $G_0(s)$ has only left half-plane zeros, and $A(s)$ is a stable all-pass transfer function, namely one such that

$$A(j\omega)A^H(j\omega) = I \tag{5.129}$$

then the application of the LTR procedure to $G_0(s)$ will result in perfect recovery of the open-loop gain characteristics at the input or output of the true plant $G(s)$. Stability margins will be worse than predicted for $G_0(s)$, however, and indeed closed-loop stability will no longer be guaranteed. On the zero locations of $A(s)$, and this can be allowed for when designing the return ratio which is to be recovered. Algorithms for obtaining state-space realizations of $A(s)$ and $G_0(s)$ are discussed by Stein and Athans (1987).

A third strategy for dealing with non-minimum-phase plant is to continue to use the LTR procedure, but to recover something other than the return ratio of an optimal state-feedback regulator or of a Kalman filter. If Section 5.4 is reviewed, it will be seen that recovery of the return ratio $-K_c(sI-A)^{-1}B$ at the plant input does not depend on K_c being an optimal state-feedback matrix. Any stabilizing state-feedback matrix, such as one obtained by some pole-placement algorithm, could be used and recovered just as well. Also, it can be shown (Stein and Athans, 1987) that any return ratio which shares the right half-plane pole and zero structure of the plant can be recovered by the LTR process. Thus the LQG/LTR approach can be replaced by a more general 'state-feedback/LTR' or 'state-observer/LTR' approach, which can be applied to non-minimum-phase plant. The usefulness of such an approach is questionable, however. The original motives for introducing the LQG/LTR technique were to allow the good robustness and performance properties of the linear quadratic regulator or the Kalman filter to be obtained, and to simplify the LQG design procedure. If both of these are now discarded, then LTR remains of interest only if some powerful methodology is brought in to replace the discarded half of LQG/LTR.

SUMMARY

In this chapter we have introduced the linear quadratic Gaussian (LQG) control problem, and outlined its solution. We have then shown how the general LQG problem can be restricted (LQG/LTR) in such a way that it yields a useful and fairly straightforward method of feedback design. The use of LQG methods guarantees internal stability, so the designer can shift attention from 'phase compensation' to 'gain shaping' in order to achieve desirable performance and robustness properties.

Adjusting the various covariance and cost-weighting matrices which appear in the LQG problem statement has been shown to be equivalent to adjusting the trade-off between the principal gains of various closed-loop transfer functions. In particular, the LQG/LTR procedure, applied at the output of the plant, has been shown to trade off the principal gains of the sensitivity $S(s)$ against those of $T(s) = I - S(s)$. Furthermore, this interpretation continues to hold for plants with right half-plane zeros, when loop transfer recovery cannot be guaranteed to work.

In this chapter we have assumed that the plant can be described by a rational transfer-function matrix, or equivalently, by a state-space model of finite state dimension.

Notes

1. We emphasize that the separation principle holds only when all three assumptions are satisfied: linear model, quadratic cost (J) and Gaussian stochastic processes (w and v).

2. This was suggested by Matt Wette, a graduate student at the University of California, Santa Barbara. It can be shown that this approximation achieves the smallest possible 'gap' between v_j and any real vector. ('Gap' here means the sine of the angle between two vectors – see Golub and Van Loan (1983).)

3. The state estimate $\hat{x}$ is given in terms of the plant output (for $u \equiv 0$) by $\hat{x}(s) = (sI - A)^{-1} K_f F_f^{-1}(s) y(s)$. Thus equation (5.76) shows that the filter action can always be interpreted as creating a white process $F_f^{-1}(s)y(s)$ (the 'innovation process'), and then applying this white process to $(sI - A)^{-1} K_f$ to obtain the optimal estimate $\hat{x}$.

4. It is also possible to use the original plant model, without augmentation, during the recovery step. This gives approximate, but often sufficient, LTR. It is of course essential to use the full augmented model when implementing the compensator.

5. The notation $K_f = \mathrm{LQE}(A, \Gamma, C, W, V)$ and $K_c = \mathrm{LQR}(A, B, M^T QM, R)$ is consistent with that used in the *Control System Toolbox* software, which is available for use with PC-Matlab and Pro-Matlab (see Chapter 8).

EXERCISES

5.1   (Alternative derivation of the LTR result.) Let u_1 be the signal at the input to the matrix B of the LQG compensator (see Figure 5.2), and let u be the signal at the plant input. Show that

$$\hat{x}(s) = \phi(s)[B\{C\phi(s)B\}^{-1} - K_f\{I + C\phi(s)K_f\}^{-1}]C\phi(s)Bu_1$$
$$- \phi(s)K_f\{I + C\phi(s)K_f\}^{-1}C\phi(s)Bu$$

where $\phi(s) = (sI - A)^{-1}$. Hence show that LTR is obtained at the plant input if

$$K_f\{I + C\phi(s)K_f\}^{-1} = B\{C\phi(s)B\}^{-1}$$

Show that this condition is satisfied asymptotically if $K_f \to q^{1/2}BX$, as $q \to \infty$, for any non-singular X.

5.2   In Section 5.6 it is shown that applying the LTR procedure at the plant output is equivalent to trading off the sensitivity $S(j\omega)$ against $I - S(j\omega)$. Show that applying the procedure at the plant input is equivalent to trading off $[I + K(j\omega)G(j\omega)]^{-1}$ against $[I + K(j\omega G(j\omega)]^{-1}K(j\omega)G(j\omega)$.
(*Hint*: Consider a disturbance acting at the *input* of the plant, and a weighting function $W = R^{-1/2}Q^{1/2}G$.)

5.3   Repeat the LQG/LTR design performed on the AIRC model (see the Appendix) in Section 5.8, but aiming for a closed-loop bandwidth of about $3\,\text{rad}\,\text{s}^{-1}$. Compare your design with the one obtained as the solution of Exercise 4.1.

5.4   Use the LQG/LTR method to design a feedback compensator for the TGEN model (see the Appendix), assuming a control structure with one degree of freedom, and the same specifications as in Exercise 4.2.

5.5   Use the LQG/LTR method to design a feedback compensator for the RPV model given in the Appendix. The specification is that $\bar{\sigma}[S(j\omega)]$ is to be minimized, where $S(j\omega) = [I + G(j\omega)K(j\omega)]^{-1}$ is the sensitivity, subject to the constraints

$$\bar{\sigma}[I - S(j\omega)] < 2, \quad \text{for all } \omega$$

and

$$\bar{\sigma}[I - S(j\omega)] < \frac{10}{\omega}, \quad \omega > 100$$

Examine the step responses of your design.
(Note that the RPV model is open-loop unstable. You are not expected to obtain an exact solution to the minimization problem.)

References

Anderson B.D.O. and Moore J.B. (1972). *Linear Optimal Control.* Englewood Cliffs NJ: Prentice-Hall.

Åström K.J. and Wittenmark B. (1984). *Computer-Controlled Systems.* Englewood Cliffs NJ: Prentice-Hall.

Banks S.P. (1986). *Control Systems Engineering.* Englewood Cliffs NJ: Prentice-Hall.

Davis M.H.A. and Vinter R.B. (1985). *Stochastic Modelling and Control.* London: Chapman & Hall.

Doyle J.C. (1978). Guaranteed margins for LQG regulators. *IEEE Transactions on Automatic Control,* **AC-23**, 756–7.

Doyle J.C. and Stein G. (1979). Robustness with observers. *IEEE Transactions on Automatic Control,* **AC-24**, 607–11.

Doyle J.C. and Stein G. (1981). Multivariable feedback design: Concepts for a classical/modern synthesis. *IEEE Transactions on Automatic Control,* **AC-26**, 4–16.

Franklin G.F. and Powell J.D. (1980). *Digital Control of Dynamic Systems.* Reading MA: Addison-Wesley.

Golub G.H. and Van Loan C.F. (1983). *Matrix Computations.* Baltimore MD: Johns Hopkins University Press.

IEEE (1971). Special Issue on the LQG Problem. *IEEE Transactions on Automatic Control,* **AC-16**, 527–869.

Kalman R.E. (1964). When is a linear control system optimal? *Journal of Basic Engineering (Trans. ASME D),* **86**, 51–60.

Kwakernaak H. (1969). Optimal low sensitivity linear feedback systems. *Automatica,* **5**, 279–86.

Kwakernaak H. and Sivan R. (1972). *Linear Optimal Control Systems.* New York: Wiley.

MacFarlane A.G.J. (1963). An eigenvector solution of the optimal linear regulator, *Journal of Electronics and Control,* **14**, 643–54.

MacFarlane A.G.J. (1970). The return-difference and return-ratio matrices and their use in the analysis and design of multivariable feedback control systems. *Proceedings of the Institution of Electrical Engineers,* **117**, 2037–49.

Potter J.E. (1966). Matrix quadratic solutions. *SIAM Journal on Applied Mathematics,* **14**, 496–501.

Safonov M.G. and Athans M. (1977). Gain and phase margins of multiloop LQG regulators. *IEEE Transactions on Automatic Control,* **AC-22**, 173–9.

Safonov M.G., Laub A.J. and Hartmann G.L. (1981). Feedback properties of multi-variable systems: The role and use of the return difference matrix. *IEEE Transactions on Automatic Control*, **AC-26**, 47–65.

Stein G. and Athans M. (1987). The LQG/LTR procedure for multivariable feedback control design. *IEEE Transactions on Automatic Control*, **AC-32**, 105–14.

CHAPTER 6

The Youla Parametrization and H_∞ Optimal Control

6.1   Introduction
6.2   A motivating example: sensitivity minimization
6.3   The H_∞ problem formulation
6.4   The Youla (or Q) parametrization
6.5   Solution of the H_∞ problem
6.6   The Hankel approximation problem

6.7   An algorithm for solving general H_∞ problems
6.8   Design example
6.9   Review and comments
Summary
Exercises
References

6.1   Introduction

This chapter is primarily concerned with H_∞ (pronounced as 'H-infinity') optimal control. This relatively new approach to feedback design is explained, and a complete derivation of a solution for a special case (the '1-block' problem) is given in Sections 6.4 to 6.6. A better algorithm, however, and one which can be applied to general H_∞ problems, is given in Section 6.7 (without any derivation). Our derivation introduces the celebrated 'Youla parametrization' of all stabilizing feedback controllers for a given plant, which we explore in some detail in Section 6.4. It also introduces the Hankel-approximation problem, in Section 6.6. This has important applications, particularly to model approximation.

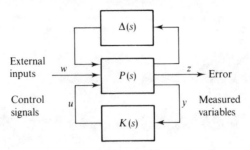

Figure 6.1 Standard representation of inaccurately known plant under feedback control.

In Chapter 3 we saw that one can obtain an accurate representation of the uncertainty inherent in a plant model in the form of Figure 6.1. Recall that in this representation 'external inputs' (w) is a vector of all the signals entering the system, and 'error' (z) is a vector of all the signals required to characterize the behaviour of the closed-loop system. Both these vectors may contain elements which are abstract in the sense that they may be defined mathematically, but do not represent signals which actually exist at any point in the system. u is the vector of control signals, and y is the vector of measured outputs.

In this chapter we shall use the representation shown in Figure 6.2, which is the same as Figure 6.1 but with the additive perturbation $\Delta(s)$ not shown. $P(s)$ is derived from the *nominal* plant model, just as in Chapter 3. However, it may also include weighting functions which depend on the design problem that is being solved.

Suppose that $P(s)$ is partitioned as

$$P(s) = \begin{bmatrix} P_{11}(s) & P_{12}(s) \\ P_{21}(s) & P_{22}(s) \end{bmatrix} \tag{6.1}$$

so that

$$z = P_{11}w + P_{12}u, \qquad y = P_{21}w + P_{22}u \tag{6.2}$$

Then we can eliminate u and y, using $u = Ky$, to obtain

$$z = [P_{11} + P_{12}K(I - P_{22}K)^{-1}P_{21}]w \tag{6.3}$$

The expression enclosed in brackets in equation (6.3) will occur frequently in the rest of this chapter, so it will be convenient to have a shorthand notation for it. It is becoming conventional to denote it by $F_l(P, K)$, so that we can rewrite (6.3) as[1]

$$z = F_l(P, K)w \tag{6.4}$$

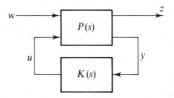

Figure 6.2 As Figure 6.1, but with perturbation Δ removed.

By suitably defining w and z (or, equivalently, P), it is possible to put a number of practical design problems into the form

minimize $\|F_1(P, K)\|_\infty$

where the minimization is over all realizable controllers $K(s)$ which stabilize the closed loop, and $\|.\|_\infty$ is as defined in Section 3.8.

This is known as the H_∞-**optimization problem**. In this chapter we shall show how this problem arises in a few examples, and how to solve it using state-space algorithms.

On the way to the solution of the H_∞ problem we shall examine the structure of all stabilizing controllers for a given plant, and show that the tricks we had to use in Chapter 5 to obtain good results with LQG methods were in no way special to the use of those methods: all stabilizing controllers can be thought of as observer/state-feedback combinations. Furthermore, all the closed-loop properties which are usually of interest, such as the sensitivity function $S(s)$, the closed-loop transfer function $T(s)$ or, more generally, $F_1(P(s), K(s))$, depend in a very simple manner on a single (albeit transfer-function matrix valued) parameter. These results are of fundamental importance in feedback theory, and their interest is not confined to the context of H_∞ design.

In Chapter 3 we defined 'H_∞' to be the set of transfer functions of asymptotically stable, realizable systems. In this chapter we shall often write $X \in H_\infty$ to mean that X is an asymptotically stable, realizable system. The term 'H_∞ problem' arises from the fact that we are minimizing $\|F_1(P, K)\|_\infty$ over all $F_1(P, K)$, such that $F_1(P, K) \in H_\infty$ and the feedback combination of P and K is internally stable. A consistent term for the LQG problem, which is sometimes used, is 'H_2 problem', since that requires the minimization of $\|F_1(P, K)\|_2$ over all $F_1(P, K) \in H_2$, again with the constraint of internal stability.

6.2   A motivating example: sensitivity minimization

Let us begin by looking at a relatively simple example. Consider the familiar feedback loop shown in Figure 6.3. To make everything as simple as possible,

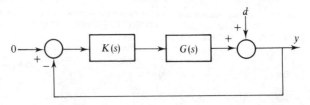

Figure 6.3   Feedback configuration for motivating example.

we assume that all the signals are scalars, and that G is stable. Suppose that we do not know what disturbances to expect, but we do know that their spectrum is essentially confined to the frequency range $0 \leqslant \omega \leqslant \omega_b$. A plausible design goal is to find a controller K which minimizes the worst-case excursion of y resulting from any disturbance d (and stabilizes the loop, of course), and this could be achieved by solving the problem

$$\text{minimize } \| S \|_\infty \tag{6.5}$$

or, equivalently,

$$\text{minimize } \sup_\omega |S(j\omega)| \tag{6.6}$$

where $S = (I + GK)^{-1}$ is the sensitivity function.[2] In fact this goal can be improved by limiting the minimization to the frequency range in which disturbances occur, since we know that making the sensitivity small outside this range will lead to unnecessary noise amplification and poor stability margins. We can do this by finding a (stable, minimum-phase) transfer function $W(s)$ with the properties

$$|W(j\omega)| \approx 1, \quad 0 \leqslant \omega \leqslant \omega_b \tag{6.7}$$

and

$$|W(j\omega)| \ll 1, \quad \omega > \omega_b \tag{6.8}$$

and then posing the problem

$$\text{minimize } \sup_\omega |W(j\omega)S(j\omega)| \tag{6.9}$$

We now make a crucial observation. Suppose that we define

$$Q = K(I + GK)^{-1} \tag{6.10}$$

then we can rewrite S in terms of Q and G as

$$S = I - GQ \tag{6.11}$$

Furthermore, it is easy to check that

$$(I + KG)^{-1} K = Q \tag{6.12}$$

$$(I + KG)^{-1} = I - QG \tag{6.13}$$

and

$$(I + GK)^{-1} G = (I - GQ)G \tag{6.14}$$

Now, the four matrices defined in (6.11)–(6.14) are just the matrices H_{ij} which appeared in Definition 2.7. Recall that the closed-loop system is internally stable if and only if each of these is stable. Since Q appears in each of these expressions in a particularly simple manner, and since we have assumed G to be stable, we can see immediately that the closed loop will be stable whenever Q is stable, because then each of (6.11)–(6.14) contains simply a product of stable matrices. What is more, the closed loop will be stable only if Q is stable, for if Q were unstable then (6.12) would be unstable, and the whole system would be unstable. We can also see from (6.10) that K will be proper (realizable) only if Q is proper (because with K proper, $Q \to K$ as $|s| \to \infty$), and from (6.15) below we shall be able to deduce that Q will be proper only if K is proper. Combining these observations with (6.10), we see that all realizable stabilizing controllers are given by

$$K = (I - QG)^{-1} Q \tag{6.15}$$

where the 'parameter' Q ranges over all proper, stable transfer functions. This is the celebrated **Youla parametrization** of all stabilizing controllers. We shall see later that this parametrization can be generalized to unstable plant.

We can now reformulate our optimization problem with the aid of (6.11) as

$$\text{minimize} \sup_{\substack{\text{stable } Q \quad \omega}} |W(I - GQ)(j\omega)| \tag{6.16}$$

Note that the use of the Youla parametrization has allowed us to formulate the design problem as an almost unconstrained optimization problem (the only constraint being that Q be proper and stable) with a relatively simple criterion function (Q appears linearly inside the modulus symbols).

This simple example displays many of the features which occur in more general H_∞ problems. First, the Youla parametrization plays essentially the same role in simplifying the optimization problem as it does in more general problems. Secondly, if we write $J = WS$ for the function whose magnitude is being minimized, and J^* for this function when the optimal controller is in place, then it can be shown that J^* is a constant. This is not generally true, but it is true that $|J^*(j\omega)|$ is constant, namely that J^* is an all-pass function. This implies that the choice of the weighting function W is crucial in the formulation of problems which are sensible engineering problems. Features such as the sharpness of the 'cut-off' of $|W(j\omega)|$ will determine the sharpness with which $|S(j\omega)|$ rises (above ω_b), since $|S|$ will have the same shape as $|W^{-1}|$.

We also remark on those features which are special to this example. First, the optimal solution is not actually realizable because it would require an improper controller. This can be avoided by use of a more elaborate criterion function – for example, penalizing control energy as well as sensitivity. Secondly, the problem actually has no solution if the plant has no right half-plane zeros, since the (weighted) sensitivity can then be made arbitrarily small. This simply shows that in our quest for a simple example we have oversimplified the problem, and posed one which is not realistic. Thirdly, the problem can be solved without any new theory, other than the Youla parametrization (Zames, 1981). This is emphatically not the case for other H_∞ problems.

6.3 The H_∞ problem formulation

6.3.1 Examples of H_∞ problems

SENSITIVITY MINIMIZATION

As a first example, let us simply put the sensitivity minimization example treated in Section 6.2 into the general form presented in Section 6.1. To do this, we use a familiar manipulation to write

$$W(I + GK)^{-1} = W[I - GK(I + GK)^{-1}] \qquad (6.17)$$

Since we require that

$$F_l(P, K) = W(I + GK)^{-1} \qquad (6.18)$$

we deduce, by comparing (6.17) with (6.3), that we need

$$P_{11} = W, \qquad P_{12} = -WG, \qquad P_{21} = I, \qquad P_{22} = -G \qquad (6.19)$$

The reader can check that this is equivalent to choosing $w = d$ and $z = Wy$, where d and y are the disturbance and output variables, respectively, as shown in Figure 6.3.

Note that this reformulation of the problem remains valid if we attempt to minimize $\|WS\|_\infty$ for a multivariable plant. In this case P_{11} and P_{21} will always be square, whereas P_{12} and P_{22} will have the same dimensions as G. This is important because the H_∞ theory becomes considerably more complicated if P_{12} has more rows than columns (or if P_{21} has more columns than rows), although the algorithms required to compute H_∞ solutions remain similar.

ROBUSTNESS TO ADDITIVE PERTURBATIONS

Consider the diagram shown in Figure 6.4: the real plant is modelled as a nominal plant G in parallel with an unstructured additive perturbation Δ. The only things known about Δ are that there is a frequency-dependent upper bound on its gain:

$$\bar{\sigma}[\Delta(j\omega)] < |r(j\omega)|, \quad \text{for each } \omega \tag{6.20}$$

where $r(s)$ is some (scalar) transfer function; and that the number of right half-plane poles of $G + \Delta$ is known and remains constant as Δ varies. A controller K is to be found which will stabilize the loop for all possible perturbations of the plant. The stability of the loop is the same as the stability of the loop shown in Figure 6.5 (note that we have used a positive-feedback convention in Figure 6.4). If K stabilizes the nominal plant G, so that $K(I - GK)^{-1}$ is stable, then the small-gain theorem (see Chapter 3) tells us that this loop will remain stable if $\|\Delta K(I - GK)^{-1}\|_\infty < 1$. Now,

$$\|\Delta K(I - GK)^{-1}\|_\infty = \|rr^{-1}\Delta K(I - GK)^{-1}\|_\infty$$

$$\leqslant \|r^{-1}\Delta\|_\infty \|rK(I - GK)^{-1}\|_\infty$$

$$< \|rK(I - GK)^{-1}\|_\infty \tag{6.21}$$

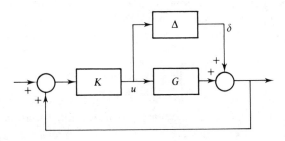

Figure 6.4 Feedback around a plant with additive uncertainty.

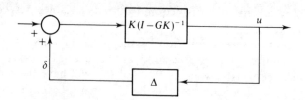

Figure 6.5 Feedback loop with the same stability as that in Figure 6.4.

Thus robust stability is assured if $\|rK(I-GK)^{-1}\|_\infty \leqslant 1$. Furthermore, it can be shown that if this condition is not met, then there exists a Δ in the expected set of perturbations which will destabilize the system. So to establish whether it is possible to achieve robust stability of the plant G in the presence of all possible perturbations Δ, and if so to maximize the stability margin, we should solve the H_∞ problem

$$\text{minimize } \|rK(I-GK)^{-1}\|_\infty \tag{6.22}$$

(over all stabilizing K).

To put this into standard form, we require that

$$F_l(P, K) = rK(I-GK)^{-1} \tag{6.23}$$

which we obtain by setting

$$P_{11} = 0_{l, m}, \qquad P_{12} = rI_l, \qquad P_{21} = I_m, \qquad P_{22} = G \tag{6.24}$$

where we have assumed that G has l inputs and m outputs.

MIXED PERFORMANCE AND ROBUSTNESS OBJECTIVE

Suppose now that we wish to obtain good disturbance-rejection performance, and to maintain stability in the presence of unstructured multiplicative output perturbations. That is, we wish to keep both the sensitivity S and the closed-loop transfer function $I-S$ small in magnitude. We cannot do both simultaneously, of course, but must emphasize one or other objective at each frequency. We can do this by making

$$F_l(P, K) = \begin{bmatrix} W_1 S \\ W_2(I-S) \end{bmatrix} \tag{6.25}$$

where W_1 and W_2 are frequency-dependent weighting matrices; W_2 would typically be chosen by considering frequency-dependent bounds on $\bar{\sigma}(\Delta)$, as discussed in Chapter 3. Defining G and K as in Figure 6.3, we see that we should choose

$$P_{11} = \begin{bmatrix} W_1 \\ 0 \end{bmatrix}, \qquad P_{12} = \begin{bmatrix} -W_1 G \\ W_2 G \end{bmatrix}, \qquad P_{21} = I, \qquad P_{22} = -G$$

(6.26)

6.3.2 Performance robustness: an unsolved problem

It is known that some important design problems cannot be formulated as H_∞ problems. In particular, the problem of achieving performance robustness, namely the maintenance of performance objectives – not merely stability – in the presence of unmodelled perturbations, is one such problem. We saw in Chapter 3 that the solution to this problem requires the minimization of the so-called structured singular value $\mu(M)$ of a transfer function matrix M, and that this can be very different from $\bar{\sigma}[M]$. However, we also saw in Chapter 3 that

$$\mu(M) \leqslant \inf_{D \in \underline{D}} \bar{\sigma}[DMD^{-1}]$$

(6.27)

where $\underline{D}$ is a set of diagonal, real positive matrices whose structure matches that of the plant perturbation. Specifically, let us suppose that the plant perturbation has the structure

$$\Delta = \text{diag}\{\Delta_1, \Delta_2, \ldots, \Delta_n\}$$

(6.28)

Then $\mathbf{D}$ has the structure

$$\mathbf{D} = \text{diag}\{d_1 I, d_2 I, \ldots, d_n I\}$$

(6.29)

with the dimensions of the blocks in (6.29) matching those in (6.28). Furthermore, the equality holds in (6.27) if $n \leqslant 3$, and numerical experiments suggest that (6.27) gives a reasonably tight bound even if $n > 3$.

This suggests that the performance robustness problem could be solved by solving the problem

$$\text{minimize} \ \|DF_1(P, K)D^{-1}\|_\infty$$
$$\text{K, D}$$

(6.30)

by iteratively solving for D and K. With D fixed this is the standard H_∞ problem, and with K fixed the minimization over D is convex, so at worst it can be performed by search techniques. This approximate solution has been implemented by Doyle (1985), who reports good results, but who has also discovered a simple example for which such an iterative scheme will fail to find the optimum K and D.

6.4   The Youla (or Q) parametrization

6.4.1   Fractional representations

In Chapter 2 we introduced matrix-fraction descriptions (MFDs) of rational transfer functions, of the form

$$G(s) = N(s)D^{-1}(s) = \tilde{D}^{-1}(s)\tilde{N}(s) \tag{6.31}$$

where N, D, $\tilde{N}$ and $\tilde{D}$, are polynomial matrices, with N and D right coprime, and $\tilde{N}$ and $\tilde{D}$ left coprime. We now generalize these ideas to **fractional representations**: every proper transfer function (matrix) $G(s)$ can be written in the form

$$G(s) = U(s)V^{-1}(s) = \tilde{V}^{-1}(s)\tilde{U}(s) \tag{6.32}$$

where U, V, $\tilde{U}$ and $\tilde{V}$, are stable transfer functions, with U and V right coprime, and $\tilde{U}$ and $\tilde{V}$ left coprime. Here right coprimeness means that if U and V have a common stable right factor X, so that $U = WX$ and $V = ZX$, then X^{-1} is stable, or equivalently, X has all its zeros in the left half-plane. (There is a corresponding definition for left coprimeness.)

EXAMPLE 6.1

$$G(s) = \begin{bmatrix} \dfrac{s-2}{s+1} & \dfrac{s+3}{s-4} \end{bmatrix}$$

$$= \begin{bmatrix} \dfrac{s-4}{s+4} \end{bmatrix}^{-1} \begin{bmatrix} \dfrac{(s-2)(s-4)}{(s+1)(s+4)} & \dfrac{s+3}{s+4} \end{bmatrix}$$

$$= \begin{bmatrix} \dfrac{s-2}{s+1} & \dfrac{s+3}{s+4} \end{bmatrix} \begin{bmatrix} 1 & 0 \\ 0 & \dfrac{s-4}{s+4} \end{bmatrix}^{-1}$$

As an aside, we remark that the mathematical basis of both MFDs and fractional representations is the theory of **rings**. The set of polynomial matrices is a ring, and the units of this ring are the unimodular matrices. The set of stable transfer functions is also a ring, the units of this ring being the minimum-phase transfer functions. Roughly speaking, the idea of a unit is that it does not change the properties of an object significantly when it multiplies it. Thus we use MFDs when we are not concerned with transformations that leave the pole–zero properties of a transfer function unchanged, and we use fractional representations when we are not concerned with transformations which leave the right half-plane pole–zero properties of a transfer function unchanged. Since our principal concern in this section is with stability, it is appropriate to use fractional representations.

A useful characterization of coprimeness is given by:

Theorem 6.1 (Bezout's theorem): U and V are right coprime if and only if there exist stable X and Y such that

$$XU + YV = I \qquad (6.33)$$

A corresponding theorem characterizes left coprimeness.

If G has a stabilizable and detectable realization (A, B, C, D), then we can obtain right and left fractional representations for it as follows. Let F be any matrix such that $A + BF$ is asymptotically stable – that is, F is the solution to any state-feedback problem. (Note that we assume that $\dot{x} = Ax + Bu$ and $u = +Fx$, whereas in Chapter 5 we assumed that $u = -K_c x$. We make this change of sign in order to be consistent with most of the literature on H_∞.) Let u, x and y denote the input, state and output vectors of G, as usual, and define $v = u - Fx$. Then

$$\dot{x} = (A + BF)x + Bv \qquad (6.34)$$

$$u = Fx + v \qquad (6.35)$$

$$y = (C + DF)x + Dv \qquad (6.36)$$

so

$$u(s) = [F(sI - A - BF)^{-1}B + I]v(s) = M(s)v(s), \quad M \text{ stable} \qquad (6.37)$$

and

$$y(s) = [(C + DF)(sI - A - BF)^{-1}B + D]v(s) = N(s)v(s), \quad N \text{ stable} \qquad (6.38)$$

so that $y(s) = N(s)M^{-1}(s)u(s)$, and hence $G = NM^{-1}$.

To get the other representation, choose any H such that $A + HC$ is stable (that is, solve any observer problem). Consider the observer equations

$$\dot{\zeta} = (A + HC)\zeta - Hy + (B + HD)u \qquad (6.39)$$

$$\eta = C\zeta + Du - y \qquad (6.40)$$

Then

$$\eta(s) = -[C(sI - A - HC)^{-1}H + I]y(s) = \tilde{M}(s)y(s), \quad \tilde{M} \text{ stable} \qquad (6.41)$$

and

$$\eta(s) = [C(sI - A - HC)^{-1}(B + HD) + D]u(s) = \tilde{N}(s)u(s), \quad \tilde{N} \text{ stable}$$
(6.42)

so that $y(s) = \tilde{M}^{-1}(s)\tilde{N}(s)u(s)$, and hence $G = \tilde{M}^{-1}\tilde{N}$.

We shall be able to show later that these fractional representations are in fact coprime. We shall also see that under certain conditions we can choose F (or H) such that N (or $\tilde{N}$) has the all-pass (or inner) property, namely that $N^H(j\omega)N(j\omega) = I$, which is very useful for the solution of the H_∞ problem.

6.4.2 Parametrization of all stabilizing controllers

From now on we assume a positive-feedback convention (as in Figure 6.4). We saw in Chapter 2 that internal stability of the closed loop is equivalent to the matrix

$$\begin{bmatrix} I & -K \\ -G & I \end{bmatrix}^{-1}$$

being asymptotically stable.

> **Theorem 6.2:** Let G and K have the fractional representations $G = NM^{-1} = \tilde{M}^{-1}\tilde{N}$ and $K = UV^{-1} = \tilde{V}^{-1}\tilde{U}$. Then the feedback loop is stable if and only if both
>
> (1) $\qquad \begin{bmatrix} M & U \\ N & V \end{bmatrix}^{-1}$
>
> and
>
> (2) $\qquad \begin{bmatrix} \tilde{V} & -\tilde{U} \\ -\tilde{N} & \tilde{M} \end{bmatrix}^{-1}$
>
> are stable.
>
> *Proof:*
> (1)
>
> $$\begin{bmatrix} I & K \\ G & I \end{bmatrix} = \begin{bmatrix} I & UV^{-1} \\ NM^{-1} & I \end{bmatrix}$$
>
> $$= \begin{bmatrix} M & U \\ N & V \end{bmatrix} \begin{bmatrix} M^{-1} & 0 \\ 0 & V^{-1} \end{bmatrix}$$
> (6.43)

so

$$\begin{bmatrix} I & K \\ G & I \end{bmatrix}^{-1} = \begin{bmatrix} M & 0 \\ 0 & V \end{bmatrix} \begin{bmatrix} M & U \\ N & V \end{bmatrix}^{-1} \qquad (6.44)$$

Now, using the fact that the pairs (N, M) and (U, V) are coprime, and Theorem 6.1, it is easy to show that

$$\begin{bmatrix} M & 0 \\ 0 & V \end{bmatrix} \text{ and } \begin{bmatrix} M & U \\ N & V \end{bmatrix} \text{ are coprime.}$$

Hence $\begin{bmatrix} I & K \\ G & I \end{bmatrix}^{-1}$ is stable only if $\begin{bmatrix} M & U \\ N & V \end{bmatrix}^{-1}$ is stable. Also $\begin{bmatrix} M & 0 \\ 0 & V \end{bmatrix}$ is stable, so $\begin{bmatrix} M & U \\ N & V \end{bmatrix}^{-1}$ is stable only if $\begin{bmatrix} I & K \\ G & I \end{bmatrix}^{-1}$ is stable. The only thing to be shown now is that $\begin{bmatrix} I & -K \\ -G & I \end{bmatrix}^{-1}$ is stable if and only if $\begin{bmatrix} I & K \\ G & I \end{bmatrix}^{-1}$ is stable. This is most easily done by considering the transmission zeros of $\begin{bmatrix} I & -K \\ -G & I \end{bmatrix}$ and $\begin{bmatrix} I & K \\ G & I \end{bmatrix}$, noting that they are the same, since the determinants of both matrices are given by $\det(I - GK)$.[3]

This completes the proof of (1); (2) is proved similarly. ∎

Theorem 6.3: With the same notation as in Theorem 6.2, if the feedback loop is stabilizable, then $M, N, U, V, \tilde{M}, \tilde{N}, \tilde{U}$ and $\tilde{V}$ can be chosen such that

$$\begin{bmatrix} \tilde{V} & -\tilde{U} \\ -\tilde{N} & \tilde{M} \end{bmatrix} \begin{bmatrix} M & U \\ N & V \end{bmatrix} = \begin{bmatrix} I & 0 \\ 0 & I \end{bmatrix} \qquad (6.45)$$

and $K = UV^{-1} = \tilde{V}^{-1}\tilde{U}$ is a stabilizing controller.

Proof: Choose $M, N, \tilde{M}$ and $\tilde{N}$, to be as in (6.37), (6.38), (6.41) and (6.42), respectively, so that M has the state-space realization $(A+BF, B, F, I)$, N has the realization $(A+BF, B, C+DF, D)$, and so on.

We can obtain a stabilizing controller by using the state-feedback matrix F and the observer-gain matrix H in a combined state-feedback/observer structure, as shown in Figure 6.6. (Such controllers are often called **observer-based**.)

We have already shown, in Chapter 5, that such a controller is stabilizing (except that there we assumed that $D=0$; the proof follows in the same way in the general case).

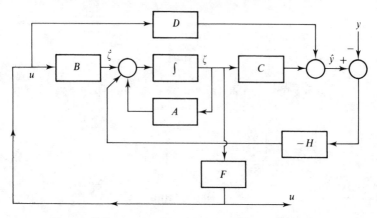

Figure 6.6 An observer-based stabilizing controller.

From Figure 6.6 we obtain the equations

$$\dot\zeta = A\zeta + Bu + H(\hat y - y) \tag{6.46}$$

and

$$u = F\zeta \tag{6.47}$$

which can be combined to give

$$\dot\zeta = (A + BF + HC + HDF)\zeta - Hy \tag{6.48}$$

Now, let $\mu = y - C\zeta - Du$. Then, since $\hat y = C\zeta + Du$, (6.46) becomes

$$\dot\zeta = A\zeta + Bu - H\mu \tag{6.49}$$

$$= (A + BF)\zeta - H\mu \tag{6.50}$$

But

$$y = C\zeta + Du + \mu \tag{6.51}$$

$$= (C + DF)\zeta + \mu \tag{6.52}$$

so we have

$$y(s) = [-(C + DF)(sI - A - BF)^{-1}H + I]\mu(s) = V(s)\mu(s),$$
$$V \text{ stable} \tag{6.53}$$

and

$$u(s) = -F(sI - A - BF)^{-1}H\mu(s) = U(s)\mu(s), \quad U \text{ stable}$$
(6.54)

so that $u(s) = U(s)V^{-1}(s)y(s)$, and hence $K = UV^{-1}$.

By a dual construction, it can be shown that if

$$\tilde{V}(s) = -F(sI - A - HC)^{-1}(B + HD) + I$$
(6.55)

and

$$\tilde{U}(s) = -F(sI - A - HC)^{-1}H$$
(6.56)

then $K = \tilde{V}^{-1}\tilde{U}$, and $\tilde{U}$ and $\tilde{V}$ are stable by construction.

Now, from (6.37), (6.38), (6.53) and (6.54), we see that

$$\begin{bmatrix} M & U \\ N & V \end{bmatrix}$$

has the state-space realization

$$\left(A + BF, \; [B \; -H], \; \begin{bmatrix} F \\ C + DF \end{bmatrix}, \; \begin{bmatrix} I & 0 \\ D & I \end{bmatrix} \right)$$

and from (6.41), (6.42), (6.55) and (6.56), we see that

$$\begin{bmatrix} \tilde{V} & -\tilde{U} \\ -\tilde{N} & \tilde{M} \end{bmatrix}$$

has the realization

$$\left(A + HC, \; [-(B + HD) \; H], \; \begin{bmatrix} F \\ C \end{bmatrix}, \; \begin{bmatrix} I & 0 \\ -D & I \end{bmatrix} \right)$$

It is straightforward to show that, if a transfer-function matrix X has the realization (A, B, C, D) and D^{-1} exists, then X^{-1} has the realization $(A - BD^{-1}C, -BD^{-1}, D^{-1}C, D^{-1})$. From this we see that

$$\begin{bmatrix} M & U \\ N & V \end{bmatrix}^{-1}$$

has the realization

$$\left(A+BF-[B \quad -H] \begin{bmatrix} I & 0 \\ -D & I \end{bmatrix} \begin{bmatrix} F \\ C+DF \end{bmatrix} \right.,$$

$$\left. -[B \quad -H] \begin{bmatrix} I & 0 \\ -D & I \end{bmatrix}, \begin{bmatrix} I & 0 \\ -D & I \end{bmatrix} \begin{bmatrix} F \\ C+DF \end{bmatrix}, \begin{bmatrix} I & 0 \\ -D & I \end{bmatrix} \right)$$

$$= \left(A+BF-[B \quad -H] \begin{bmatrix} F \\ C \end{bmatrix}, [-(B+HD) \quad H], \right.$$

$$\left. \begin{bmatrix} F \\ C \end{bmatrix}, \begin{bmatrix} I & 0 \\ -D & I \end{bmatrix} \right)$$

$$= \left(A+HC, [-(B+HD) \quad H], \begin{bmatrix} F \\ C \end{bmatrix}, \begin{bmatrix} I & 0 \\ -D & I \end{bmatrix} \right)$$

which is the realization given above for

$$\begin{bmatrix} \tilde{V} & -\tilde{U} \\ -\tilde{N} & \tilde{M} \end{bmatrix}$$

Thus (6.45) is verified for the particular fractional representations we have constructed. K is stabilizing by construction, so Theorem 6.3 is proved. ∎

Note that (6.45) can be combined with Theorem 6.1 to prove that all four fractional representations which we have constructed for G and K are coprime. For example, to show that U and V are coprime, apply Theorem 6.1 with $X = -\tilde{N}$ and $Y = \tilde{M}$.

A most important consequence of Theorem 6.2 is that *any* controller $K = UV^{-1} = \tilde{V}^{-1}\tilde{U}$ for which (6.45) holds is a stabilizing controller. We make use of this now in proving the central result of this section:

Theorem 6.4:   Let $K_0 = U_0 V_0^{-1} = \tilde{V}_0^{-1}\tilde{U}_0$ be such that (6.45) holds. For any $Q \in H_\infty$ (that is, for any realizable, stable Q of compatible dimensions), define

$$U = U_0 + MQ, \qquad V = V_0 + NQ \qquad\qquad (6.57)$$

$$\tilde{U} = \tilde{U}_0 + Q\tilde{M}, \qquad \tilde{V} = \tilde{V}_0 + Q\tilde{N} \qquad\qquad (6.58)$$

(1)   Then $UV^{-1} = \tilde{V}^{-1}\tilde{U}$, and $K = UV^{-1} = \tilde{V}^{-1}\tilde{U}$ is a stabilizing controller for $G = NM^{-1} = \tilde{M}^{-1}\tilde{N}$.

(2)   Furthermore, *any* stabilizing controller has fractional representations (6.57) and (6.58).

Proof:

(1)

$$\begin{bmatrix} \tilde{V} & -\tilde{U} \\ -\tilde{N} & \tilde{M} \end{bmatrix}\begin{bmatrix} M & U \\ N & V \end{bmatrix} = \begin{bmatrix} \tilde{V}_0 + Q\tilde{N} & -\tilde{U}_0 - Q\tilde{M} \\ -\tilde{N} & \tilde{M} \end{bmatrix}$$
$$\times \begin{bmatrix} M & U_0 + MQ \\ N & V_0 + NQ \end{bmatrix} \tag{6.59}$$

Since (6.45) holds for U_0, V_0, $\tilde{U}_0$ and $\tilde{V}_0$, we have

$$\tilde{N}M - \tilde{M}N = 0 \quad \text{and} \quad \tilde{N}U_0 - \tilde{M}V_0 = \tilde{V}_0 M - \tilde{U}_0 N = I \tag{6.60}$$

Performing the multiplication in (6.59) and using (6.60), we obtain

$$\begin{bmatrix} \tilde{V} & -\tilde{U} \\ -\tilde{N} & \tilde{M} \end{bmatrix}\begin{bmatrix} M & U \\ N & V \end{bmatrix} = \begin{bmatrix} I & 0 \\ 0 & I \end{bmatrix} \tag{6.61}$$

From the (1, 2) block we get $\tilde{V}U = \tilde{U}V$, so $UV^{-1} = \tilde{V}^{-1}\tilde{U}$. From Theorem 6.2 we see that $K = UV^{-1} = \tilde{V}^{-1}\tilde{U}$ is stabilizing.

(2) Suppose that K is realizable, stabilizes the feedback loop, and has the fractional representation $K = UV^{-1} = \tilde{V}^{-1}\tilde{U}$.

Consider

$$\left[\begin{bmatrix} \tilde{V} & -\tilde{U} \\ -\tilde{N} & \tilde{M} \end{bmatrix}\begin{bmatrix} M & U \\ N & V \end{bmatrix}\right]^{-1}$$
$$= \begin{bmatrix} (\tilde{V}M - \tilde{U}N)^{-1} & 0 \\ 0 & (-\tilde{N}U + \tilde{M}V)^{-1} \end{bmatrix} \tag{6.62}$$

From Theorem 6.2 we know that this is the product of two stable matrices, and so is itself stable. Hence $(\tilde{M}V - \tilde{N}U)^{-1}$ is stable. Let

$$Z = \tilde{M}V - \tilde{N}U \tag{6.63}$$

and

$$Q = M^{-1}(UZ^{-1} - U_0) \tag{6.64}$$

(note that $MQ \in H_\infty$). Then

$$V_0 + NQ = V_0 + NM^{-1}(UZ^{-1} - U_0) \tag{6.65}$$

$$= V_0 + \tilde{M}^{-1}\tilde{N}(UZ^{-1} - U_0) \tag{6.66}$$

$$= \tilde{M}^{-1}[\tilde{M}V_0 - \tilde{N}U_0 + \tilde{N}UZ^{-1}] \tag{6.67}$$

$$= \tilde{M}^{-1}[I + \tilde{N}UZ^{-1}] \tag{6.68}$$

(since U_0 and V_0 satisfy equation (6.45))

$$= \tilde{M}^{-1}[Z + \tilde{N}U]Z^{-1} \tag{6.69}$$

$$= VZ^{-1} \tag{6.70}$$

(from (6.63)). Note that from this we can deduce that $NQ \in H_\infty$.
From equation (6.64) we have

$$U = (U_0 + MQ)Z \tag{6.71}$$

and combining this with (6.70) we get

$$K = UV^{-1} = (U_0 + MQ)(V_0 + NQ)^{-1} \tag{6.72}$$

so that $((U_0 + MQ), (V_0 + NQ))$ is a fractional representation of K.
If we now set $\tilde{U} = \tilde{U}_0 + Q\tilde{M}$ and $\tilde{V} = \tilde{V}_0 + Q\tilde{N}$, and proceed as in the proof of part (1), it will be apparent that $\tilde{V}^{-1}\tilde{U} = UV^{-1}$.
The only thing left to do is to prove that $Q \in H_\infty$. This follows easily, since we have shown that both $MQ \in H_\infty$ and $NQ \in H_\infty$, and N and M are assumed to be coprime.   ∎

Theorem 6.4 gives a surprising and very powerful result. It says that, once we know one stabilizing controller for a plant, we can easily generate the family of all stabilizing controllers, by means of fractional representations. We shall soon see that everything can be expressed in terms of state-space models, so that practical algorithms based on this result can be obtained.
But first we shall explore some implications of Theorem 6.4. We begin by obtaining an expression for the controller's transfer function K in terms of K_0 and Q:

$$K = (U_0 + MQ)(V_0 + NQ)^{-1}$$

$$= (K_0 + MQV_0^{-1})(I + NQV_0^{-1})^{-1} \tag{6.73}$$

$$= (K_0 + MQV_0^{-1})[I - NQ(I + V_0^{-1}NQ)^{-1}V_0^{-1}] \tag{6.74}$$

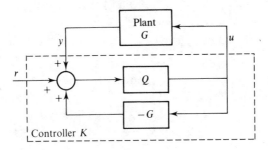

Figure 6.7 Controller structure if the plant is stable.

(by the matrix-inversion lemma)

$$= K_0 - K_0 NQ(I + V_0^{-1}NQ)^{-1}V_0^{-1} + MQV_0^{-1}(I + NQV_0^{-1})^{-1} \tag{6.75}$$

$$= K_0 - (K_0 NQ - MQ)(I + V_0^{-1}NQ)^{-1}V_0^{-1} \tag{6.76}$$

(since $V_0^{-1}(I + NQV_0^{-1})^{-1} = (I + V_0^{-1}NQ)^{-1}V_0^{-1}$)

$$= K_0 - (U_0 V_0^{-1}N - M)Q(I + V_0^{-1}NQ)^{-1}V_0^{-1} \tag{6.77}$$

But

$$U_0 V_0^{-1}N - M = \tilde{V}_0^{-1}\tilde{U}_0 N - M = \tilde{V}_0^{-1}(\tilde{U}_0 N - \tilde{V}_0 M) = -\tilde{V}_0^{-1} \tag{6.78}$$

(by equation (6.45)). So

$$K = K_0 + \tilde{V}_0^{-1}Q(I + V_0^{-1}NQ)^{-1}V_0^{-1} \tag{6.79}$$

If we suppose temporarily that the plant G is stable, then we can take $N = G$, $M = I$ and $K_0 = 0$; the latter has representations $U_0 = \tilde{U}_0 = 0$ and $V_0 = \tilde{V}_0 = I$, for example. In this case (6.79) simplifies to $K = Q(I + GQ)^{-1}$, from which we obtain $Q = K(I - GK)^{-1}$, which is precisely the 'Youla parameter' Q which we used in the sensitivity-minimization example in Section 6.2 (except for a sign change, the result of using a negative-feedback convention in the example, and a positive-feedback convention here). From the expression for K we can see that K can be obtained as the feedback connection of Q and $-G$, as shown in Figure 6.7. This shows that the controller is cancelling the plant's dynamics by putting $-G$ in parallel with the plant's transfer function G, and inserting arbitrary stable dynamics, so that $u = Qr$, where r is the 'reference input'.

But (6.79) shows that things get more complicated if the plant is unstable. Why is this? In particular, why will the simple scheme of Figure 6.7

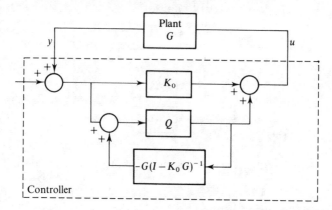

Figure 6.8 Controller structure if the plant is unstable, but can be stabilized by a stable controller K_0.

fail? The reason is clearly that the instability in G only 'disappears' from certain transmission paths – in particular, from the path between r and u. Using the controller shown in Figure 6.7, we would have, for example, that $y = GQr$, which would be unstable.

But suppose that we first stabilized the plant using a (feedback) controller K_0, and then applied the simple scheme of Figure 6.7? What structure would our controller have then, in terms of the original plant? It turns out that *if K_0 can be chosen to be stable*, it would have precisely the structure given by (6.79)! We shall show this shortly, but we first note its implication for design methods. One might suspect that following a two-step design procedure, stabilizing the plant in the first step and then modifying the dynamics of the stabilized plant in the second, would not allow all possible designs to be achieved: the first step might in some way limit the design freedom in the second step. Theorem 6.4 (and its consequences) shows that this is not the case if the plant can be stabilized by a stable controller.

Figure 6.8 shows the controller which results from such a two-step procedure. The stable controller K_0 is used to stabilize the plant. This yields a modified system, with transfer function (from u to y):

$$H = G(I - K_0 G)^{-1} \tag{6.80}$$

The scheme of Figure 6.7 is now used to construct a controller for H which consists of the feedback connection of $Q \in H_\infty$ and $-H$. The transfer function of the complete controller is

$$K_1 = K_0 + Q(I + HQ)^{-1} \tag{6.81}$$

and

$$H = NM^{-1}(I - \tilde{V}_0^{-1}\tilde{U}_0 NM^{-1})^{-1} \qquad \text{(6.82)}$$

$$= N(M - \tilde{V}_0^{-1}\tilde{U}_0 N)^{-1} \qquad \text{(6.83)}$$

$$= N\tilde{V}_0 \qquad \text{(6.84)}$$

(by 6.78). Hence we have

$$K_1 = K_0 + Q(I + N\tilde{V}_0 Q)^{-1}, \quad Q \in H_\infty \qquad \text{(6.85)}$$

Now consider the controller K defined by (6.79), which we rewrite as

$$K = K_0 + \tilde{V}_0^{-1} Q V_0^{-1}(I + NQV_0^{-1}), \quad Q \in H_\infty \qquad \text{(6.86)}$$

Remember that we have assumed that $K_0 \in H_\infty$; but this implies that V_0^{-1} and $\tilde{V}_0^{-1} \in H_\infty$, so that

$$R = \tilde{V}_0^{-1} Q V_0^{-1} \in H_\infty, \quad \text{if and only if } Q \in H_\infty \qquad \text{(6.87)}$$

(The 'only if' comes from the fact that $V_0 \in H_\infty$ and $\tilde{V}_0 \in H_\infty$.) Given any $Q \in H_\infty$ in (6.86) we can obtain the corresponding $R \in H_\infty$ from (6.87), and given any $R \in H_\infty$, we can obtain the corresponding $Q \in H_\infty$ from

$$Q = \tilde{V}_0 R V_0 \qquad \text{(6.88)}$$

We can therefore rewrite (6.86) as

$$K = K_0 + R(I + N\tilde{V}_0 R)^{-1} \quad (R \in H_\infty) \qquad \text{(6.89)}$$

which has precisely the same form as (6.85). Thus we have shown the equivalence of (6.79) to a controller obtained by a two-step method, provided K_0 can be chosen to be stable.

Some plants can be stabilized only by an unstable controller K_0, in which case the scheme shown in Figure 6.8 will fail because the instability in K_0 may show up at some nodes in the system. For this general case, Figure 6.9 shows a block-diagram representation of (6.79). Here the dashed-line rectangle encloses that part of the controller which remains fixed while Q is allowed to vary over H_∞ to obtain all possible stabilizing controllers.

6.4.3   All stabilizing controllers are observer-based

Here we show that all the controllers described by (6.79), and hence all stabilizing controllers, have observer-based state-space realizations. We have

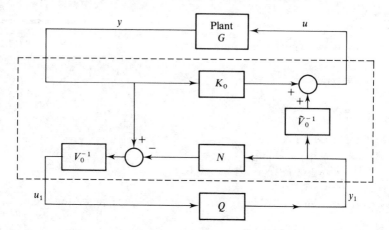

Figure 6.9   General structure of stabilizing controllers ($Q \in H_\infty$).

already obtained realizations of N, K_0, V_0 and $\tilde{V}_0$ (in equations (6.38), (6.47), (6.48), (6.53) and (6.55)). We begin by assembling these together to obtain a realization for the fixed part of the controller, enclosed by the dashed-line rectangle in Figure 6.9. Let this fixed part have transfer function J, so that the controller can be redrawn as in Figure 6.10. Then, in the notation introduced in Section 6.1,

$$K = F_1(J, Q) \tag{6.90}$$

and (from Figure 6.9)

$$J = \begin{bmatrix} K_0 & \tilde{V}_0^{-1} \\ V_0^{-1} & -V_0^{-1}N \end{bmatrix} \tag{6.91}$$

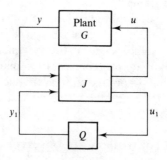

Figure 6.10   As Figure 6.9, but with the system enclosed by the dashed-line rectangle represented by J.

Using the formula for the realization of an inverse system (see Section 6.4.2), it is easy to show that V_0^{-1} and $\tilde{V}_0^{-1}$ have the following realizations:

$$V_0^{-1}: (A+BF+HC+HDF, H, C+DF, I) \tag{6.92}$$

$$\tilde{V}_0^{-1}: (A+BF+HC+HDF, B+HD, F, I) \tag{6.93}$$

and hence $V_0^{-1}N$ has the realization

$$V_0^{-1}N: \left(\begin{bmatrix} A+BF+HC+HDF & H(C+DF) \\ 0 & A+BF \end{bmatrix}, \begin{bmatrix} HD \\ B \end{bmatrix}, \right.$$

$$\left. [C+DF \quad C+DF], D \right) \tag{6.94}$$

The change of state coordinates

$$z = \begin{bmatrix} I & I \\ 0 & I \end{bmatrix} x$$

shows that an alternative realization is

$$V_0^{-1}N: \left(\begin{bmatrix} A+BF+HC+HDF & 0 \\ 0 & A+BF \end{bmatrix}, \begin{bmatrix} B+HD \\ B \end{bmatrix}, \right.$$

$$\left. [C+DF \quad 0], D \right) \tag{6.95}$$

which is unobservable, and can therefore be simplified to

$$V_0^{-1}N: (A+BF+HC+HDF, B+HD, C+DF, D) \tag{6.96}$$

Putting together equations (6.47), (6.48), (6.92), (6.93) and (6.96), we obtain a realization for J:

$$J: \left(A+BF+HC+HDF, [-H \quad B+HD], \begin{bmatrix} F \\ -(C+DF) \end{bmatrix}, \right.$$

$$\left. \begin{bmatrix} 0 & I \\ I & -D \end{bmatrix} \right) \tag{6.97}$$

If we let the inputs to J be the vector $[y^T \ y_1^T]^T$ and its outputs be $[u^T \ u_1^T]^T$, as shown in Figure 6.10, then the state-space equations corresponding to (6.97)

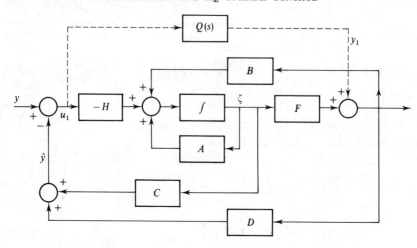

Figure 6.11 Alternative representation of all stabilizing controllers ($Q \in H_\infty$).

are

$$\dot{\zeta} = (A + BF + HC + HDF)\zeta - Hy + (B + HD)y_1 \tag{6.98}$$

$$u = F\zeta + y_1 \tag{6.99}$$

$$u_1 = -(C + DF)\zeta + y - Dy_1 \tag{6.100}$$

It is straightforward to check that (6.98) and (6.99) are the equations of an observer-based controller, as shown in Figure 6.6, which has been modified by adding an additional signal y_1 to the controller output u. The additional term $(B + HD)y_1$ in (6.98) ensures that the observer is driven by the same control signal as the plant. Equation (6.100) reveals that the signal u_1 is just the difference between the estimated and the actual plant output (the **innovations signal**, if the observer is a Kalman filter). But $y_1 = Qu_1$, so the structure of every stabilizing controller has the form shown in Figure 6.11: an observer/state-feedback structure, with an arbitrary, stable, proper transfer function Q inserted as shown by the dashed line. (Compare Figure 6.11 with Figure 5.2.)

Yet another interpretation of the structure of stabilizing controllers can be obtained as follows. Let x_q be the state vector of a realization of Q. Then we can supplement (6.98)–(6.100) with the equations

$$\dot{x}_q = A_q x_q + B_q u_1 \tag{6.101}$$

$$y_1 = C_q x_q + D_q u_1 \tag{6.102}$$

To simplify the algebra, we shall assume that the plant is strictly proper, namely that $D=0$. However, we shall retain $D_q \neq 0$.

If we eliminate u_1 and $\bar{y}_1$ from (6.98)–(6.102), setting $D=0$, we obtain

$$\dot{\zeta} = (A + BF + HC - BD_qC)\zeta + BC_qx_q + (BD_q - H)y \qquad (6.103)$$

$$\dot{x}_q = A_qx_q - B_qC\zeta + B_qy \qquad (6.104)$$

$$u = (F - D_qC)\zeta + C_qx_q + D_qy \qquad (6.105)$$

Now, if we let $x_e = \begin{bmatrix} \zeta \\ x_q \end{bmatrix}$, $A_e = \begin{bmatrix} A & 0 \\ 0 & A_q \end{bmatrix}$, $B_e = \begin{bmatrix} B \\ 0 \end{bmatrix}$

$$C_e = [C \ \ 0], \qquad F_e = [F \ \ C_q], \qquad H_e = \begin{bmatrix} H \\ -B_q \end{bmatrix}$$

then (6.103)–(6.105) can be rewritten as

$$\dot{x}_e = (A_e + B_eF_e + H_eC_e - B_eD_qC_e)x_e + (B_eD_q - H_e)y \qquad (6.106)$$

$$u = (F_e - D_qC_e)x_e + D_qy \qquad (6.107)$$

This gives a state-space realization of the complete controller K (complete in that it includes the dynamics of both J and Q). Note that K is strictly proper if and only if $D_q = 0$, namely if and only if Q is strictly proper.

If $D_q = 0$ then (6.106) and (6.107) are the equations of an observer-based controller for a plant with realization $(A_e, B_e, C_e, 0)$. But

$$C_e(sI - A_e)^{-1}B_e = C(sI - A)^{-1}B = G(s) \qquad (6.108)$$

so (6.106) and (6.107) are actually the equations of an observer-based controller for the original plant, with its dynamics augmented in such a way that its transfer function is unchanged. Hence we conclude that all strictly proper stabilizing controllers have the structure of an observer-based controller for an augmented model of the plant.

If $D_q \neq 0$, then we retain the observer-based structure, with the addition of a signal path through D_q, as shown in Figure 6.12. Note that this controller is stabilizing with *any* D_q, provided F_e and H_e give a stable observer/state-feedback combination. (*Exercise*: Why?)

6.4.4 Parametrization of closed-loop transfer functions

To investigate those closed-loop transfer functions which appear in H_∞ problems, we need to return to Figure 6.2 and exploit the structure which the

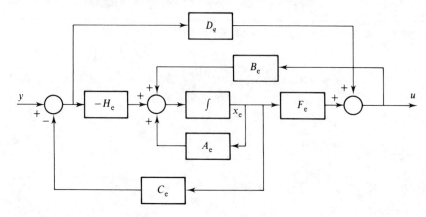

Figure 6.12 State-space representation of all stabilizing controllers for strictly proper plant.

controller possesses. Figure 6.13 shows the original scheme of Figure 6.2, suitably redrawn. Note that $P_{22} = G$ (since we are using a positive-feedback convention; it would be $P_{22} = -G$ otherwise). Let us combine P and J into the transfer function T, as shown in Figure 6.14. (*Warning*: In earlier chapters we have used the symbol 'T' to denote something else.) We have

$$z = F_1(P, K)w \tag{6.109}$$

$$= F_1(T, Q)w \tag{6.110}$$

$$= [T_{11} + T_{12}Q(I - T_{22}Q)^{-1}T_{21}]w \tag{6.111}$$

We shall show that $T_{22} = 0$, so that (6.111) simplifies to

$$z = [T_{11} + T_{12}QT_{21}]w \tag{6.112}$$

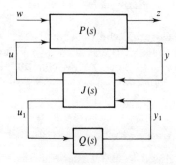

Figure 6.13 As Figure 6.2, but with the controller represented as in Figure 6.10.

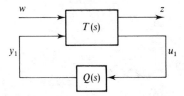

Figure 6.14 As Figure 6.13, but with the feedback connection of P and J represented by T.

which shows that, even in the general case, closed-loop transfer functions are affine in the Youla parameter Q. Note (from Figure 6.9 or 6.11) that u_1 has the same dimension as y, and y_1 has the same dimension as u. Consequently T_{11}, T_{12} and T_{21} have the same dimensions as P_{11}, P_{12} and P_{21}, respectively.

To show that $T_{22} = 0$, we use Theorem 6.4 to obtain

$$K(I - P_{22}K)^{-1} = (U_0 + MQ)(V_0 + NQ)^{-1}$$
$$\times [I - \tilde{M}^{-1}\tilde{N}(U_0 + MQ)(V_0 + NQ)^{-1}]^{-1} \quad \textbf{(6.113)}$$

$$= (U_0 + MQ)[(V_0 + NQ) - \tilde{M}^{-1}\tilde{N}(U_0 + MQ)]^{-1} \quad \textbf{(6.114)}$$

$$= (U_0 + MQ)[(\tilde{M}V_0 - \tilde{N}U_0) + (\tilde{M}N - \tilde{N}M)Q]^{-1}\tilde{M} \quad \textbf{(6.115)}$$

$$= -(U_0 + MQ)\tilde{M} \quad \textbf{(6.116)}$$

in which the last step follows from (6.60). From this we get

$$F_l(P, K) = P_{11} - P_{12}(U_0 + MQ)\tilde{M}P_{21} \quad \textbf{(6.117)}$$

$$= (P_{11} - P_{12}U_0\tilde{M}P_{21}) - P_{12}MQ\tilde{M}P_{21} \quad \textbf{(6.118)}$$

Comparing this with (6.111), we obtain the correspondences

$$T_{11} = P_{11} - P_{12}U_0\tilde{M}P_{21}$$

$$T_{12} = -P_{12}M, \qquad T_{21} = \tilde{M}P_{21}, \qquad T_{22} = 0 \quad \textbf{(6.119)}$$

Another way to see that $T_{22} = 0$ is to redraw Figure 6.11 in the form of Figure 6.15, with $Q(s)$ omitted. Now T_{22} is just the transfer function between y_1 and u_1. A standard result of observer theory, and one which we also came across in Chapter 5, is that this transfer function is zero since the state

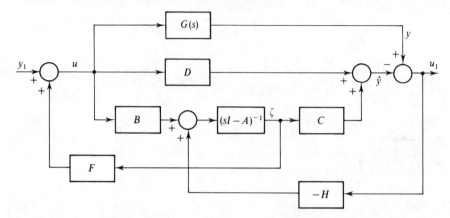

Figure 6.15 Plant with stabilizing controller as in Figure 6.11, but with Q omitted.

estimate ζ is uncontrollable from y_1, and the plant's state is unobservable from u_1 (see Exercise 6.5).

This proves the main result of this section.

In order to be able to solve H_∞ problems using state-space algorithms, we need state-space models of T_{11}, T_{12} and T_{21}. We shall assume that a (minimal) realization for P is

$$P: \left(A, [B_1\ B_2], \begin{bmatrix} C_1 \\ C_2 \end{bmatrix}, \begin{bmatrix} D_{11} & D_{12} \\ D_{21} & D_{22} \end{bmatrix} \right) \tag{6.120}$$

so that $P_{ij}(s) = C_i(sI-A)^{-1}B_j + D_{ij}, \quad i,j = 1, 2$ (6.121)

It can be shown that, if F is a stabilizing state-feedback matrix for the plant (that is, if $A + B_2 F$ is stable), and H is a stabilizing observer-gain matrix for the plant (i.e. if $A + HC_2$ is stable), then realizations of T_{11}, T_{12} and T_{21} are given by

$$T_{11}: \left(\begin{bmatrix} A+B_2F & -B_2F \\ 0 & A+HC_2 \end{bmatrix}, \begin{bmatrix} B_1 \\ B_1+HD_{21} \end{bmatrix}, \right.$$

$$\left. [C_1+D_{12}F \quad -D_{12}F], D_{11} \right) \tag{6.122}$$

$$T_{12}: (A+B_2F, B_2, C_1+D_{12}F, D_{12}) \tag{6.123}$$

$$T_{21}: (A+HC_2, B_1+HD_{21}, C_2, D_{21}) \tag{6.124}$$

Note that these are all stable.

This concludes our rather long section on the Youla parametrization. We can now use the results we have developed, particularly (6.112), to solve the H_∞ problem.

6.5   Solution of the H_∞ problem

6.5.1   Equivalence to the model-matching problem

In Section 6.1 we formulated the general H_∞ problem as

$$\text{minimize } \| F_1(P, K) \|_\infty$$

over stabilizing compensators K.
   Combining Theorem 6.4 with (6.112), we can rewrite this as

$$\underset{Q \in H_\infty}{\text{minimize}} \| T_{11} + T_{12} Q T_{21} \|_\infty$$

This is sometimes known as the **model-matching problem** because to solve it we need to choose Q such that $T_{12} Q T_{21}$ 'matches' the 'model' $-T_{11}$ as well as possible.
   If T_{12} and T_{21} are square, it looks at first sight as if we should be able to set $Q = -T_{12}^{-1} T_{11} T_{21}^{-1}$, and the problem would be solved. In practice this is never possible because the resulting Q is neither proper nor stable. For illustration, consider the sensitivity-minimization example presented in Section 6.2. We saw in Section 6.3.1 that this problem corresponds to choosing

$$P_{11} = W, \qquad P_{12} = -WG, \qquad P_{21} = I$$

Since G is assumed to be stable, we can take $T_{11} = P_{11}$, $T_{12} = P_{12}$ and $T_{21} = P_{21}$ in this case (see Exercise 6.3) which gives

$$-T_{12}^{-1} T_{11} T_{21}^{-1} = G^{-1}$$

This is unstable, if G is assumed to have a right half-plane zero. Even if this assumption is dropped, G^{-1} is not proper if we assume that G is strictly proper.
   To obtain a solution of the model-matching problem, we shall show in the next section that it is equivalent to yet another problem, the so-called 'Hankel approximation problem', for which a complete solution algorithm exists and will be given in Section 6.6.
   Once a solution to this problem is obtained, in the form of a state-space realization of Q, the required controller can be obtained (in state-space form) using equations (6.103)–(6.105).

6.5.2   Equivalence to the Hankel approximation problem

In this and subsequent sections we shall use $X^*(s)$ to denote $X^T(-s)$. We shall usually drop the argument (s), as elsewhere in this chapter.

Let us assume, for the time being, that T_{12} and T_{21} are square and all-pass (or **inner**), namely that

$$T_{12}T_{12}^* = I \quad \text{and} \quad T_{21}^* T_{21} = I \tag{6.125}$$

It is easy to show that, if X and Y are all-pass, and $\sigma(.)$ denotes any singular value of $(.)$, then $\sigma(XAY) = \sigma(A)$, and that, for any A, $\sigma(A^*) = \sigma(A)$. Hence we have

$$\| T_{11} + T_{12}QT_{21} \|_\infty = \| T_{12}(T_{12}^* T_{11} T_{21}^* + Q)T_{21} \|_\infty \tag{6.126}$$

$$= \| T_{12}^* T_{11} T_{21}^* + Q \|_\infty \tag{6.127}$$

$$= \| T_{21} T_{11}^* T_{12} + Q^* \|_\infty \tag{6.128}$$

Since $Q \in H_\infty$ and $T_{ij} \in H_\infty$, Q^* has only unstable poles, while $T_{21} T_{11}^* T_{12}$ potentially has both stable and unstable poles.

It can be shown, by a rather complicated analysis (Limebeer and Hung, 1987), that $T_{21} T_{11}^* T_{12}$ in fact has only stable poles. If we define

$$R = T_{21} T_{11}^* T_{12} \tag{6.129}$$

then

$$\min_{Q \in H_\infty} \| T_{11} + T_{12}QT_{21} \|_\infty = \min_{Q \in H_\infty} \| R + Q^* \|_\infty \tag{6.130}$$

which converts the model-matching problem into the problem of approximating a stable transfer function (R) by an unstable one $(-Q^*)$. This is known as the **Hankel approximation problem**, or the **Nehari extension problem**.

We made two assumptions about T_{12} and T_{21} to obtain (6.130): that they are square and all-pass. Somewhat surprisingly, the second assumption is not at all restrictive because it is usually possible to choose F and H in (6.123) and (6.124) so that the assumption is satisfied (see Exercise 6.8). It is always possible to find a suitable F and H if P_{12} and P_{21} are square, and P is stabilizable.

The validity of the first assumption, however, that T_{12} and T_{21} are square, is determined entirely by the problem statement, and is violated by some important classes of problem – for example, the mixed performance and robustness problem described in Section 6.3.1.

6.5.3   1-block, 2-block and 4-block problems

If T_{12} has more columns than rows, and T_{21} has more rows than columns, then there is no problem, because T_{12}^* and T_{21}^* can still be found such that equations (6.125) hold. The whole of the development given above is still valid.

More commonly, though, T_{12} has more rows than columns, or T_{21} has more columns than rows, and things then become more complicated. It is no longer possible for (6.125) to hold; however, it is always possible to find transfer functions $T_{12\perp}$ and $T_{21\perp}$ such that $[T_{12}\ T_{12\perp}]$ and $[T_{21}^T\ T_{21\perp}^T]^T$ are square and all-pass. Hence we have that

$$\begin{bmatrix} T_{12}^* \\ T_{12\perp}^* \end{bmatrix}[T_{12}\ T_{12\perp}]=I \quad \text{and} \quad \begin{bmatrix} T_{21} \\ T_{21\perp} \end{bmatrix}[T_{21}^*\ T_{21\perp}^*]=I \qquad \textbf{(6.131)}$$

Proceeding as above, we obtain

$$\|T_{11}+T_{12}QT_{21}\|_\infty = \left\| \begin{bmatrix} T_{12}^* \\ T_{12\perp}^* \end{bmatrix}[T_{11}+T_{12}QT_{21}][T_{21}^*\ T_{21\perp}^*] \right\|_\infty$$

$$\textbf{(6.132)}$$

$$= \left\| \begin{bmatrix} T_{12}^* \\ T_{12\perp}^* \end{bmatrix} T_{11}[T_{21}^*\ T_{21\perp}^*]+\begin{bmatrix} Q & 0 \\ 0 & 0 \end{bmatrix} \right\|_\infty \quad \textbf{(6.133)}$$

$$= \left\| \begin{bmatrix} R_{11}+Q & R_{12} \\ R_{21} & R_{22} \end{bmatrix} \right\|_\infty \qquad \textbf{(6.134)}$$

where

$$R_{11}=T_{12}^* T_{11} T_{21}^*, \qquad R_{12}=T_{12}^* T_{11} T_{21\perp}^*$$

$$R_{21}=T_{12\perp}^* T_{11} T_{21}^*, \qquad R_{22}=T_{12\perp}^* T_{11} T_{21\perp}^*$$

If T_{21} is square (or has more rows than columns), then $T_{21\perp}$ does not exist, and (6.134) simplifies to

$$\|T_{11}+T_{12}QT_{21}\|_\infty = \left\| \begin{bmatrix} R_{11}+Q \\ R_{21} \end{bmatrix} \right\|_\infty \qquad \textbf{(6.135)}$$

and if T_{12} is square (or has more columns than rows) then $T_{12\perp}$ does not exist, and (6.134) simplifies to

$$\|T_{11}+T_{12}QT_{21}\|_\infty = \|[R_{11}+Q\ \ R_{12}]\|_\infty \qquad \textbf{(6.136)}$$

In the literature on H_∞ control theory, problems which result in the formulation (6.130) are often called **1-block problems**, while those which result in

(6.135) or (6.136) are called **2-block problems**, and those which result in (6.134) are called **4-block problems**.

It is possible to develop solutions to all these problems along similar lines to that taken above for the 1-block problem (Doyle, 1984), but this approach would lead to unnecessarily complicated derivations for the 2-block and 4-block problems, and to very inefficient solution algorithms. It is better to use an alternative algorithm, which is given in Section 6.7 without any derivation. That algorithm is in fact to be preferred for all three kinds of H_∞ problem, although a workable algorithm is obtained for the 1-block problem by solving (6.130), using the results obtained in Section 6.6.

The minimum achievable value of $\| T_{11} + T_{12} Q T_{21} \|_\infty$ can be computed for 1-block problems, and the optimal value of Q can be obtained by a single application of the appropriate algorithm. This is not so for 2-block or 4-block problems, for which one must choose some real positive scalar γ, apply the algorithm to discover whether there exists a Q such that $\| T_{11} + T_{12} Q T_{21} \|_\infty < \gamma$, and if so obtain the corresponding compensator. An optimal controller is found by iteratively searching over values of γ. There is some reason to believe that such an iteration is inevitable in any algorithm for solving general H_∞ optimization problems.

It should be borne in mind, however, that often we do not really need to minimize $\| T_{11} + T_{12} Q T_{21} \|_\infty$, but just keep it below some particular value. This is true of the robustness problem treated in Section 6.3.1, for example. If we simply wish to obtain $\| T_{11} + T_{12} Q T_{21} \|_\infty < m$, say, then we need only take $\gamma = m$, and check whether a solution exists (and if so, find it). In this case no iteration is involved.

6.6   The Hankel approximation problem

6.6.1   The Hankel norm

In Section 6.5.2 we saw how to convert 1-block H_∞ problems into so-called Hankel approximation problems which take the form

$$\underset{Y \in H_\infty}{\text{minimize}} \| G + Y^* \|_\infty, \quad G \in H_\infty \tag{6.137}$$

In this section we shall show how to solve such problems, again using state-space algorithms. Our treatment will therefore be limited to rational G, and we shall assume that we have a realization $G: (A, B, C, D)$ available as 'input' to our algorithms.

First we need to define the **Hankel norm** of G, which will be denoted by $\| G \|_H$. We assume that $G \in H_\infty$, and we let $P = P^T$ and $Q = Q^T$ be the solutions of the **Lyapunov equations**

$$AP + PA^T + BB^T = 0 \tag{6.138}$$

and

$$A^T Q + QA + C^T C = 0 \tag{6.139}$$

(P and Q are the **controllability** and **observability gramians**, respectively).
Then the Hankel norm of G is

$$\| G \|_H = \bar{\sigma}[(PQ)^{1/2}] \tag{6.140}$$

that is, it is the spectral norm of $(PQ)^{1/2}$. A more informative interpretation of
$\| G \|_H$ is the following. We can define the **Hankel operator** corresponding to G
as the following operator on functions $v(t)$:

$$(\Gamma_G v)(t) = \int_0^\infty C \exp[A(t + \tau)] B v(\tau) d\tau \tag{6.141}$$

Note that this is not the same as the familiar convolution operator, because of
the $+$ sign in the kernel. Suppose that we have an input $u(t)$ which is non-
zero only for $-\infty < t \leqslant 0$, and let the resulting output over the interval
$0 \leqslant t < \infty$ be $y(t)$. Then Γ_G is the operator which maps $u(t)$ into $y(t)$; that is, it
maps **past inputs** into **future outputs**. This can be shown as follows.
Let $v(t) = u(-t)(0 \leqslant t < \infty)$. Then

$$(\Gamma_G v)(t) = \int_0^\infty C \exp[A(t + \tau)] Bu(-\tau) d\tau \tag{6.142}$$

$$= \int_{-\infty}^0 C \exp[A(t - \tau)] Bu(\tau) d\tau \tag{6.143}$$

$$= y(t) \tag{6.144}$$

(since $u(t) = 0$ for $t > 0$).

The Hankel norm $\| G \|_H$ can be shown to be the largest singular value
of the Hankel operator Γ_G. We shall not explain here what is meant by a
singular value of an operator, but remark that it can be given a 'largest-gain'
interpretation in the following sense:

$$\| G \|_H = \sup \frac{\| y(t) \|_2}{\| u(t) \|_2} \quad \begin{cases} y(t) = 0, & t < 0 \\ u(t) = 0, & t > 0 \end{cases} \tag{6.145}$$

Note that only the dynamic part (A, B, C) of G influences the Hankel
norm, which is independent of the matrix D.

We shall now state a theorem which places a lower bound on the
achievable error in the Hankel approximation problem, in terms of the
Hankel norm. In the next section we shall show that a particular approxima-

tion achieves this lower bound, and hence solves the Hankel approximation problem.

> **Theorem 6.5** (Glover, 1984):   Given rational $G \in H_\infty$, then, for any $Y \in H_\infty$,
>
> $$\|G\|_H \leqslant \|G + Y^*\|_\infty \tag{6.146}$$
>
> *Proof*:   Omitted – see Glover (1984).

6.6.2   Glover's algorithm

We now present an algorithm, developed by Glover (1984), for solving the Hankel approximation problem

$$\underset{Y \in H_\infty}{\text{minimize}} \|G + Y^*\|_\infty, \quad G \in H_\infty \tag{6.147}$$

when G is square.

Step 1   Obtain a **balanced realization** of G: (A, B, C, D), namely one such that the Lyapunov equations

$$A\Sigma + \Sigma A^T + BB^T = 0 \tag{6.148}$$

and

$$A^T\Sigma + \Sigma A + C^T C = 0 \tag{6.149}$$

have a common diagonal solution

$$\Sigma = \begin{bmatrix} \sigma_1 I & 0 \\ 0 & \Sigma_1 \end{bmatrix} \tag{6.150}$$

where $\Sigma_1 = \text{diag}\{\sigma_2, \ldots, \sigma_n\}$, and $\sigma_1 > \sigma_2 \geqslant \ldots \geqslant \sigma_n > 0$. It can be shown that such a balanced realization always exists (if $G \in H_\infty$); for a method of finding it see Exercise 6.13 and Section 8.3.9.

   Note that

$$\|G\|_H = \sigma_1 \tag{6.151}$$

Step 2   Partition A, B and C conformally with Σ:

$$A = \begin{bmatrix} A_{11} & A_{12} \\ A_{21} & A_{22} \end{bmatrix}, \quad B = \begin{bmatrix} B_1 \\ B_2 \end{bmatrix}, \quad C = [C_1 \ C_2] \tag{6.152}$$

Let $\Gamma = \Sigma_1^2 - \sigma_1^2 I$ (where the unit matrix has the same dimensions as Σ_1 – not usually the same as the unit matrix appearing in (6.150)), and choose a U such that $UU^T = I$ and $B_1 = -C_1^T U$; such a U always exists, since $B_1 B_1^T = C_1^T C_1$.

Step 3 Define the following matrices:

$$\hat{A} = \Gamma^{-1}(\sigma_1^2 A_{22}^T + \Sigma_1 A_{22} \Sigma_1 - \sigma_1 C_2^T U B_2^T) \tag{6.153}$$

$$\hat{B} = \Gamma^{-1}(\Sigma_1 B_2 + \sigma_1 C_2^T U) \tag{6.154}$$

$$\hat{C} = -C_2 \Sigma_1 - \sigma_1 U B_2^T \tag{6.155}$$

$$\hat{D} = -D + \sigma_1 U \tag{6.156}$$

and let Y_{opt} have the realization $(-\hat{A}^T, -\hat{C}^T, \hat{B}^T, \hat{D}^T)$.
Then we have

Theorem 6.6:

(1) $Y_{\text{opt}} \in H_\infty$ (6.157)

(2) $\| G + Y_{\text{opt}}^* \|_\infty = \| G \|_H$ (6.158)

(3) $(G + Y_{\text{opt}}^*)(G + Y_{\text{opt}}^*)^* = \sigma_1^2 I$ (6.159)

REMARKS

(1)   Parts (1) and (2) of the theorem, together with Theorem 6.5, show that Y_{opt} is a solution to the Hankel approximation problem. Part (3) shows that the approximation error $G + Y^*$ is an all-pass function, and hence that the magnitude of the error is the same at all frequencies. In applications to H_∞ problems this has the consequence that $\bar{\sigma}[F_1(P, K_{\text{opt}})(j\omega)]$ is constant.

(2)   As remarked earlier, when solving H_∞ problems we may be satisfied by finding a $Y \in H_\infty$ such that $\| G + Y^* \|_\infty \leqslant m$, where m is some specified real number, and not be particularly interested in finding the optimal Y which minimizes $\| G + Y^* \|_\infty$. A suitable Y will exist if and only if $m \geqslant \sigma_1$, and can be found from the following realization for Y^*:

$$\Gamma_m = \Sigma^2 - m^2 I \tag{6.160}$$

$$\hat{A} = \Gamma_m^{-1}(m^2 A^T + \Sigma A \Sigma) \tag{6.161}$$

$$\hat{B} = \Gamma_m^{-1} \Sigma B \tag{6.162}$$

$$\hat{C} = -C\Sigma \tag{6.163}$$

$$\hat{D} = -D \tag{6.164}$$

Safonov *et al.* (1987) have pointed out that in this case there is no need to find a balanced realization: it is sufficient to find the controllability and observability gramians $P = P^T$ and $Q = Q^T$ which solve the Lyapunov equations

$$AP + PA^T + BB^T = 0 \tag{6.165}$$

and

$$A^T Q + QA + C^T C = 0 \tag{6.166}$$

for any realization of G, and then replace (6.160) by

$$\Gamma_m = QP - m^2 I \tag{6.167}$$

and replace $\Sigma A \Sigma$ in (6.161) by QAP, Σ in (6.162) by Q and Σ in (6.163) by P.

Proof: The proof is omitted. It is long but quite straightforward, and is given by Glover (1984).

Theorem 6.6 assumes that G is square. If it is not square, we augment G with rows or columns of zeros to make it square. Suppose that we need to augment it with columns, so that $G_{sq} = [G \ \ 0]$ is square. We then use Theorem 6.6 to obtain a square matrix $Y \in H_\infty$, such that $\|G_{sq} + Y^*\|_\infty = \|G_{sq}\|_H = \|G\|_H$. Now partition Y so that

$$\|G_{sq} + Y^*\|_\infty = \|[G + Y_1^* \quad 0 + Y_2^*]\|_\infty \tag{6.168}$$

$$= \|G\|_H$$

Then

$$\|G + Y_1^*\|_\infty \leqslant \|G\|_H \tag{6.169}$$

and hence, by Theorem 6.5,

$$\|G + Y_1^*\|_\infty = \|G\|_H \tag{6.170}$$

Therefore Y_1 is a solution to the Hankel approximation problem.

The theory and results of the Hankel approximation have a number of applications in addition to solving H_∞ problems. The principal one is to the approximation of high-order state-space models by simpler ones (Glover, 1984).

6.7 The Glover–Doyle algorithm for general H_∞ problems

The algorithm to be presented in this section is taken from Glover and Doyle (1988); a proof of its validity is outlined by Doyle *et al.* (1988). At the time of writing these results are very new, and several alternative proofs are being developed. It is too early to say which of these will eventually emerge as the shortest, clearest or most elegant.

The algorithm yields a family of stabilizing controllers K, such that

$$\| F_1(P,K) \|_\infty < \gamma \tag{6.171}$$

for some chosen value of γ, if any such controllers exist. We assume that a realization of P is given by (6.120) and, using earlier notation, that the dimensions of P_{11}, P_{12}, P_{21} and P_{22} are $p_1 \times m_1$, $p_1 \times m_2$, $p_2 \times m_1$ and $p_2 \times m_2$, respectively. Furthermore we assume that (A, B_2, C_2, D_{22}) – namely the plant – is stabilizable and detectable, so that stabilizing controllers exist, and that $\operatorname{rank}(D_{12}) = m_2$ and $\operatorname{rank}(D_{21}) = p_2$, in order to ensure the realizability of controllers which solve (6.171).

To simplify the notation, we perform non-singular transformations on u and y (see Figure 6.2):

$$\tilde{u} = S_u u, \qquad \tilde{y} = S_y y \tag{6.172}$$

and unitary transformations on w and z:

$$\tilde{w} = T_w w, \qquad \tilde{z} = T_z z \tag{6.173}$$

$(T_w T_w^{\mathrm{H}} = I = T_z T_z^{\mathrm{H}})$. This gives

$$\begin{bmatrix} \tilde{z} \\ \tilde{y} \end{bmatrix} = \begin{bmatrix} \tilde{P}_{11} & \tilde{P}_{12} \\ \tilde{P}_{21} & \tilde{P}_{22} \end{bmatrix} \begin{bmatrix} \tilde{w} \\ \tilde{u} \end{bmatrix} \tag{6.174}$$

$$= \begin{bmatrix} T_z P_{11} T_w^{\mathrm{H}} & T_z P_{12} S_u^{-1} \\ S_y P_{21} T_w^{\mathrm{H}} & S_y P_{22} S_u^{-1} \end{bmatrix} \begin{bmatrix} \tilde{w} \\ \tilde{u} \end{bmatrix} \tag{6.175}$$

and, with a feedback controller $\tilde{K} = S_u K S_y^{-1}$ in place,

$$\tilde{w} = F_1(\tilde{P}, \tilde{K}) \tilde{z} = [T_w F_1(P,K) T_z^{\mathrm{H}}] \tilde{z} \tag{6.176}$$

Since T_w and T_z are unitary, we have

$$\| F_1(\tilde{P}, \tilde{K}) \|_\infty = \| F_1(P,K) \|_\infty \tag{6.177}$$

We can therefore replace P by $\tilde{P}$, find a suitable controller $\tilde{K}$, and then obtain the required controller

$$K = S_u^{-1} \tilde{K} S_y \tag{6.178}$$

Now, the point of these transformations is that we can choose T_z and S_u such that

$$\tilde{D}_{12} = T_z D_{12} S_u^{-1} = \begin{bmatrix} 0 \\ I \end{bmatrix} \tag{6.179}$$

and T_w and S_y such that

$$\tilde{D}_{21} = S_y D_{21} T_w^{\mathrm{H}} = [0\ I] \tag{6.180}$$

One way of finding the required transformations to achieve this is to use the singular-value decompositions of D_{12} and D_{21} to obtain row and column compressions, as explained in Chapter 8. Since this can always be done (because of our assumptions about the ranks of D_{12} and D_{21}) we shall assume that these transformations have already been performed, and therefore that

$$D_{12} = [0\ I]^{\mathrm{T}} \quad \text{and} \quad D_{21} = [0\ I] \tag{6.181}$$

We assume that D_{11} is partitioned as

$$D_{11} = \begin{bmatrix} D_{1111} & D_{1112} \\ D_{1121} & D_{1122} \end{bmatrix} \tag{6.182}$$

where D_{1122} has m_2 rows and p_2 columns.

For the time being we assume that the plant is strictly proper, namely that

$$D_{22} = 0 \tag{6.183}$$

Finally we assume that

$$\mathrm{rank} \begin{bmatrix} A - j\omega I & B_2 \\ C_1 & D_{12} \end{bmatrix} = n + m_2 \tag{6.184}$$

and

$$\mathrm{rank} \begin{bmatrix} A - j\omega I & B_1 \\ C_2 & D_{21} \end{bmatrix} = n + p_2 \tag{6.185}$$

for all real ω, where A has dimensions $n \times n$.

It is useful to introduce the notation

$$D_{1*} = [D_{11} \ D_{12}] \quad \text{and} \quad D_{*1} = [D_{11}^T \ D_{21}^T]^T \tag{6.186}$$

and define

$$R = D_{1*}^T D_{1*} - \begin{bmatrix} \gamma^2 I_{m_1} & 0 \\ 0 & 0 \end{bmatrix} \tag{6.187}$$

and

$$\tilde{R} = D_{*1} D_{*1}^T - \begin{bmatrix} \gamma^2 I_{p_1} & 0 \\ 0 & 0 \end{bmatrix} \tag{6.188}$$

Now we define X_∞ and Y_∞ to be the stabilizing solutions of the following Riccati equations:

$$X_\infty (A - BR^{-1} D_{1*}^T C_1) + (A - BR^{-1} D_{1*}^T C_1)^T X_\infty$$
$$- X_\infty BR^{-1} B^T X_\infty + C_1^T (I - D_{1*} R^{-1} D_{1*}^T) C_1 = 0 \tag{6.189}$$

$$Y_\infty (A - B_1 D_{*1}^T \tilde{R}^{-1} C)^T + (A - B_1 D_{*1}^T \tilde{R}^{-1} C) Y_\infty$$
$$- Y_\infty C^T \tilde{R}^{-1} C Y_\infty + B_1 (I - D_{*1}^T \tilde{R}^{-1} D_{*1}) B_1^T = 0 \tag{6.190}$$

By 'stabilizing' solutions we mean that $X_\infty = X_\infty^T$ and $Y_\infty = Y_\infty^T$, and the matrices $A + BF$ and $A + HC$ have all their eigenvalues in the open left half-plane, where F and H are defined by (6.191) and (6.192) below.

Equations (6.189) and (6.190) have the same form as the Riccati equations encountered in Chapter 5, except that the constant and quadratic terms are generally not positive-semidefinite. For example, $BR^{-1} B^T$, which appears in the quadratic term in (6.189), may be indefinite if γ is large enough (see equation (6.187)). Because of this, stabilizing solutions to these equations may not exist; non-existence of one or both solutions indicates that there is *no* stabilizing controller which satisfies (6.171). The equations can be solved using the algorithms given in Chapter 8; more specific algorithms have been developed recently which exploit the special form of (6.189) and (6.190), but those described in Chapter 8 are sufficient in most cases.

Having obtained X_∞ and Y_∞, we define

$$F = -R^{-1}(D_{1*}^T C_1 + B^T X_\infty) \tag{6.191}$$

$$H = -(B_1 D_{*1}^T + Y_\infty C^T)\tilde{R}^{-1} \tag{6.192}$$

and partition them:

$$F = [F_{11}^T \quad F_{12}^T \quad F_2^T]^T \tag{6.193}$$

$$H = [H_{11} \quad H_{12} \quad H_2] \tag{6.194}$$

with F_{11}, F_{12} and F_2 having $m_1 - p_2, p_2$ and m_2 rows, respectively, and H_{11}, H_{12} and H_2 having $p_1 - m_2, m_2$ and p_2 columns, respectively.

Now we can state the main result (in which $\rho(.)$ denotes the spectral radius):

Theorem 6.7 (Glover and Doyle, 1988):

(1)   A stabilizing controller exists, such that $\|F_l(P,K)\|_\infty < \gamma$, if and only if

(a) $\gamma > \max\{\bar{\sigma}[D_{1111}, D_{1112}], \quad \bar{\sigma}[D_{1111}^T, D_{1121}^T]\}$ $\tag{6.195}$

and

(b)   there exist solutions $X_\infty \geqslant 0$ and $Y_\infty \geqslant 0$ of (6.189) and (6.190), respectively, such that

$$\rho(X_\infty Y_\infty) < \gamma^2 \tag{6.196}$$

(2)   If (a) and (b) above are satisfied, then all (rational) stabilizing controllers K, for which $\|F_l(P,K)\|_\infty < \gamma$, are given by

$$K = F_l(K_a, \Phi), \tag{6.197}$$

for any rational $\Phi \in H_\infty$ such that $\|\Phi\|_\infty < \gamma$ where K_a has the realization

$$K_a: \left(\hat{A}, [\hat{B}_1 \ \hat{B}_2], \begin{bmatrix} \hat{C}_1 \\ \hat{C}_2 \end{bmatrix}, \begin{bmatrix} \hat{D}_{11} & \hat{D}_{12} \\ \hat{D}_{21} & 0 \end{bmatrix} \right) \tag{6.198}$$

and

$$\hat{D}_{11} = -D_{1121}D_{1111}^T(\gamma^2 I - D_{1111}D_{1111}^T)^{-1}D_{1112} - D_{1122} \tag{6.199}$$

$\hat{D}_{12}$ and $\hat{D}_{21}$ are any matrices (of size $m_2 \times m_2$ and $p_2 \times p_2$, respectively) such that

$$\hat{D}_{12}\hat{D}_{12}^T = I - D_{1121}(\gamma^2 I - D_{1111}^T D_{1111})^{-1}D_{1121}^T \tag{6.200}$$

$$\hat{D}_{21}^T\hat{D}_{21} = I - D_{1112}^T(\gamma^2 I - D_{1111}D_{1111}^T)^{-1}D_{1112} \tag{6.201}$$

(these can be obtained by Cholesky factorization, for example – see Golub and Van Loan, 1983); the remaining matrices in (6.198) are given by

$$\hat{B}_2 = (B_2 + H_{12})\hat{D}_{12} \tag{6.202}$$

$$\hat{C}_2 = -\hat{D}_{21}(C_2 + F_{12})Z \tag{6.203}$$

$$\hat{B}_1 = -H_2 + \hat{B}_2\hat{D}_{12}^{-1}\hat{D}_{11} \tag{6.204}$$

$$\hat{C}_1 = F_2 Z + \hat{D}_{11}\hat{D}_{21}^{-1}\hat{C}_2 \tag{6.205}$$

$$\hat{A} = A + HC + \hat{B}_2\hat{D}_{12}^{-1}\hat{C}_1 \tag{6.206}$$

where

$$Z = (I - \gamma^2 Y_\infty X_\infty)^{-1} \tag{6.207}$$

One member of the family of solutions is obtained by taking $\Phi \equiv 0$, and this is called the **central** or **maximum-entropy controller**. It has the realization

$$K: (\hat{A}, \hat{B}_1, \hat{C}_1, \hat{D}_{11})$$

and so has the same state dimension as the realization used for P.

The above algorithm assumes that $D_{22} = 0$, namely that the plant is strictly proper. If this is not the case, it is sufficient to do the following. Apply the algorithm to

$$\hat{P} = P - \begin{bmatrix} 0 & 0 \\ 0 & D_{22} \end{bmatrix} \tag{6.208}$$

and obtain a controller $\hat{K}$. Then form the controller

$$K = \hat{K}(I + D_{22}\hat{K})^{-1} \tag{6.209}$$

Figure 6.16 shows the structure of P and K in terms of $\hat{P}$ and $\hat{K}$. It is straightforward to show that K stabilizes P if $\hat{K}$ stabilizes $\hat{P}$ (from Figure 6.16, for example), and that

$$F_l(P, K) = F_l(\hat{P}, \hat{K}) \tag{6.210}$$

The only slight complication is that any controller $\hat{K}(s)$ for which $\det(I + D_{22}\hat{K}(\infty)) = 0$ is not admissible.

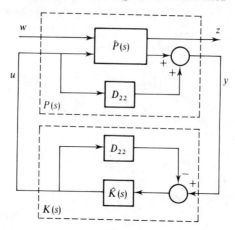

Figure 6.16   Controller for a plant which is not strictly proper.

To find a controller which minimizes $\| F_1(P, K) \|_\infty$ it is necessary to use the algorithm iteratively, reducing γ until the limiting value γ_0 is reached, such that $\rho(X_\infty Y_\infty) = \gamma_0^2$, or until one or other of the two Riccati equations (6.189) and (6.190) fails to have a positive semi-definite solution.

It can be shown that, if γ is increased to very large values, then the central controller generated by the Glover–Doyle algorithm converges to the LQG controller which minimizes $\| F_1(P, K) \|_2$. There are many connections between the H_∞ theory and LQG (H_2) theory, some of which are highlighted by Doyle *et al.* (1988). The most obvious similarities are that both approaches require the solution of two Riccati equations, and both yield a controller with the same state dimension as the original problem description – namely the state dimension of the plant, together with any augmenting systems.

6.8   Design example

6.8.1   Design specification

We apply the H_∞ design method to the AIRC model used in earlier chapters and defined in the Appendix. The design specification is essentially the same as in Chapter 5: a closed-loop bandwidth of about $10\,\text{rad}\,\text{s}^{-1}$, with reasonably damped responses, and zero sensitivity at zero frequency. Since we have the benefit of having already found some controller designs for this plant, we can be more specific and aim to improve on the LQG/LTR design we obtained in the previous chapter.

In particular, we shall see whether we can keep $\max(\| S \|_\infty, \| T \|_\infty)$ below the level achieved with the LQG/LTR controller (where S is the sensitivity and T the closed-loop transfer function), while reducing $\bar{\sigma}(S)$ more quickly below $10 \, \text{rad} \, \text{s}^{-1}$, and reducing $\bar{\sigma}(T)$ more quickly above $10 \, \text{rad} \, \text{s}^{-1}$. In order to do this, we shall solve the mixed performance and robustness problem defined in Section 6.3.1; that is, we take

$$F_1(P, K) = \begin{bmatrix} W_1 S \\ W_2 T \end{bmatrix} \tag{6.211}$$

We choose the weights to be

$$W_1(s) = w_1(s)I_3 \quad \text{and} \quad W_2(s) = w_2(s)I_3 \tag{6.212}$$

where $w_1(s)$ and $w_2(s)$ are scalar transfer functions.
The functions

$$w_1(s) = \frac{(s+6)^2}{s(s+0.6)} \tag{6.213}$$

and

$$w_2(s) = \frac{(s+10)(s+50)}{500} \tag{6.214}$$

penalize S and T, respectively, slightly more severely than the weights which are implicit in the trade-off obtained with the LQG/LTR design (see

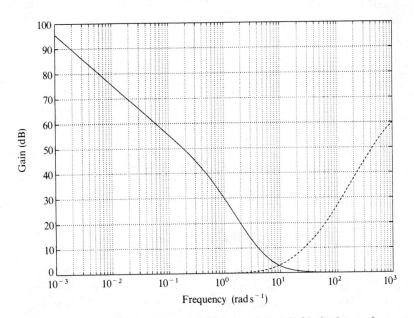

Figure 6.17 Weightings on S (solid curve) and T (dashed curve).

Figure 5.17), and both have a gain close to unity near $10 \, \text{rad} \, \text{s}^{-1}$. We shall have to modify these functions slightly in order to shift the pole of $w_1(s)$ away from the origin, and to make $w_2(s)$ proper; details of the required modifications will be given below. The gains of the *modified* weighting functions are shown in Figure 6.17.

6.8.2 Application of the Glover–Doyle algorithm

We saw in Section 6.3.1 that we require

$$P = \begin{bmatrix} P_{11} & P_{12} \\ P_{21} & P_{22} \end{bmatrix} = \left[\begin{array}{c|c} W_1 & -W_1 G \\ 0 & W_2 G \\ \hline I & -G \end{array} \right] \qquad (6.215)$$

where G is the transfer function of the plant, so that, in the notation of Section 6.7, we have $m_1 = 3$, $m_2 = 3$, $p_1 = 6$ and $p_2 = 3$. (We have a '2-block' problem, since P_{21} is square.) If the plant has the realization $G: (A, B, C, D)$, and each of the weighting functions has the realization $W_i: (A_i, B_i, C_i, D_i)$ $(i = 1, 2)$, then it is straightforward to show that P has the realization $P: (A_p, B_p, C_p, D_p)$, where

$$A_p = \begin{bmatrix} A & 0 & 0 \\ B_1 C & A_1 & 0 \\ B_2 C & 0 & A_2 \end{bmatrix}, \qquad B_p = \begin{bmatrix} 0 & B \\ -B_1 & B_1 D \\ 0 & B_2 D \end{bmatrix} \qquad (6.216)$$

$$C_p = \begin{bmatrix} -D_1 C & -C_1 & 0 \\ D_2 C & 0 & C_2 \\ -C & 0 & 0 \end{bmatrix}, \qquad D_p = \begin{bmatrix} D_1 & -D_1 D \\ 0 & D_2 D \\ I_3 & -D \end{bmatrix}$$

The partitions in (6.216) correspond to the partitions between P_{11}, P_{12}, P_{21} and P_{22}. (This is not the only possible realization of P, of course, but it is the easiest to obtain.)

If each of the scalar weighting functions $w_i(s)$ has the realization w_i: $(\alpha_i, \beta_i, \gamma_i, \delta_i)$ then realizations of W_i are given by

$$\begin{aligned} A_i &= \text{diag}\{\alpha_i\}, & B_i &= \text{diag}\{\beta_i\} \\ C_i &= \text{diag}\{\gamma_i\}, & D_i &= \delta_i I_3 \end{aligned} \qquad (6.217)$$

where the matrices α_i, β_i and γ_i are repeated three times on each principal diagonal, since our plant has three outputs. Clearly w_1 requires a two-state realization, and we shall have to add two poles to w_2 in order to make it realizable, so it too will require two states. Each of A_1 and A_2 is therefore a 6×6 matrix, and, since the plant has five states, A_p has dimensions 17×17. (The other dimensions are: $B_p \in \mathbb{R}^{17 \times 6}$, $C_p \in \mathbb{R}^{9 \times 17}$ and $D_p \in \mathbb{R}^{9 \times 6}$.)

Inserting poles at -1000 makes a negligible difference to w_2 over the frequency range of interest to us, so in place of (6.214) we shall use

$$w_2(s) = \frac{2000(s+10)(s+50)}{(s+1000)^2} \tag{6.218}$$

the gain being chosen to given $w_2(0) = 1$.

Now we need to consider the various rank conditions which are required by the Glover–Doyle algorithm. First of all, we require that rank $[-(D_1 D)^T (D_2 D)^T]^T = m_2 = 3$, and rank $(I_3) = p_2 = 3$. The second of these is satisfied, but the first is not, since $D = 0$ for our plant. We therefore need to modify D so that the rank condition is met, but in such a way that the behaviour of the plant is changed very little over significant frequencies. Since $\underline{\sigma}(G(\mathrm{j}100)) \approx 5 \times 10^{-5}$, then replacing D by

$$\tilde{D} = 10^{-5} I_3 \tag{6.219}$$

should have no perceptible effect on the plant's behaviour below $100\,\mathrm{rad\,s^{-1}}$. We therefore replace D_p, as defined in (6.216), by

$$D_p = \left[\begin{array}{c|c} D_1 & -D_1\tilde{D} \\ 0 & B_1\tilde{D} \\ \hline I_3 & -D \end{array} \right] \tag{6.220}$$

Note that we leave $-D$ unchanged in the lower right corner of D_p, since the algorithm is actually simplified if P_{22} is strictly proper (condition (6.183)). Rank condition (6.185) becomes, at $\omega = 0$,

$$\mathrm{rank} \begin{bmatrix} A & 0 & 0 & 0 \\ B_1 C & A_1 & 0 & -B_1 \\ B_2 C & 0 & A_2 & 0 \\ -C & 0 & 0 & I \end{bmatrix} = n + p_2 = 20 \tag{6.221}$$

It is easy to see that this condition is violated, since both A and A_1 have eigenvalues at zero. We therefore use

$$A_p = \begin{bmatrix} \tilde{A} & 0 & 0 \\ B_1 C & \tilde{A}_1 & 0 \\ B_2 C & 0 & A_2 \end{bmatrix} \tag{6.222}$$

in place of the earlier definition (6.216), and define $\tilde{A}$ and $\tilde{A}_1$ as follows. Setting $\tilde{A} = A$, except for $\tilde{A}(1, 1) = \lambda$, replaces the eigenvalue at 0 by one at λ and leaves all the other eigenvalues of A unchanged. Choosing $\lambda = -0.001$ has very little effect on the behaviour of the plant, except at very low frequencies. Examination of the principal gains of both the original plant $G: (A, B, C, D)$, and the modified version $\tilde{G}: (\tilde{A}, B, C, \tilde{D})$ shows that almost no difference is perceptible between 0.001 and $100 \, \text{rad} \, \text{s}^{-1}$.

The origin eigenvalues of A_1 are shifted by changing $w_1(s)$; clearly we need to insert a pole at some very low frequency instead of zero. If we wish to force three 'integrators' into the controller, rather than two (similar to what we did in the earlier designs), then this frequency should be lower than $|\lambda|$, so that the controller is forced to inject gain down to this frequency (since the plant gain levels out below $|\lambda|$). We therefore replace (6.213) by

$$w_1(s) = \frac{(s+6)^2}{(s+0.00006)(s+0.6)} \tag{6.223}$$

and use the corresponding $\tilde{A}_1$.

Rank condition (6.184) is satisfied at $\omega = 0$, even without the changes made above. Conditions (6.184) and (6.185) are equivalent to the conditions that the systems with realizations (A, B_2, C_1, D_{12}) and (A, B_1, C_2, D_{21}) (using notation as in (6.184) and (6.185)) have no transmission zeros, and no pole–zero cancellations, on the imaginary axis. One way of checking this is to plot the principal gains of these systems across a wide range of frequencies, and see whether any of them approaches zero (since a transfer function loses rank at a transmission zero). Note that the first of these systems is not square, and so is extremely unlikely to have any transmission zeros (see Chapter 2). Such checks reveal that there are no further problems with these rank conditions.

We now have a problem description which can be submitted to the Glover–Doyle algorithm. We shall not give a detailed description of the computations required, except to note that, since $m_1 = p_2$, the matrices D_{1111} and D_{1121} (in (6.182)) do not exist, and therefore (6.199)–(6.201) simplify to

$$\hat{D}_{11} = -D_{1122}, \qquad \hat{D}_{12} = I_3, \qquad \hat{D}_{21} = I_3 \tag{6.224}$$

respectively.

6.8.3 Adjustment of γ and weights

Before using the algorithm, we need to choose a suitable value of γ such that $\| F_l(P, K) \|_\infty < \gamma$. From the LQG/LTR design we know that $\max(\| S \|_\infty, \| T \|_\infty) \leqslant 4 \, \text{dB}$, approximately, should be feasible. From Figure 6.17 we see that both $|w_1(j\omega)|$ and $|w_2(j\omega)|$ take values close to $3 \, \text{dB}$

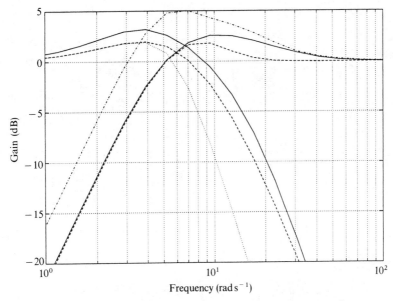

Figure 6.18 Principal gains of S and T with $\gamma = 10^{-6}$.

near $\omega = 10$, which is where $\bar{\sigma}(S)$ and $\bar{\sigma}(T)$ are likely to have their peak values. We can therefore estimate that a value in the vicinity of $7\,\text{dB}$, or 2.2, should be possible. This estimate may be optimistic since our weighting $w_2(s)$ is intended to force a faster 'roll-off' above $\omega = 10$ than was achieved by the LQG/LTR design, and this may force larger peaks of S or T.

We begin by choosing a very large value of γ, 10^6. As remarked in Section 6.7, the limiting solution of (6.171), as γ is increased indefinitely, is in fact the 'LQG' solution which minimizes $\| F_1(P, K) \|_2$. It is useful to start with this solution, if only to 'debug' the problem definition. The algorithm yields a positive-semidefinite stabilizing X_∞ (solution of (6.189)), as expected, but, rather surprisingly at first, $Y_\infty = 0$. In fact this is not unusual for 2-block problems, and depends on the locations of the transmission zeros of certain subsystems. We take the (17-state) central or maximum-entropy controller defined by 6.199 and (6.204–6), and this gives principal gains of S and T as shown in Figure 6.18. This already looks like a useful design, with low- and high-frequency behaviour much as expected, and the bandwidth about right. The peak value of $\bar{\sigma}(S)$ is $5\,\text{dB}$, which occurs at $7\,\text{rad s}^{-1}$; since $|w_1(j7)| \approx 5\,\text{dB}$, this design certainly gives $\| F_1(P, K) \|_\infty \geqslant 10\,\text{dB} = 3.16$. Note that $\bar{\sigma}(F_1(P, K))$ does not reach its peak value at $\omega = 7$, however; for example at $\omega = 1$ we have $\bar{\sigma}(S) = -17\,\text{dB}$ and $|w_1(j)| = 30\,\text{dB}$, so $\| F_1(P, K) \|_\infty \geqslant 13\,\text{dB} = 4.47$.

Starting with this initial design, we reduce γ until we find the limiting value. The evidence so far suggests we should look in the range $1 < \gamma < 10$; in fact condition (6.195) indicates that we can restrict the search to $\sqrt{2} < \gamma < 10$,

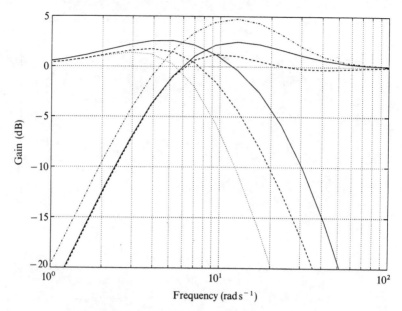

Figure 6.19   Principal gains of S and T with $\gamma = 3.5$.

since $\bar{\sigma}(D_{1112}) = \sqrt{2}$ for our problem. Since we always obtain the solution $Y_\infty = 0$, condition (6.196), namely $\rho(X_\infty Y_\infty) < \gamma^2$, is always satisfied. The limiting value of γ is therefore indicated by (6.189) failing to have a positive-semidefinite stabilizing solution. As the limiting value is approached, the solution X_∞ becomes unbounded, and when it is exceeded (γ too small) only indefinite solutions can be obtained.

A little iteration 'by hand' showed that the smallest value which can be achieved is approximately[4] $\gamma = 3.5$. With this value of γ, the principal gains of S and T are as shown in Figure 6.19. The effect of reducing γ is to increase the bandwidth a little, but $\|S\|_\infty$ and $\|T\|_\infty$ have hardly been affected. This has occurred, of course, because $\bar{\sigma}(W_1 S)$ and $\bar{\sigma}(W_2 T)$ achieve their peak values at frequencies other than those at which $\bar{\sigma}(S)$ and $\bar{\sigma}(T)$ reach *their* peak values.

The results obtained so far indicate that the weights $w_1(s)$ and $w_2(s)$ do not quite reflect our intentions. They have certainly forced $\bar{\sigma}(S)$ and $\bar{\sigma}(T)$ to reduce rapidly for $\omega < 10$ and $\omega > 10$, respectively, but they have not given enough emphasis to the reduction of $\|S\|_\infty$ and $\|T\|_\infty$. At this stage it is necessary to go back to the beginning and adjust the weights to reflect the designer's intentions more accurately. With well-conceived software this need not be very time-consuming, but we shall limit ourselves to just one further iteration. Figure 6.19 shows that the current weights place more emphasis on keeping $\bar{\sigma}(T)$ small than on keeping $\bar{\sigma}(S)$ small, so if we just increase $w_1(s)$ by a constant gain, without changing its dynamics, we should be able to decrease $\|S\|_\infty$ relative to $\|T\|_\infty$, and thus perhaps reduce $\max(\|S\|_\infty, \|T\|_\infty)$.

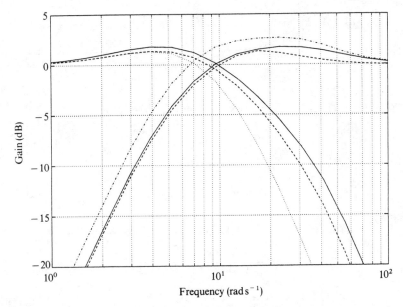

Figure 6.20 Principal gains of S and T with w_1 increased by a factor of 4, and $\gamma = 7.8$.

With W_1 increased to $W_1(s) = 4w_1(s)I$, the limiting value of γ is about 7.8, and the resulting gains of S and T are as shown in Figure 6.20. Both $\|S\|_\infty$ and $\|T\|_\infty$ have been reduced significantly, to about 3 and 2 dB, respectively. As expected, $\bar{\sigma}(S)$ has been reduced at frequencies lower than $10\,\mathrm{rad\,s}^{-1}$, while $\bar{\sigma}(T)$ has been increased at higher frequencies. Figure 6.21 shows the step responses obtained with this design. Note that the steep 'roll-off' of $\bar{\sigma}(T)$ which has been achieved (compared with previous designs) has resulted in the initial slopes of the responses being smaller. This indicates that the control signals required are considerably smaller than in the earlier designs.

6.8.4 Comparison with previous designs

When comparing the H_∞ procedure described for this example with the LQG procedure used in the design example in Section 5.8, it should be remembered that we have described one particular way of using the H_∞ approach, and one particular way of using the LQG approach. Many of the differences which are apparent arise from the choice of these particular variants of the methods, and not from the use of the H_∞ rather than the LQG problem formulation.

The only difference which is a result of choosing the H_∞ procedure is that it is easier to control the peak values of $\bar{\sigma}(S)$ and $\bar{\sigma}(T)$, whereas the LQG approach manipulates all the principal gains simultaneously. This is a

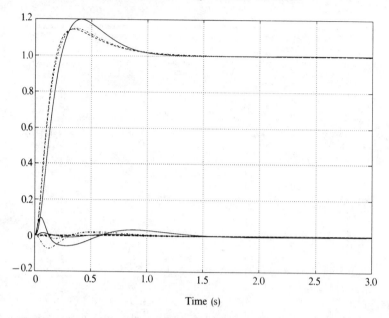

Figure 6.21 Closed-loop step responses to step demand on output 1 (solid curves), output 2 (dashed curves) and output 3 (dash–dotted curves).

significant difference, of course, and the prime justification for using the H_∞ approach is the claim that it is precisely these peak values which one usually needs to control.

However, most of the apparent differences between the two design examples result from the use of a closed-loop approach here, and an open-loop approach in Section 5.8. This accounts for the amount of attention we had to devote to the choice of weighting functions, since these really encapsulate all the design decisions. One advantage of the closed-loop approach over the LQG/LTR approach is that it makes it easier to control the behaviour of $\bar{\sigma}(S)$ and $\bar{\sigma}(T)$: we were able to achieve steeper reductions of these – particularly of $\bar{\sigma}(T)$ – than in the LQG/LTR design, *and* to do so with smaller peak values of $\bar{\sigma}(S)$ and $\bar{\sigma}(T)$.

A major disadvantage of the closed-loop approach, however, is one which has not yet become apparent. If we examine the transmission zeros of the controller, or the eigenvalues of the closed-loop system, we discover that the controller cancels the plant's resonant poles at $-0.018 \pm 0.182j$; this may be unacceptable, as discussed in Chapter 4. As usual, this happens because a closed-loop specification has been imposed without regard to the plant's characteristics. Although LQG controllers – such as the one obtained in Section 6.8.3 with $\gamma = 10^6$ – generally do introduce such cancellations, the LQG/LTR procedure avoided doing so, since the open-loop return ratio which was recovered was allowed to exhibit the same resonance as the plant.

Undesirable cancellations can be avoided, even with closed-loop approaches, by using more elaborate problem formulations which result in restraints on the principal gains of transfer functions such as SG (see Section 4.8.2). Alternatively, McFarlane and Glover (1988) have recently proposed an H_∞ design procedure which relies on open-loop gain shaping, and therefore offers the prospect of avoiding such cancellations.

6.9   Review and comments

6.9.1   The Youla parametrization

The development of the Youla parametrization and its consequences – the structure of all stabilizing controllers, for example – has introduced a new theory of fundamental importance into the subject of linear-feedback systems. It is rather surprising, therefore, to find that the parametrization itself was already known in the 1950s, when it was applied in studies of optimal control (Raggazini and Franklin, 1958, Chapter 7). Its deeper implications went unnoticed at the time, however. It was revived in the 1970s by Kucera (1974, p. 116) and Youla *et al.* (1976), its current name being taken from the latter. The general form of the parametrization, as given by Theorem 6.4, was developed by Desoer *et al.* (1980)[5] with the aid of coprime fractional representations. (See also Callier and Desoer (1982).) The state-space interpretations were developed by Doyle (1984); a crucial ingredient of this work was the paper by Nett *et al.* (1984) which first showed how to obtain state-space realizations of fractional representations.

Whereas Kucera (1974) and Youla *et al.* (1976) employed the parametrization for quadratic (H_2) optimization, Zames (1981) introduced it into the solution of H_∞ problems.

6.9.2   Alternative approaches to H_∞ optimal control

In this chapter we have presented effective algorithms for solving H_∞ design problems, and in order to do this, and to link with earlier chapters, we have placed much emphasis on state-space descriptions. One effect of this has been to obscure the fact that much of the *theory* of H_∞ optimization is best understood in the context of complex analysis. That is, it is essentially a frequency-domain theory. Most of the new terminology, such as 'H_∞' itself, comes from that field (see Rudin (1966), for example), and all the early theoretical contributions to the development of the theory (for example Zames, 1981; Zames and Francis, 1983; Francis and Zames, 1984; Francis *et al.*, 1984) were set in that context.

Above all, the complex-analysis setting gives an essential basis for the development of algorithms which can be applied to systems containing time delays, or which need to be represented by irrational transfer functions. Such algorithms are still being developed.

We shall now outline very briefly how complex analysis comes into the feedback design problem. Consider SISO systems first. If the plant has a zero at $s = z_i$ in the right half-plane, then we know that we cannot attempt to cancel that zero by a pole of the controller, if we are to maintain internal stability. Thus we must have

$$T(z_i) = \frac{G(z_i)K(z_i)}{1 - G(z_i)K(z_i)} = 0 \tag{6.225}$$

and hence the sensitivity function $S(s) = 1 - T(s)$ must satisfy

$$S(z_i) = 1 \tag{6.226}$$

If we attempt to minimize $|S(j\omega)|$, as Zames (1981) did, we must do so subject to the constraint (6.226), which acts at each right half-plane zero of the plant. We also require $S(s)$ to be analytic in the right half-plane in order to obtain closed-loop stability. The problem is therefore equivalent to solving an interpolation problem in which the interpolating function is constrained to be analytic in a half-plane, as well as being optimal. This is a standard problem in complex analysis, known as the **Neyanlinna–Pick** problem.

For multivariable systems we have

$$\det T(z_i) = 0 \tag{6.227}$$

and hence

$$\bar{\sigma}[S(z_i)] \geqslant 1 \tag{6.228}$$

We can proceed by exploiting either the theory of matrix-valued Nevanlinna–Pick problems, as have Chang and Pearson (1984), or operator theory, as have Francis et al. (1984) and Francis (1987).

The development given in Sections 6.4 and 6.5 of this chapter follows Doyle (1984) and Chu (1985). The link between Sections 6.4 and 6.5, namely the equivalence of the H_∞ problem to the 'model-matching' problem, was first established by Safonov and Verma (1985).

SUMMARY

This chapter has covered a wide range of topics, all of them intimately connected with the relatively new H_∞ theory of feedback design.

Coprime factorizations of multivariable linear systems have been introduced, and methods of obtaining state-space realizations of such factorizations presented. These factorizations have been used to explore and exhibit the structure of all stabilizing controllers for a system, and to obtain a simple parametrization of all closed-loop transfer functions.

A derivation of the solution of 1-block H_∞ problems has been given, and the Glover–Doyle algorithm has been presented (without derivation) for the solution of general H_∞ problems. The topic of Hankel approximation has been introduced, and it has been shown to have applications in H_∞ optimal control theory and in model approximation.

A design example has been presented.

As was the case in Chapter 5, most of the material in this chapter has again assumed that the plant can be described by a state-space model with finite state dimension.

Notes

1. $F_l(P, K)$ is called a **linear fractional transformation** of P and K. The subscript stands for 'lower', and indicates that K is below P in Figure 6.2. Had we drawn the feedback path containing K above P, then we would have needed the 'upper' transformation

$$F_u(P, K) = P_{22} + P_{21} K (I - P_{11} K)^{-1} P_{12}$$

Of course everything here is a function of s. We shall sometimes write $F_l(P(s), K(s))$, but we shall usually omit the argument.

2. We saw in Chapter 3 that this formulation corresponds to minimizing

$$\| y \|_2^2 = \int_0^\infty y^2(t) dt$$

if the only thing known about the disturbance d is that $\| d \|_2 \leqslant c$, where c is some positive number.

3. From Schur's formula:

$$\det \begin{bmatrix} A & B \\ C & D \end{bmatrix} = \det A \det (D - CA^{-1}B)$$

4. The accuracy with which the limiting value of γ can be determined depends on the sophistication of both the search strategy and the Riccati-equation solver. In this case both were low. The search was performed manually. The Riccati equations were solved by using the function 'LQR', which is provided with the PC-Matlab and Pro-Matlab *Control System Toolbox* software, with the checking for definiteness of the matrices Q and R suppressed. This function is not intended for solving Riccati equations with indefinite terms, but it usually works. It is likely to fail sooner than a more tailor-made algorithm, however, as the limiting value of γ is approached.

5. In works written or co-written by Desoer the parametrization is invariably known as the 'Q parametrization'.

EXERCISES

6.1   If a plant with transfer function G is stable, find an expression for the closed-loop transfer function

$$T = GK(I + GK)^{-1}$$

in terms of the Youla parameter Q.

   If the plant has more inputs than outputs, show that the 'perfect tracking' condition $T = I$ can be approached arbitrarily closely over any desired frequency range (for almost all plants). Why can it not be achieved exactly? What additional condition is required if the number of inputs equals the number of outputs?

6.2   Figure 6.22 shows the structure of **internal-model control** (Garcia and Morari, 1982). The plant G is assumed to be stable. Show that, if $C(s)$ is stable and $G_0(s) = G(s)$, then the closed loop is stable. Show also that any loop compensator $K(s)$ can be interpreted as having the internal-model control structure (assuming that $G_0 = G$ and G is stable), find the corresponding $C(s)$, and hence deduce that $C(s)$ is the same as the Youla parameter $Q(s)$.

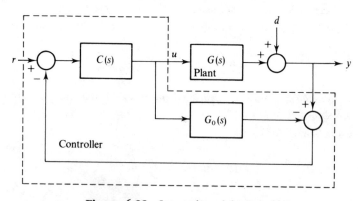

Figure 6.22   Internal-model control.

6.3   If P is stable in Section 6.4.4, show that one can take $T_{11} = P_{11}$, $T_{12} = P_{12}$ and $T_{21} = P_{21}$.

6.4   Suppose that P (in Section 6.4.4) has fractional representations

$$P = N_p M_p^{-1} = \tilde{M}_p^{-1} \tilde{N}_p$$

and that

$$N_p = \begin{bmatrix} N_{11} & N_{12} \\ N_{21} & N_{22} \end{bmatrix}, \qquad \tilde{N}_p = \begin{bmatrix} \tilde{N}_{11} & \tilde{N}_{12} \\ \tilde{N}_{21} & \tilde{N}_{22} \end{bmatrix}$$

the partitioning corresponding to that of P. Given the realizations (6.123) and (6.124) for T_{12} and T_{21}, show that $T_{12} = N_{12}$ and $T_{21} = \tilde{N}_{21}$.

(*Hint*: If F is a stabilizing state feedback for the plant P_{22}, then $[0 \; F^\mathrm{T}]^\mathrm{T}$ is a stabilizing state feedback for P if the realization (6.120) is assumed.)

6.5   Show that the state-space equations corresponding to Figure 6.15 are

$$\begin{bmatrix} \dot{x} \\ \dot{\zeta} \end{bmatrix} = \begin{bmatrix} A & BF \\ -HC & A+BF+HC \end{bmatrix} \begin{bmatrix} x \\ \zeta \end{bmatrix} + \begin{bmatrix} B \\ B \end{bmatrix} y_1$$

$$u = [C \; -C] \begin{bmatrix} x \\ \zeta \end{bmatrix}$$

By defining a new state variable $e = x - \zeta$, so that

$$\begin{bmatrix} x \\ e \end{bmatrix} = \begin{bmatrix} I & 0 \\ I & -I \end{bmatrix} \begin{bmatrix} x \\ \zeta \end{bmatrix}$$

show that the transfer function from y_1 to u_1 is zero.

6.6   If $X(s)$ has realization (A, B, C, D), show that $X^*(s) = X^\mathrm{T}(-s)$ has realization $(-A^\mathrm{T}, -C^\mathrm{T}, B^\mathrm{T}, D^\mathrm{T})$.

6.7   Consider the following feedback system:

Plant: $y = G(u + d)$

Controller: $u = K(r + y)$

Errors: $e_1 = W_1 u, \quad e_2 = W_2(r + y)$

It is known that $\|r\|_2 \leqslant 1$ and $\|d\|_2 \leqslant 1$, and it is desired to design K so as to minimize

$$\left\| \begin{bmatrix} e_1 \\ e_2 \end{bmatrix} \right\|_2$$

Show that this can be formulated as a standard H_∞ problem with

$$P_{11} = \begin{bmatrix} 0 & 0 \\ W_2 & W_2 G \end{bmatrix}, \qquad P_{12} = \begin{bmatrix} W_1 \\ W_2 G \end{bmatrix}$$

$$P_{21} = [I \ G], \qquad\qquad P_{22} = G$$

If G is stable (and W_1 and W_2 are chosen to be stable, of course), show that the resulting optimization problem reduces to

$$\min_{Q \in H_\infty} \left\| \begin{bmatrix} R_{11} + Q \\ R_{21} \end{bmatrix} \right\|_\infty$$

even though P_{21} is not square.

(*Hint*: Show that, in this case, $R_{12} = R_{22} = 0$. The result actually holds if G is unstable too (Chu, 1985).)

6.8   Suppose that G has minimal realization $G: (A, B, C, D)$. If $\tilde{N}$ and $\tilde{M}$ have realizations

$$\tilde{N}: (A + HC, B + HD, C, D)$$

and

$$\tilde{M}: (A + HC, -H, C, -I)$$

where

$$H = -XC^\mathsf{T}$$

and X is the stabilizing solution of the Riccati equation

$$AX + XA^\mathsf{T} - XC^\mathsf{T}CX = 0$$

show that $G = \tilde{M}^{-1}\tilde{N}$ and $\tilde{M}\tilde{M}^* = I$ (namely, that $\tilde{M}$ is all-pass). Use the following lemma, due to Glover (1984):

Lemma 6-1:   If $G(s)$ has minimal realization $G: (A, B, C, D)$, then G is all-pass, namely $GG^* = I$, if

(1)

$$DD^\mathsf{T} = I$$

(2)   There exist X and Y, such that

$$CX + DB^\mathsf{T} = 0 \quad \text{and} \quad B^\mathsf{T}Y + D^\mathsf{T}C = 0$$

where $X = X^\mathsf{T}$ and $Y = Y^\mathsf{T}$,

$$AX + XA^\mathsf{T} + BB^\mathsf{T} = 0 \quad \text{and} \quad A^\mathsf{T}Y + YA + C^\mathsf{T}C = 0$$

6.9 Suppose that G has at least as many columns as rows, minimal realization (A, B, C, D), and $\det(DD^T) \neq 0$. Let $\tilde{N}$ and $\tilde{M}$ have the realizations

$$\tilde{N}: (A + HC, B + HD, (DD^T)^{-1/2} C, (DD^T)^{-1/2} D)$$

and

$$\tilde{M}: (A + HC, H, (DD^T)^{-1/2} C, (DD^T)^{-1/2})$$

where

$$H = -XC^T$$

and X is the stabilizing solution of the Riccati equation

$$AX + XA^T - (XC^T + BD^T)(DD^T)^{-1}(CX + DB^T) + BB^T = 0$$

Show that $G = \tilde{M}^{-1} \tilde{N}$ and $\tilde{N} \tilde{N}^* = I$ (namely, that $\tilde{N}$ is all-pass). (*Hint*: Use Lemma 6.1 above. Note that a suitable solution X exists only if $G(j\omega) G^T(-j\omega) > 0$ for all ω.)

6.10 Work through the Glover–Doyle algorithm of Section 6.7 assuming that $D_{11} = 0$, and observe the consequent simplification.

6.11 Show that, if

$$P = \hat{P} + \begin{bmatrix} 0 & 0 \\ 0 & D_{22} \end{bmatrix}$$

and

$$K = \hat{K}(I + D_{22}\hat{K})^{-1}$$

then

$$F_1(P, K) = F_1(\hat{P}, \hat{K})$$

(that is, show that (6.210) holds).
 Find a state-space realization of K, given a realization of $\hat{K}$.

6.12 Show that the Hankel norm $\|G\|_H$ is an **input/output invariant**, namely that it remains unaltered if a different realization $(TAT^{-1}, TB, CT^{-1}, D)$ is taken for G.

6.13 Let (A, B, C, D) be a minimal realization of $G \in H_\infty$, and let $P = P^T$ and $Q = Q^T$ be the solutions of the Lyapunov equations

$$AP + PA^T + BB^T = 0$$
$$A^T Q + QA + C^T C = 0$$

It can be shown that $P > 0$ and $Q > 0$.

Let $Q = R^T R$. Show that there exists U such that $U^H U = I$ and

$$RPR^T = U\Sigma^2 U^H$$

where $\Sigma = \text{diag}\{\sigma_1, \sigma_2, \ldots, \sigma_n\}$, $\sigma_1 \geqslant \sigma_2 \geqslant \ldots \geqslant \sigma_n > 0$.

Hence show that the realization $G: (TAT^{-1}, TB, CT^{-1}, D)$, where

$$T = \Sigma^{-1/2} U^H R$$

is balanced in the sense defined in Section 6.6.2.

6.14 (a) Solve the robust stabilization problem defined in Section 6.3.1, using the general H_∞ problem formulation. Assume that $r(s) = \text{constant}$, and G is square. Show that this requires the solution of Riccati equations of the form

$$A^T X + XA - XBB^T X = 0$$

   (b) Obtain the solution if the plant is stable, and interpret it.

   (c) Obtain the solution if the plant transfer function is

$$G(s) = \frac{0.1}{s - 0.1} + \frac{s - 10}{s + 10} - 1$$

and

$$r(s) = 0.25$$

6.15 Repeat Exercise 5.3, using the H_∞ approach.

6.16 Repeat Exercise 5.4, using the H_∞ approach.

6.17 Repeat Exercise 5.5, using the H_∞ approach.

6.18 In the design example of Section 6.8, controller poles occur at -6×10^{-5}, instead of the origin. Consider methods of obtaining a state-space realization of the controller in which these poles have been shifted to the origin.

(*Note*: Section 8.3.6 shows how to obtain a decomposition $K = K_- + K_+$, in which all the poles of K_- lie to the left of some real number ε, and all the poles of K_+ lie to the right of ε.)

In practice, would it be necessary to have the controller poles exactly at the origin, or could they be left at -6×10^{-5}?

References

Callier F.M. and Desoer C.A. (1982). *Multivariable Feedback Systems*. Berlin: Springer-Verlag.

Chang B.C. and Pearson J.B. (1984). Optimal disturbance reduction in linear multivariable systems. *IEEE Transactions on Automatic Control*, **AC-29**, 880–7.

Chu C.-C. (1985). H_∞ optimization and robust multivariable control. *PhD Thesis*, University of Minnesota.

Desoer C.A., Liu R.W., Murray J. and Saeks R. (1980). Feedback system design: The fractional representation approach to analysis and synthesis. *IEEE Transactions on Automatic Control*, **AC-25**, 399–412.

Doyle J.C. (1984). Lecture Notes, ONR/Honeywell Workshop on Advances in Multivariable Control.

Doyle J.C. (1985). Structured uncertainty in control system design. In *Proc. 24th IEEE Conf. on Decision and Control*, Fort Lauderdale FL, Dec. 1985, pp. 260–5.

Doyle J., Glover K., Khargonekar P. and Francis B. (1988). State-space solutions to standard H_2 and H_∞ control problems. In *Proc. American Control Conf.*, Atlanta GA, pp. 1691–6.

Francis B.A., Helton J.W. and Zames G. (1984). H_∞ optimal feedback controllers for linear multivariable systems. *IEEE Transactions on Automatic Control*, **AC-29**, 888–900.

Francis B.A. and Zames G. (1984). On H_∞ optimal sensitivity theory for SISO feedback systems. *IEEE Transactions on Automatic Control*, **AC-29**, 9–16.

Francis B.A. (1987). *A Course in H_∞ Control Theory*. Lecture Notes in Control and Information Sciences, Vol. 88. Berlin: Springer-Verlag.

Garcia E.C. and Morari M. (1982). Internal Model Control: 1. A unifying review and some new results. *Industrial and Engineering Chemistry Process Design and Development*, **21**, 308–23.

Glover K. (1984). All optimal Hankel norm approximations of linear multivariable systems and their L_∞ error bounds. *International Journal of Control*, **39**, 1115–93.

Glover K. and Doyle J.C. (1988). State-space formulae for all stabilizing controllers that satisfy an H_∞ norm bound and relations to risk sensitivity. *Systems and Control Letters*, **11**, 167–72.

Golub G.H. and van Loan C.F. (1983). *Matrix Computations*. Baltimore: John Hopkins UP

Kucera V. (1974). *Discrete Linear Control: The Polynomial Equation Approach*. New York: Wiley.

Limebeer D.J.N. and Hung Y.S. (1987). An analysis of the pole–zero cancellations in H_∞ optimal control problems of the first kind. *SIAM Journal on Control and Optimization*, **25**, 1457–93.

McFarlane D. and Glover K. (1988). An H_∞ design procedure using robust stabilization of normalized coprime factors. In *Proc. 27th IEEE Conf. on Decision and Control*, Austin TX, pp. 1343–8.

Nett C.N., Jacobson C.A. and Balas M.J. (1984). A connection between state-space and doubly coprime fractional representations. *IEEE Transactions on Automatic Control*, **AC-29**, 831–2.

Raggazini J.R. and Franklin G.F. (1958). *Sampled-Data Control Systems*. New York: McGraw-Hill.

Rudin W. (1966). *Real and Complex Analysis*. New York: McGraw-Hill.

Safonov M.G. and Verma M.S. (1985). L_∞ sensitivity optimization and Hankel approximation. *IEEE Transactions on Automatic Control*, **AC-30**, 279–80.

Safonov M.G., Jonckheere E.A., Verma M. and Limebeer D.J.N. (1987). Synthesis of positive real multivariable feedback systems. *International Journal of Control*, **45**, 817–42.

Vidyasagar M. (1985). *Control System Synthesis: A Factorization Approach*. Boston MA: MIT Press.

Youla D.C., Jabr H.A. and Bongiorno J.J. (1976). Modern Wiener–Hopf design of optimal controllers, part II: The multivariable case. *IEEE Transactions on Automatic Control*, **AC-21**, 319–38.

Zames G. (1981). Feedback and optimal sensitivity: Model reference transformations, multiplicative semi-norms, and approximate inverses. *IEEE Transactions on Automatic Control*, **AC-26**, 301–20.

Zames G. and Francis B.A. (1983). Feedback, minimax sensitivity, and optimal robustness. *IEEE Transactions on Automatic Control*, **AC-28**, 585–601.

CHAPTER 7

Design by Parameter Optimization

7.1 Introduction
7.2 Edmunds' algorithm
7.3 The method of inequalities
7.4 Multi-objective optimization

7.5 Conclusion
Summary
Exercises
References

7.1 Introduction

It frequently happens that the structure of a controller is constrained in some way. In the process industries, for example, there are considerable costs associated with each non-zero element of the controller's transfer-function matrix. Every such element represents a considerable amount of hardware, possibly extensive additional wiring, additional testing and maintenance procedures, and additional training for the plant's operating personnel. There are therefore strong pressures to keep the number of 'cross-couplings' as low as possible. In aerospace applications, considerations of reliability may dictate an analog realization of the controller, with the smallest possible number of states.

Such constraints are not respected by any of the design techniques we have described in earlier chapters, except for the direct and inverse Nyquist-array techniques when driven entirely manually (so that the designer has explicit control over the controller's structure), and the QFT approach. The designer using the approximate commutative controller approach can control the McMillan degree of the controller, but not the distribution of the

325

dynamics over the controller's elements: in general, each element of the controller's transfer function will have the same complexity. With the LQG and H_∞ approaches he loses control of the McMillan degree as well, and cannot enforce the open-loop stability of the controller. He must therefore either split the design problem into sub-problems which do not have structure constraints, and apply these design techniques to the sub-problems, or abandon the theoretically based techniques and adopt an *ad hoc* approach.

In this chapter we shall outline three such *ad hoc* approaches to the problem. They all consist of choosing the controller structure somehow, and then optimizing the free parameters within that structure. By modifying the problem formulation, the method of choosing the structure and the optimization algorithm, one can invent an endless variety of 'design techniques'. The three we have chosen to present have been selected because they combine simplicity with reasonable effectiveness.

7.2   Edmunds' algorithm

Edmunds (1979) proposed an algorithm for optimizing controller parameters, to make the closed-loop transfer function $T = GK(I + GK)^{-1}$ approach a target transfer function as closely as possible over a specified frequency range. If a quadratic error function is minimized pointwise over frequency, and some approximations are made, the problem can be posed as a linear least-squares problem to which a standard solution is known. This makes it straightforward to code the algorithm. Another attraction (in the context of this book) is that the resulting design technique is a frequency-domain one and can therefore be used particularly naturally in conjunction with the Nyquist-like techniques discussed in Chapter 4.

However, the algorithm is rather crude for the task at hand and needs to be given a sensible target to aim for, together with a compatible controller structure, if it is to succeed. It is particularly suitable, therefore, for 'tuning' parameter values which have already been obtained by some other method.

7.2.1   The algorithm

We now describe the algorithm. Suppose, as usual, that our plant is represented by the transfer function matrix G, with l columns (inputs) and m rows (outputs), and that we are to design a controller with transfer-function matrix K (with m columns and l rows). Let the closed-loop transfer function actually achieved by a controller K be $T = GK(I + GK)^{-1}$, and let the 'target' transfer function which we would like to achieve be T_t. Corresponding to T_t, there is a 'target' controller K_t such that

$$GK_t = T_t(I - T_t)^{-1} \tag{7.1}$$

We define an error function

$$E = T_t - T \tag{7.2}$$

Then it can be shown, with a little manipulation, that

$$(I - T)(GK_t - GK)(I - T_t) = E \tag{7.3}$$

If we suppose that $\|E\|$ is sufficiently small, which will be the case if K is sufficiently close to K_t, then, by replacing $I - T$ by $I - T_t$ in (7.3), we obtain

$$(I - T_t)(GK_t - GK)(I - T_t) \approx E \tag{7.4}$$

since

$$(I - T)(GK_t - GK)(I - T_t) = (I - T_t)(GK_t - GK)(I - T_t) + O(\|E\|^2) \tag{7.5}$$

Now let us write

$$K(s) = \frac{1}{d(s)} N(s) \tag{7.6}$$

where $d(s)$ is a common-denominator polynomial which is assumed to be known, and $N(s)$ is a matrix of polynomials of known degrees but with unknown coefficients. Finally, we define

$$B(s) = I - T_t(s) \tag{7.7}$$

$$A(s) = \frac{1}{d(s)} B(s) G(s) \tag{7.8}$$

and

$$Y(s) = B(s) G(s) K_t(s) B(s) \tag{7.9}$$

Then (7.4) becomes

$$Y(s) \approx A(s) N(s) B(s) + E(s) \tag{7.10}$$

The noteworthy features here are that the unknown coefficients in $N(s)$ appear linearly in this expression, that $A(s)$, $B(s)$ and $Y(s)$ are all known and can be evaluated at particular values of s when required and, hence, that the problem of finding $N(s)$ which minimizes

$$\|E\|_2^2 = \int_{-\infty}^{\infty} \mathrm{tr}[E^{\mathrm{T}}(-j\omega)E(j\omega)]\,d\omega$$

is a linear least-squares problem if the approximate equality in (7.10) is replaced by exact equality.

To put (7.10) into the more familiar standard form in which linear least-squares problems are usually seen, we need to 'stack' the columns of Y, N and E on top of each other. For this purpose we define their columns by

$$Y(s) = [y_1(s) \ldots y_m(s)] \tag{7.11}$$

$$N(s) = [n_1(s) \ldots n_m(s)] \tag{7.12}$$

$$E(s) = [e_1(s) \ldots e_m(s)] \tag{7.13}$$

We also need to use the $\otimes$ notation for the **Kronecker** or **tensor product** of two matrices: if P has p rows and q columns, and Q has r rows and s columns, then $P \otimes Q$ is the $pr \times qs$ matrix:

$$P \otimes Q = \begin{bmatrix} p_{11}Q & p_{12}Q & \cdots & p_{1s}Q \\ p_{21}Q & p_{22}Q & \cdots & p_{2s}Q \\ \vdots & \vdots & & \vdots \\ p_{r1}Q & p_{r2}Q & \cdots & p_{rs}Q \end{bmatrix} \tag{7.14}$$

In this notation, (7.10) can be rewritten as

$$\begin{bmatrix} y_1(s) \\ y_2(s) \\ \vdots \\ y_m(s) \end{bmatrix} \approx [B^{\mathrm{T}}(s) \otimes A(s)] \begin{bmatrix} n_1(s) \\ n_2(s) \\ \vdots \\ n_m(s) \end{bmatrix} + \begin{bmatrix} e_1(s) \\ e_2(s) \\ \vdots \\ e_m(s) \end{bmatrix} \tag{7.15}$$

Remember that $n_i(s)$ represents a vector of polynomials:

$$n_i(s) = [n_{1i}(s) \ldots n_{li}(s)]^{\mathrm{T}} \tag{7.16}$$

and suppose that

$$n_{ij}(s) = v_{ij}^0 s^p + v_{ij}^1 s^{p-1} + \ldots + v_{ij}^p \tag{7.17}$$

for some positive integer p, assuming for notational convenience that each n_{ij} has the same degree. (But this is not a real restriction, since $v_{ij}^x = 0$ is allowed, and can be forced if desired.) Then $\{v_{ij}^x\}$ is the set of controller parameters to be optimized; if each n_{ij} has degree p, there are $lm(p+1)$ of them. We need to introduce one more new notation: let $\Sigma(s)$ be the matrix (with lm rows and $lm(p+1)$ columns)

$$\Sigma(s) = \begin{bmatrix} s^p & s^{p-1} \ldots 1 & & & \mathbf{0} \\ & & s^p & s^{p-1} \ldots 1 & \\ & \mathbf{0} & & & \ddots \\ & & & & s^p & s^{p-1} \ldots 1 \end{bmatrix} \tag{7.18}$$

then

$$\begin{bmatrix} n_1(s) \\ \vdots \\ n_m(s) \end{bmatrix} = \Sigma(s)v \tag{7.19}$$

where

$$v = [v_{11}^0 \ v_{11}^1 \ldots v_{ml}^p]^T \tag{7.20}$$

So if we let

$$X(s) = [B^T(s) \otimes A(s)]\Sigma(s) \tag{7.21}$$

$$\eta(s) = [y_1^T(s) \ldots y_m^T(s)]^T \tag{7.22}$$

and

$$\varepsilon(s) = [e_1^T(s) \ldots e_m^T(s)]^T \tag{7.23}$$

then (7.15) becomes

$$\eta(s) \approx X(s)v + \varepsilon(s) \tag{7.24}$$

which is in a standard form: $\eta(s)$ is a known vector, $X(s)$ is a known matrix, v is a vector of unknown parameters and $\varepsilon(s)$ is a vector of 'errors'.

To obtain a practical algorithm, we need to evaluate $\eta(s)$ and $X(s)$ at a number of points on the imaginary axis, say $\{s = j\omega_i: i = 1, 2, \ldots, \mu\}$, and approximate $\|E\|_2$ (which is the same as $\|\varepsilon\|_2$) by

$$\|\varepsilon\|_2^2 \approx \sum_{i=1}^{\mu} \varepsilon^T(-j\omega_i)\varepsilon(j\omega_i) \tag{7.25}$$

Assembling data from all these points, we obtain

$$
\begin{bmatrix} \eta(j\omega_1) \\ \vdots \\ \eta(j\omega_\mu) \end{bmatrix} \approx \begin{bmatrix} X(j\omega_1) \\ \vdots \\ X(j\omega_\mu) \end{bmatrix} v + \begin{bmatrix} \varepsilon(j\omega_1) \\ \vdots \\ \varepsilon(j\omega_\mu) \end{bmatrix}
\tag{7.26}
$$

The standard least-squares solution to this would be (Lawson and Hanson, 1974)

$$
\hat{v} = \left\{ [X^\mathrm{T}(-j\omega_1) \ldots X^\mathrm{T}(-j\omega_\mu)] \begin{bmatrix} X(j\omega_1) \\ \vdots \\ X(j\omega_\mu) \end{bmatrix} \right\}^{-1}
$$

$$
\times \left\{ [X^\mathrm{T}(-j\omega_1) \ldots X^\mathrm{T}(-j\omega_\mu)] \begin{bmatrix} \eta(j\omega_1) \\ \vdots \\ \eta(j\omega_\mu) \end{bmatrix} \right\}
\tag{7.27}
$$

but in general this would give *complex* parameter values. We therefore have to depart slightly from the standard problem, in order to obtain real parameters. To do this, we use the following lemma:

Lemma 7.1: If $Y = X\theta + E$, the value of θ which minimizes $\|E\|_2$, given X and Y, and subject to the constraint $\mathrm{Im}\{\theta\} = 0$, is

$$
\hat{\theta} = [\mathrm{Re}\{X^\mathrm{H}X\}]^{-1} \mathrm{Re}\{X^\mathrm{H}Y\}
\tag{7.28}
$$

With the aid of this lemma we obtain the optimal *real* parameters:

$$
\hat{v} = \left(\mathrm{Re} \left\{ [X^\mathrm{T}(-j\omega_1) \ldots X^\mathrm{T}(-j\omega_\mu)] \begin{bmatrix} X(j\omega_1) \\ \vdots \\ X(j\omega_\mu) \end{bmatrix} \right\} \right)^{-1}
$$

$$
\times \mathrm{Re} \left\{ [X^\mathrm{T}(-j\omega_1) \ldots X^\mathrm{T}(-j\omega_\mu)] \begin{bmatrix} \eta(j\omega_1) \\ \vdots \\ \eta(j\omega_\mu) \end{bmatrix} \right\}
\tag{7.29}
$$

The validity of this algorithm depends on the validity of (7.4), which in turn depends on the size of $\|E\|_2$ obtained with the parameter vector $\hat{v}$. There are, however, ways of extending its validity. Let the controller obtained from (7.29) be K_i, and suppose that we somehow obtain another controller

$$K_{i+1} = K_i + (\Delta K)_i \qquad (7.30)$$

Let the closed-loop transfer functions obtained with these be T_i and T_{i+1}, respectively, and let

$$E_i = T_t - T_i \qquad (7.31)$$

Then, from (7.3),

$$(I - T_i)G(K_t - K_i)(I - T_t) = E_i \qquad (7.32)$$

and

$$(I - T_{i+1})G(K_t - K_{i+1})(I - T_t) = E_{i+1} \qquad (7.33)$$

If we assume that $T_{i+1} \approx T_i$, then subtracting (7.33) from (7.32) gives

$$(I - T_i)G(\Delta K)_i(I - T_t) \approx E_i - E_{i+1} \qquad (7.34)$$

Everything in (7.34) is known, except $(\Delta K)_i$ and E_{i+1}, and (7.34) can be put into the same form as (7.10), which again leads to a linear least-squares problem (assuming that only the numerator elements of $(\Delta K)_i$ are unknown). Thus (7.34) can be solved to find the $(\Delta K)_i$ which approximately minimizes $\|E_{i+1}\|_2$, in the same way as before. This clearly leads to an iterative algorithm since (7.34) can be repeated as many times as required.

Edmunds (1979), who suggested this approach, uses

$$(I - T_i)G(\Delta K)_i \Xi_i (I - T_t) \approx E_i - E_{i+1} \qquad (7.35)$$

where

$$\Xi_i = I + (I - T_i)G(K_t - K_i) \qquad (7.36)$$

The approximation (7.35) is less crude than (7.34).

A further refinement, which is advisable, is to replace (7.30) by

$$K_{i+1} = K_i + \alpha_i(\Delta K)_i \qquad (7.37)$$

where α_i is a scalar. The reason is that, although a reduced error can probably be obtained by changing K_i in the 'direction' of $(\Delta K)_i$, the size of the step dictated by (7.30) is almost certainly inappropriate because of the approximations involved in (7.34) or (7.35). One can either repeatedly halve α_i

(starting with $\alpha_i = 1$) until $\|E_{i+1}\| < \|E_i\|$, or perform a more sophisticated line search to find the value of α_i which minimizes $\|E_{i+1}\|$.

7.2.2 Comments on Edmunds' algorithm

The main advantage of Edmunds' algorithm is its flexibility. Selected elements of the controller K, and even particular coefficients, can be constrained to be zero: corresponding elements of v and columns of $X(s)$ are simply omitted in (7.24). It is also straightforward to emphasize the attainment of certain elements of T_t, at the expense of others, by minimizing the weighted error

$$\|\varepsilon\|_W^2 = \sum_{i=1}^{\mu} \varepsilon^T(-j\omega_i) W \varepsilon(j\omega_i) \tag{7.38}$$

where $W = \text{diag}\{w_j\}$ ($w_j > 0$, $j = 1, 2, \ldots, lm$), instead of (7.25). This again is a standard problem, with a standard solution:

> **Lemma 7.2:** If $Y = X\theta + E$, the value of θ which minimizes $\|E\|_W = E^H W E$, given X, Y and $W > 0$, subject to the constraint $\text{Im}\{\theta\} = 0$, is
>
> $$\hat{\theta} = [\text{Re}\{X^H W X\}]^{-1} \text{Re}\{X^H W Y\} \tag{7.39}$$

This leads to the replacement of (7.29) by

$$\hat{v} = \left(\text{Re}\left\{ [X^T(-j\omega_1) \ldots X^T(-j\omega_\mu)] \begin{bmatrix} WX(j\omega_1) \\ \vdots \\ WX(j\omega_\mu) \end{bmatrix} \right\} \right)^{-1}$$

$$\times \text{Re}\left\{ [X^T(-j\omega_1) \ldots X^T(-j\omega_\mu)] \begin{bmatrix} W\eta(j\omega_1) \\ \vdots \\ W\eta(j\omega_\mu) \end{bmatrix} \right\} \tag{7.40}$$

The particular value of this possibility is that one usually has little idea of how to choose the off-diagonal elements of T_t. A frequently adopted strategy, therefore, is to set them all to zero, and to attach very small weight to achieving them. This allows an initial controller design to be obtained. The weights on the off-diagonal terms can then be increased, if it is important to reduce the interaction (off-diagonal) terms in T.

It is immediately apparent from (7.38) and (7.40) that one can apply different weights at different frequencies, which can be useful on occasions. In particular, it is possible for the designed controller to give an unstable closed loop, even if T is very close to T_t. This is often because the low-frequency behaviour of GK is very different from that of GK_t (but not affecting T much if $\underline{\sigma}(GK)$ is large), perhaps to the extent of having inappropriate signs of zero-frequency gains in K, which leads to violation of the generalized Nyquist stability theorem. Such a problem may sometimes be corrected by placing more weight on the lower frequencies in (7.40).

A further advantage of the algorithm is that, although we developed it by writing $K(s)$ with a common denominator (see equation (7.6)), it is possible to assign a different denominator polynomial to each element of K.

If T_t is block-diagonal, then B (in equation (7.7)), and hence $B^T \otimes A$ (in equation (7.21)), is also block-diagonal, and the problem decomposes into several smaller sub-problems. If T_t is actually diagonal (which is usually the case), then each column of $N(s)$ can be optimized independently, and we have m separate problems in each of which there are $l(p+1)$ parameters to be optimized.

It is imperative that the actual algorithm employed solves (7.26) by using a numerically stable procedure and does not use (7.29), since in the neighbourhood of the true solution the matrix

$$
\mathrm{Re}\left\{ [X^T(-j\omega_1) \ldots X^T(-j\omega_\mu)] \begin{bmatrix} X(j\omega_1) \\ \vdots \\ X(j\omega_\mu) \end{bmatrix} \right\}
$$

approaches singularity. A numerically stable algorithm is obtained as follows. In the notation of Lemma 7.1, let $X = X_{\mathrm{Re}} + jX_{\mathrm{Im}}$, $Y = Y_{\mathrm{Re}} + jY_{\mathrm{Im}}$, then $\hat{\theta}$, defined by (7.28), is also obtained as the least-squares solution of the equation

$$
\begin{bmatrix} Y_{\mathrm{Re}} \\ Y_{\mathrm{Im}} \end{bmatrix} = \begin{bmatrix} X_{\mathrm{Re}} \\ X_{\mathrm{Im}} \end{bmatrix} \theta + \begin{bmatrix} E_{\mathrm{Re}} \\ E_{\mathrm{Im}} \end{bmatrix}
$$

This can be solved in a numerically stable way by using the algorithm described in Section 8.2.3. There is also scope for using recursive techniques to update the solution each time data $X(j\omega_k)$ and $\eta(j\omega_k)$ are calculated at a new frequency ω_k.

As is usually the case with non-linear least-squares optimization (which is what Edmunds' algorithm is really doing), success depends on having a sufficiently good initial design K_0 to use in (7.37). If the design obtained from (7.29) is not good enough, it may be necessary to obtain an initial design by some other technique, omit (7.29), and repeatedly solve (7.34) and (7.37). Of course, if T_t is too incompatible with the properties of G and the assumed structure of K, then no reasonable solution exists at all. One should therefore ensure that the 'roll-off' rates of T_t are compatible with those of G, that any right half-plane zeros of G are preserved in T_t, and so on.

A difficult task facing the designer who adopts this approach to parameter optimization is that he must choose the controller's denominator polynomials. It is possible to optimize the denominator polynomials as well, but it is then no longer possible to approximate the dependence of the fitting errors on the controller parameters by a linear relationship, and a considerably more complicated algorithm results. This possibility has been explored by Hung and MacFarlane (1982) in connection with a design based on the reversed-frame normalization technique (see Chapter 4).

7.2.3 A refinement of the algorithm

A refined way of using Edmunds' algorithm has been developed by Nett (1988). It is designed to overcome the difficulty of specifying a target closed-loop transfer function T_t which is compatible with the plant – in particular when this has right half-plane zeros – and to make it more likely that a stabilizing controller is obtained. These improvements to Edmunds' algorithm are obtained by combining it with the Youla parametrization, described in Chapter 6.

Suppose that the plant G is asymptotically stable and has at least as many inputs as outputs. The closed-loop transfer function $T = GK(I + GK)^{-1}$ can then be written as

$$T(s) = G(s)Q(s) \tag{7.41}$$

where $Q(s)$ is the (proper, stable) Youla parameter. Let G be factored as

$$G(s) = A(s)G_0(s) \tag{7.42}$$

in which $A(s)$ is a stable *and* all-pass transfer function whose zeros are the right half-plane zeros of $G(s)$, and $G_0(s)$ is a square, stable, minimum-phase transfer function. That is, we have

$$A(s)A^*(s) = I \tag{7.43}$$

(where $A^*(s) = A^T(-s)$), and $G_0^{-1}(s)$ has all its poles in the open left half-plane. Now consider

$$Q_t(s) = G_0^{-1}(s)A^\dagger(0)T_d(s) \tag{7.44}$$

where $T_d(s)$ is an ideal desired closed-loop transfer function, $A^\dagger$ denotes the pseudo-inverse of A, and

$$A(0)A^\dagger(0) = I \tag{7.45}$$

Of course, $T_d(s)$ is chosen to be stable, so (7.44) defines a particular Youla parameter, provided the high-frequency 'roll-off' rates of $T_d(s)$ are chosen to be large enough to make $Q_t(s)$ proper. This choice of the Youla parameter gives the closed-loop transfer function

$$T_t(s) = G(s)Q_t(s) \tag{7.46}$$

$$= A(s)A^\dagger(0)T_d(s) \tag{7.47}$$

Although this is different from the original desired transfer function $T_d(s)$, it has exactly the same principal gains, since $A(s)$ is all-pass. That is, we have

$$\sigma_i[T_t(j\omega)] = \sigma_i[T_d(j\omega)] \tag{7.48}$$

so that specified robustness properties are preserved at the plant output. Also, $T_t(s)$ is achievable with the given plant since the required right half-plane zeros are built into it, as can be seen from (7.47). The presence of the factor $A^\dagger(0)$ ensures that

$$T_t(0) = T_d(0) \tag{7.49}$$

so that steady-state characteristics (most commonly, $T_d(0) = I$) are preserved.

Now $T_t(s)$ is used as the target closed-loop transfer function in Edmunds' algorithm, in place of $T_d(s)$. The corresponding target controller is given by

$$K_t(s) = Q_t(s)[I - G(s)Q_t(s)]^{-1} \tag{7.50}$$

$$= G_0^{-1}(s)A^\dagger(0)T_d(s)[I - A(s)A^\dagger(0)T_d(s)]^{-1} \tag{7.51}$$

The point of using Edmunds' algorithm, of course, is to achieve a closed-loop response which is close to $T_t(s)$, while using a controller which is significantly simpler than $K_t(s)$. Since $K_t(j\omega)$ can be evaluated at any set of frequencies, a useful procedure is to display the elements of $K_t(j\omega)$ in Nyquist or Bode form and to determine the dynamic complexity of each element of the controller $K(s)$, and its pole positions, using these displays as a guide. Edmunds' algorithm should then be able to tune the numerator coefficients of $K(s)$ to obtain a good approximation to $K_t(s)$. Note that this approach retains the freedom to give each element of $K(s)$ a different dynamic structure, and to constrain certain elements of $K(s)$ to be zero. It also gives the designer the opportunity to use his knowledge of the plant, instrumentation and actuators to make sensible compromises between what is demanded by $K_t(s)$ and what can be achieved in practice.

In order to implement this refinement of Edmunds' algorithm, it is necessary to have a way of finding the factorization (7.42). The central idea of an algorithm for doing this can be found in Exercise 6.9, which gives a way of finding state-space realizations for coprime factors $\bar{M}$ and $\bar{N}$ of G, such that

$G = \tilde{M}^{-1}\tilde{N}$ and $\tilde{N}\tilde{N}^* = I$. Here we need to obtain coprime factors M and N, such that $G = NM^{-1}$ and $NN^* = I$, for then we can take $A(s) = N(s)$ and $G_0(s) = M^{-1}(s)$, since both $M(s)$ and $M^{-1}(s)$ are stable (recall that we assume G to be stable).

Suppose now that G is square and has a minimal realization (A, B, C, D). A dual procedure to that given in Exercise 6.9 will then give coprime factors M and N, such that $NN^* = I$. Specifically, these factors have realizations

$$M: (A + BF, \ B(D^T D)^{-1/2}, \ F, \ (D^T D)^{-1/2}) \tag{7.52}$$

and

$$N: (A + BF, \ B(D^T D)^{-1/2}, \ C + DF, \ D(D^T D)^{-1/2}) \tag{7.53}$$

where

$$F = -(D^T D)^{-1}(B^T X + D^T C) \tag{7.54}$$

$A + BF$ has all its eigenvalues in the open left half-plane, and X is a solution of the Riccati equation

$$A^T X + XA - (XB + C^T D)(D^T D)^{-1}(B^T X + D^T C) + C^T C = 0 \tag{7.55}$$

For these realizations to exist D must be non-singular, which will often not be the case. But it is usually possible to make a small modification to D which renders it non-singular, but which makes very little difference to the plant's behaviour over the range of frequencies of interest.

An approach to feedback design which bears some resemblance to Edmunds' algorithm has been proposed recently by Boyd *et al.* (1988). Since any closed-loop transfer function is linear in the Youla parameter (see Section 6.4.4), it is also linear in the numerator elements of this parameter. So if the denominator elements of the Youla parameter are fixed somehow, and the closed-loop transfer function is 'tuned' in the least-squares sense, then a linear least-squares problem again results. One can now guarantee that the controller obtained is stabilizing, since the poles of the Youla parameter can be chosen to lie in the left half-plane. But the transformation from the Youla parameter to the corresponding controller is a complex one, and it is not easy to impose any desired structure on the controller by using this approach.

7.2.4 Design example

To illustrate the use of Edmunds' algorithm we shall apply it to the AIRC model used in earlier chapters and described in the Appendix. We shall aim to achieve the target closed-loop transfer function

$$T_t(s) = \text{diag}\left\{ \left(\frac{3}{s+3} \right)^2, \left(\frac{3}{s+3} \right)^2, \left(\frac{10}{s+10} \right)^2 \right\} \tag{7.56}$$

This specification calls for a faster response of the pitch angle than of the altitude or forward speed, and is more realistic than our earlier requirement of a bandwidth of 10 rad s^{-1} in response to all reference signals. The diagonal structure of $T_t(s)$ implies a complete lack of interaction between outputs.

The target closed-loop behaviour $T_t(s)$ corresponds to the controller

$$K_t(s) = G^{-1}(s) T_t(s)[I - T_t(s)]^{-1} \tag{7.57}$$

and the frequency response of this is displayed, in Bode magnitude form, in Figure 7.1. This shows that quite simple controller dynamics are sufficient to achieve the target closed-loop behaviour.

Integrators are required in columns 2 and 3 of $K(s)$, namely on the speed and pitch-angle errors, but not in column 1, namely on the altitude error. From Figure 7.1 the pole positions required for the various elements of $K(s)$ are as follows:

(1, 1): -8 (1, 2): $0, -4$ (1, 3): $0, -30$

(2, 1): -6 (2, 2): $0, -6$ (2, 3): $0, -30$

(3, 1): -6 (3, 2): $0, -6$ (3, 3): $0, -30$

Placing the pole of $k_{11}(s)$ at -6, and the poles of $k_{12}(s)$ at 0 and -6, assigns the same set of poles to each element in any one column of $K(s)$ and thus allows a simpler (five-state) realization to be obtained.

The following controller structure was therefore chosen:

$$k_{i1}(s) = \frac{v_{i1}^0 s + v_{i1}^1}{s+6} \tag{7.58}$$

$$k_{i2}(s) = \frac{v_{i2}^0 s^2 + v_{i2}^1 s + v_{i2}^2}{s(s+6)} \tag{7.59}$$

$$k_{i3}(s) = \frac{v_{i3}^0 s^2 + v_{i3}^1 s + v_{i3}^2}{s(s+30)} \tag{7.60}$$

Applying Edmunds' algorithm (modified to allow distinct denominators) to this, with equal weight placed on the achievement of each element of $T_t(j\omega)$ at each frequency ω_i, and with a set of 50 frequencies spaced logarithmically

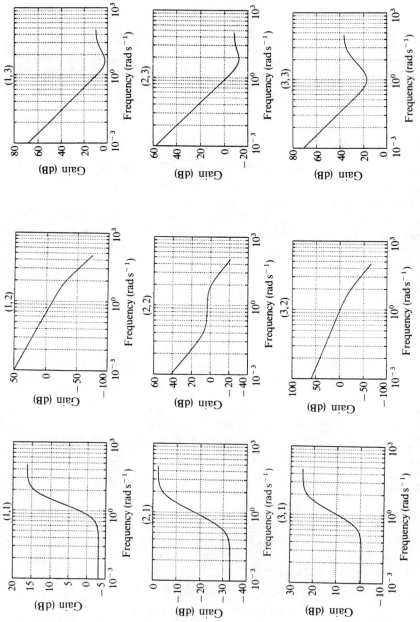

Figure 7.1 Ideal gain–frequency controller characteristics for each element of $K_t(s)$.

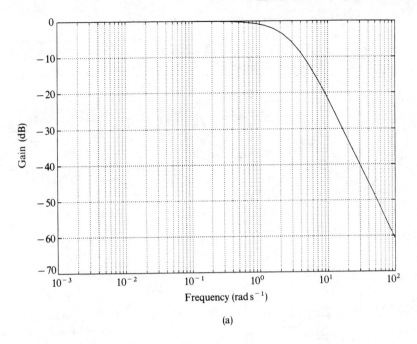

(a)

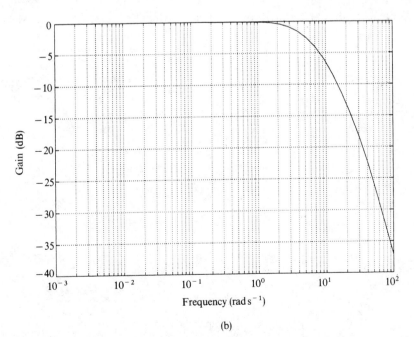

(b)

Figure 7.2 Gain of (a) $T_{11}(j\omega)$ or $T_{22}(j\omega)$, and (b) $T_{33}(j\omega)$.

Table 7.1 Numerator coefficients for tuned controller.

		Column	
Row	1	2	3
1	$v_{11}^0 = -6.5183$ $v_{11}^1 = -4.1806$	$v_{12}^0 = 8 \times 10^{-16}$ $v_{12}^1 = 6 \times 10^{-15}$ $v_{12}^2 = 1.9101$	$v_{13}^0 = -5.2967$ $v_{13}^1 = 6.5509$ $v_{13}^2 = 77.930$
2	$v_{21}^0 = -0.7822$ $v_{21}^1 = 0.1328$	$v_{22}^0 = 9 \times 10^{-16}$ $v_{22}^1 = 9.0000$ $v_{22}^2 = 0.7134$	$v_{23}^0 = -0.6153$ $v_{23}^1 = 0.6702$ $v_{23}^2 = 22.989$
3	$v_{31}^0 = -17.300$ $v_{31}^1 = -5.6199$	$v_{32}^0 = -1 \times 10^{-15}$ $v_{32}^1 = 9 \times 10^{-15}$ $v_{32}^2 = 5.3316$	$v_{33}^0 = -99.88$ $v_{33}^1 = -62.41$ $v_{33}^2 = 104.81$

between 0.001 and 100 rad s^{-1}, yields the numerator coefficients given in Table 7.1. A noteworthy feature of this is that the numerators of $k_{12}(s)$ and $k_{32}(s)$ have no dynamics, and $k_{22}(s)$ has only one zero.

This controller achieves the target closed-loop transfer function almost exactly. Figure 7.2 shows the Bode magnitude plots of the diagonal elements of the achieved closed-loop transfer function $T(s)$. Figure 7.3 shows

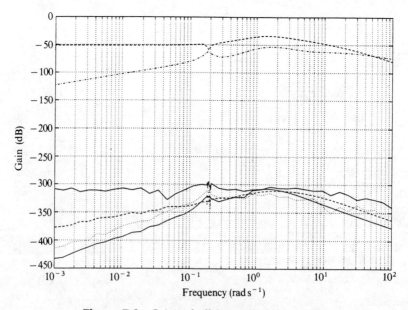

Figure 7.3 Gains of off-diagonal elements of *T*.

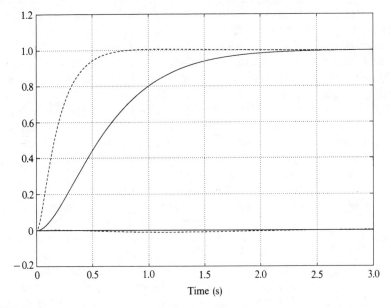

Figure 7.4 Responses to step demand on outputs 1 and 2 (solid curves) and output 3 (dashed curves).

the magnitudes of the off-diagonal elements, and it can be seen that these are all very small. Figure 7.4 shows responses to step demands on each of the outputs: the response of each output is exactly as specified, and there is virtually no interaction. The only problem is that, as in the design examples in earlier chapters, the controller introduces zeros to cancel the resonant poles at $-0.018 \pm 0.182j$.

The use of Edmunds' algorithm has allowed us to achieve a precise, albeit simple specification, using a considerably simpler controller than the ones which resulted from the LQG or H_∞ design techniques.

7.3 The method of inequalities

If one observes operators controlling a process plant manually, it becomes apparent that in many cases their objective is to hold a number of plant variables between permitted bounds, and that they are indifferent to the detailed behaviour of the variables, provided they remain within those bounds. The design specifications can therefore be written as a set of inequalities:

$$|e_i(t)| \leqslant \varepsilon_i, \quad t \geqslant 0, \quad i = 1, 2, \ldots, m \tag{7.61}$$

where each $e_i(t)$ is some function of time which represents the behaviour of one plant variable.

Zakian (Zakian and Al-Naib, 1973; Zakian, 1979) was apparently the first to suggest that many feedback-control design problems should be posed as the *satisfaction* of a set of inequalities, rather than the *minimization* of some objective function with inequalities acting as side-constraints. The shift of emphasis from an objective function to a set of inequalities gives a more accurate formal representation of many design problems, and leads to an iterative design procedure in which the designer changes the 'trade-off' between conflicting constraints by adjusting the inequalities, rather than some objective function. This is attractive, because it is usually much easier to understand the physical implications of changes in constraining inequalities than of changes in an objective function.

Optimization is still used in Zakian's approach, as will become clear shortly, but the shift of emphasis from minimization to satisfaction means that the usual ideas on choice of optimization algorithms have to be revised: speed of convergence in the neighbourhood of a minimum becomes much less important than the likelihood of finding at least one feasible point – namely one at which all the inequalities are satisfied, and the extent of information given to the user about causes of failure – which can be used to modify the inequalities, thus making success more likely at the next design iteration.

Inequality constraints arise from physical limitations as well as from desired objectives. For example, magnitudes and/or rates of change of control signals are usually limited, which results in a set of inequalities of the form

$$|u_i(t)| \leqslant \eta_i, \qquad |\dot{u}_i(t)| \leqslant v_i \tag{7.62}$$

Similar inequalities can arise not from real physical limits, but from the fact that one's model of the plant is valid only for certain ranges of variables.

If p is a vector of controller parameters which are to be designed, with a prespecified controller structure, then inequalities of the form (7.61) or (7.62) assume the form

$$\phi_i(p) \leqslant 0 \tag{7.63}$$

where each $\phi_i(p)$ is a functional. For example, to get (7.67) we define

$$\phi_i(p) = \max_{0 \leqslant t < \infty} (|e_i(t, p)|) - \varepsilon_i \tag{7.64}$$

where $e_i(t, p)$ now denotes the function $e_i(t)$ which is obtained with controller parameter p. If we wish to impose closed-loop stability as a requirement, we can define

$$\phi_i(p) = \max \text{Re} \{\text{closed-loop poles obtained with } p\} + \varepsilon \tag{7.65}$$

for some positive real ε. To constrain the value of one of the parameters (to avoid excessive gain) we use a simple function of p:

$$\phi_i(p) = |p_j| - p_{max} \tag{7.66}$$

and so on.

Very many design objectives and constraints can be put in the form (7.63), even though it may not be straightforward to express $\phi_i(p)$ mathematically. In particular, if we have an 'envelope constraint', such as

$$a(t) \leqslant y(t, p) \leqslant b(t) \tag{7.67}$$

where $a(t)$ and $b(t)$ are functions of time, we can represent it by (7.63) if we define

$$\phi_i(p) = \max_t \max \, [a(t) - y(t, p), \, y(t, p) - b(t)] \tag{7.68}$$

This is referred to as an 'infinite-dimensional constraint', since it can be regarded as the 'limit' of the set of inequalities

$$a(t_1) \leqslant y(t_1, p) \leqslant b(t_1)$$
$$\vdots \tag{7.69}$$
$$a(t_N) \leqslant y(t_N, p) \leqslant b(t_N)$$

as N is increased indefinitely. Frequency-domain constraints can also be represented by (7.63). Problems of robust design can be represented similarly: suppose that the variable y depends on a plant parameter vector δ as well as the controller parameter vector p, and we have a 'robust performance' requirement that

$$a(t) \leqslant y(t, p, \delta) \leqslant b(t), \quad \text{for all } \delta \in \Delta \tag{7.70}$$

Then we define

$$\phi_i(p) = \max_{\delta \in \Delta} \max_{0 \leqslant t < \infty} \max \, [a(t) - y(t, p, \delta), \, y(t, p, \delta) - b(t)] \tag{7.71}$$

Let S_i be the set of parameter vectors for which the ith functional inequality is satisfied:

$$S_i = \{p: \phi_i(p) \leqslant 0\} \tag{7.72}$$

Then the **admissible** or **feasible set** of parameter vectors, for which all the inequalities hold, is the intersection

$$S = \bigcap_i S_i \tag{7.73}$$

Clearly p is an admissible parameter vector if

$$\max_i \phi_i(p) \leqslant 0 \tag{7.74}$$

which shows that the search for an admissible p can be pursued by optimization, in particular by solving

$$\min_p \max_i \phi_i(p) \tag{7.75}$$

Zakian suggests an algorithm for this minimization which he calls the **moving-boundaries algorithm**. This operates as follows. We let p^k be the value of the parameter vector at the kth step, and define

$$C_i^k = \begin{cases} 0 & \text{if } \phi_i(p^k) \leqslant 0 \\ \phi_i(p^k) & \text{otherwise} \end{cases} \tag{7.76}$$

$$S_i^k = \{p: \phi_i(p) \leqslant C_i^k\} \tag{7.77}$$

and

$$S^k = \bigcap_i S_i^k \tag{7.78}$$

S^k being the set of parameter vectors for which some but not all of the inequality constraints are satisfied; it contains both p^k and the admissible set S. If we find a new parameter vector $\tilde{p}^k$, such that $\tilde{C}_i^k \leqslant C_i^k$ for each i (where $\tilde{C}_i^k$ is defined similarly to C_i^k), then we accept $\tilde{p}^k$ as the next value of the parameter vector; that is, we set $p^{k+1} = \tilde{p}^k$.

We then have

$$\max_i \phi_i(p^{k+1}) \leqslant \max_i \phi_i(p^k) \tag{7.79}$$

and

$$S \subset S^{k+1} \subset S^k \tag{7.80}$$

so that the boundary of the set in which the parameter vector is located has been moved towards the admissible set, or, rarely, has remained unaltered. The algorithm is terminated when each $C_i^k = 0$, at which point we have $S^k = S$. This process is illustrated in Figure 7.5.

If one or more of the C_i^k persists in remaining positive, this may be taken as an indication that the set of design objectives is inconsistent, while their magnitude gives some measure of how closely it is possible to approach the objectives.

The satisfaction of infinite-dimensional constraints can be tested only approximately, of course, by a discretization such as (7.69), except in those

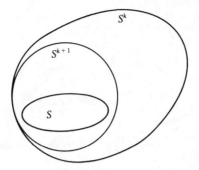

Figure 7.5 The 'moving-boundaries' process.

rare cases in which an exact value of $\phi_i(p)$ can be computed. (See Zakian (1986) for an example of such a case.)

We have not yet mentioned the difficult part of the algorithm, which is the generation of a trial parameter vector $\tilde{p}^k$, given p^k. Zakian suggests the use of an algorithm introduced by Rosenbrock (1960), and reports it to be successful, but this has to be considered a rather arbitrary choice; one could use any of a number of hill-climbing algorithms in its place. In particular, the use of relatively crude direct search methods, such as the simplex method of Nelder and Mead (1965) or one of its later developments (see Swann, 1974, or Gill *et al.*, 1981)[1] is worth investigating, particularly in view of the un-importance of speed of convergence previously mentioned in this section. Direct search methods have been used successfully by Ng (1988).

Alternatively, a standard way of dealing with a 'minimax' problem such as (7.75) is to replace it by the minimization of a smooth objective function, with non-linear constraints:

$$\min_{[p^T \, q]^T} q \tag{7.81}$$

$$\text{subject to } \phi_i(p) \leqslant q, \quad i = 1, 2, \ldots, m \tag{7.82}$$

where the minimization is over a new vector $[p^T \, q]^T$, which consists of the design parameters p and a scalar q. When formulated in this way, the problem can be solved by standard algorithms available in the major libraries of optimization software. (See Gill *et al.* (1981) for more details.)

At the opposite extreme from direct search methods are the sophisti-cated algorithms for 'semi-infinite' optimization developed by Mayne, Polak and their colleagues for solving a wide range of engineering design problems, including those posed by Zakian (Becker *et al.*, 1979; Mayne *et al.*, 1982; Polak *et al.*, 1984). These algorithms have more theoretical basis than the moving-boundaries method, and hence one can say something about their properties. For example, it can be proved that an admissible parameter vector p will be found in a finite number of steps, if it exists. However, these algorithms are far more complicated than the moving-boundaries method, their implementation being a job for experts only.

7.4   Multi-objective optimization

7.4.1   Formulation

A formulation of the control-system design problem, which has close similarities to the method of inequalities, has been proposed by Kreisselmeier and Steinhauser (1979, 1983). Instead of a set of inequalities, they translate a specific design problem into a set of performance objectives $\{J_i(p), (i = 1, 2, \ldots, m)\}$, which are chosen such that if $J_i(p_1) < J_i(p_2)$ then the design parameter (vector) p_1 is better than the design parameter p_2, as far as the objective J_i is concerned. This set of objectives can be considered to be a vector of objective functions, or a **multi-objective** function:

$$J(p) = \begin{bmatrix} J_1(p) \\ \vdots \\ J_m(p) \end{bmatrix} \tag{7.83}$$

The performance specification is satisfied if $J(p) < c$ for some positive vector c of 'thresholds' (i.e. if $J_i(p) < c_i$ for each i, and $c_i > 0$). At this stage the formulation can be seen to be completely equivalent to that of the method of inequalities, so what follows in the rest of this section can be viewed as a particular way of solving the problem posed in the previous section. As in the Method of Inequalities, the appropriate values of some of the elements of c are usually not known in advance, since they depend on qualitative judgements about trade-offs between competing objectives. These are dealt with by initially being given large values, so that the corresponding objectives play little or no role in the initial stage of design. As the design progresses the values of these elements of c are reduced until they begin to affect the design. The idea is that if this is done in a suitable interactive environment then the user can discover what compromises are necessary, and can adjust the elements of c accordingly.

7.4.2   Solution

The distinctive feature of Kreisselmeier and Steinhauser's approach is the way in which they replace the multi-objective optimization problem

$$\min_{p} \frac{J_i(p)}{c_i}, \quad i = 1, \ldots, m \tag{7.84}$$

by a smooth conventional optimization problem which, unlike (7.81)–(7.82), remains an unconstrained problem. Note that the formulation (7.84) (as opposed to a formulation such as $\min_p (J_i(p) - c_i)$) has the effect of normalizing each objective by its threshold value, so that not much attention need be paid to $J_i(p)$ if c_i is large, and if (7.84) is replaced by

$$\min_p \alpha(p) = \max_i \frac{J_i(p)}{c_i} \tag{7.85}$$

In fact, an optimization algorithm used to solve (7.85) deals with one objective at a time – the one for which the ratio $J_i(p)/c_i$ is currently the largest.

Now, (7.85) is a non-smooth optimization problem – that is, the function $\alpha(p)$ is not differentiable everywhere, but it can be approximated well by a smooth problem. Consider the function

$$\bar{\alpha}(p) = \frac{1}{\rho} \ln \left\{ \sum_{i=1}^{m} \exp \left[\rho \left(\frac{J_i(p)}{c_i} \right) \right] \right\} \tag{7.86}$$

for some $\rho > 0$: $\bar{\alpha}(p)$ is a differentiable function of the parameter vector p, and can be rewritten as

$$\bar{\alpha}(p) = \alpha(p) + \frac{1}{\rho} \ln \left\{ \sum_{i=1}^{m} \exp \left[\rho \left(\frac{J_i(p)}{c_i} - \alpha(p) \right) \right] \right\} \tag{7.87}$$

Now,

$$\frac{J_i(p)}{c_i} - \alpha(p) \leqslant 0 \tag{7.88}$$

so

$$\bar{\alpha}(p) \leqslant \alpha(p) + \frac{\ln m}{\rho} \tag{7.89}$$

Also,

$$\frac{J_k(p)}{c_k} - \alpha(p) = 0, \quad \text{for some } k \tag{7.90}$$

and hence

$$\bar{\alpha}(p) > \alpha(p) + \frac{1}{\rho} \ln[1 + \delta], \quad \delta \geqslant 0 \tag{7.91}$$

$$\geqslant \alpha(p)$$

So we have

$$\alpha(p) \leqslant \bar{\alpha}(p) \leqslant \alpha(p) + \frac{\ln m}{\rho} \qquad (7.92)$$

and we see that $\alpha(p)$ is approximated very well by $\bar{\alpha}(p)$ if $\rho \gg \ln m$.

To summarize, we started with the multi-objective optimization problem (7.84), and replaced it by the minimax problem (7.85). Finally, we approximate this by the smooth, unconstrained problem

$$\min_{p} \bar{\alpha}(p) \qquad (7.93)$$

where $\bar{\alpha}(p)$ is defined by (7.86).

This problem can be solved by a standard algorithm for unconstrained optimization. The most appropriate is probably a finite-difference quasi-Newton algorithm (Gill *et al.*, 1981), since in most cases it is impossible to compute gradient information for $\bar{\alpha}(p)$. Of course, software must be written to evaluate $\bar{\alpha}(p)$ for each design problem, and such evaluations will usually require a considerable amount of computation – for example, simulation runs with the current parameters. As with the method of inequalities, there is no need to find the true minimum of $\bar{\alpha}(p)$ for a given c; the optimization can be terminated as soon as $\bar{\alpha}(p)$ is sufficiently small. In fact, if (7.92) is to ensure a good approximation to $\alpha(p)$, the optimization should be terminated while $\bar{\alpha}(p)$ is still considerably larger than $(\ln m)/\rho$. If necessary, the elements of c can then be reduced further, and the optimization restarted.

Although the process of adjusting the threshold vector c is one which depends on the designer's skill and understanding (of the control design problem, that is, not of optimization algorithms), one should adopt a reasonably systematic procedure for making the adjustments. One possibility is the following. Let p^k and c^k denote the values of the parameters and the thresholds at the kth cycle of the optimization. If J_i is to be reduced further in the $(k+1)$th step, then set

$$c_i^{k+1} = J_i(p^k) \qquad (7.94)$$

If J_i is already satisfactory, then set

$$c_i^{k+1} = c_i^k \qquad (7.95)$$

This gives a decreasing sequence of thresholds:

$$c^1 \geqslant c^2 \geqslant \ldots \geqslant c^k \qquad (7.96)$$

and ensures that a feasible p^{k+1} can be found (since p^k is already feasible for the $(k+1)$th step). This systematic procedure may have to be abandoned

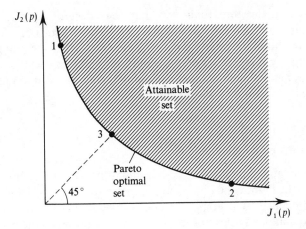

Figure 7.6 A Pareto-optimal set with two objective functions.

when trade-offs need to be made – that is, some element of c^k may need to be increased in order to allow another objective to be decreased.

Note that we can make any p feasible by choosing the elements of c to be large enough, provided all the objectives are finite. Of course, this way of making p feasible is rather artificial and should be avoided if possible, since it does not give good initial values of the parameters, and may lead to failure of the optimization in the sense that it may be impossible to bring the thresholds c_i down to acceptable values. It is far better to use any information available at the start of the design to choose parameter values which are reasonable, in the sense that most of the objectives are nearly met by them (so that most threshold values need be increased only slightly), and to increase only a few of the thresholds substantially.

From (7.86) it is clear that reducing any one of the ratios $J_i(p)/c_i$ and leaving the others unchanged results in a reduced value of $\bar{\alpha}(p)$. As a result, the optimization problem (7.93) has the desirable property that its solution is, approximately, a **Pareto optimal solution** of the multi-objective problem (7.84). This simply means that the solution is such that a reduction in any one of the objectives J_i can be achieved only by increasing at least one of the others. This is illustrated in Figure 7.6 for the case when there are only two objectives ($m = 2$). The figure shows, hatched, the space of possible values of $J_1(p)$ and $J_2(p)$ – the **attainable set**. Any point in the interior of this set has a neighbour for which the values of both $J_1(p)$ and $J_2(p)$ are reduced. Such a neighbouring point is clearly preferable to the original one. But points on the boundary of the attainable set do not have this property. Moving from one point on the boundary to another requires a trade-off: increase $J_1(p)$ and decrease $J_2(p)$, or vice versa. We should always aim to be on this boundary, which in the parlance of multiobjective optimization is called the **Pareto optimal set**.

Changing the thresholds in c has the effect of moving from one solution point on the Pareto optimal set to another. For example, in Figure 7.6 we obtain the point 1 if $c_2 \gg c_1$, the point 2 if $c_1 \gg c_2$, and the point 3 if $c_1 = c_2$. Of course, these points would be obtained exactly only if the optimization algorithm succeeded in finding the global solution to (7.84); in practice it may find only a local solution.

7.4.3   Example

Kreisselmeier and Steinhauser (1983) describe an application of their approach to the flight control of a military aircraft (the F-4C Phantom). The aircraft is modelled by a simple three-state model, and has one input – the elevator command, and three outputs – the pitch rate, the angle of attack and the normal acceleration of the centre of gravity. The states are the pitch rate, the angle of attack and the elevator deflection. The model parameters vary with the flight condition.

The controller uses only one of the outputs – the pitch rate – and has the structure shown in Figure 7.7. The choice of a reasonable structure is crucial to the success of the design technique, and it is important to note that the structure of Figure 7.7 was strongly influenced by a familiarity with both control theory and the particular model of the aircraft. Essentially, the controller consists of three simple lags in parallel. Two of these were introduced to act as approximate observers for the angle of attack and the elevator deflection, and one to provide approximate integral action. Apart from these three lags, the controller contains seven gains, one in each path through the controller. Thus there are ten parameters to be adjusted altogether.

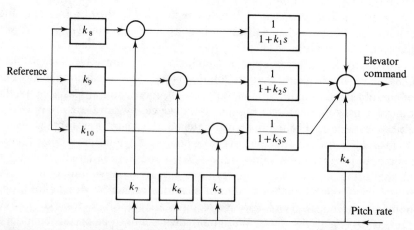

Figure 7.7 Controller structure for the flight-control example given by Kreisselmeier and Steinhauser (1983).

Many objectives were rather loosely defined for this problem. It was required to design the controller for five different flight conditions, with the aircraft parameters *and* the objectives being different for each condition (but the controller parameters were to remain the same under all flight conditions). There were thus five objective functions defined for the tracking behaviour of the pitch rate in response to reference signals, and five for the tracking behaviour of the elevator rate. Also, five objectives were defined for the behaviour of the pitch rate and elevator rate in response to disturbances. Fifteen objectives defined bounds on acceptable closed-loop pole locations, and seven defined bounds for acceptable controller gains. This makes forty-two objective functions in all!

We leave the reader to consult Kreisselmeier and Steinhauser's paper for further details of the formulation of the problem, and for an account of the resulting design. Suffice it to say here that the design technique was successful when applied to this very challenging problem.

7.5 Conclusion

The idea that optimization of parameters within a fixed structure should be used to design control systems is by no means new. It was applied as early as the 1950s, but the state of contemporary computing technology prevented its widespread exploitation. For it is only rarely that there is an obvious translation of one's real objectives into a mathematical optimization problem. In most cases one's first formulation of the problem produces unsatisfactory, even nonsensical, results, and one has to revise the formulation several, perhaps many times before useful results are obtained. This was certainly a discouraging prospect in the days when each reformulation involved, typically, taking a stack of punched cards to the local computing centre, then waiting perhaps 24 hours for the results of a run, finding that one or two cards needed changing, and repeating the process until success was achieved.

Today the position is radically different, of course, since one can sit at a terminal or a personal workstation, submit runs using macros with their own parameters, and obtain results within a few seconds or minutes, already displayed graphically and annotated. Even major changes, such as redefinitions of performance functionals, can often be effected quickly with the use of an editor. The prospect of using an iterative design procedure in such an environment is not at all daunting, and a notable feature of all three approaches we have examined in this chapter is that the formulation of the optimization problem is not seen as sacrosanct. Indeed, the user is encouraged to explore possibilities by changing the formulation frequently.

In the author's opinion, the major change in control-system design over the next few years will be the increasing use of parameter optimization. Several factors are likely to produce this change. First, current CPU technology is still capable of several doublings of computing speed. In the past few years the speed of commercially available workstations, for example, has doubled approximately annually (there have been four doublings during the writing of this book). This means that the great majority of optimization problems which we may formulate today, even those requiring a huge amount of computation in order to evaluate the performance functionals, will become solvable in a reasonable amount of time – minutes rather than hours. (But it should be emphasized that current, even slightly obsolete machines will solve many realistic design problems in only a few minutes or less.)

Secondly, the development of powerful optimization algorithms, such as those due to Mayne, Polak and their colleagues (Becker *et al.*, 1979; Mayne *et al.*, 1982; Polak *et al.*, 1984), will undoubtedly continue. This will make success with difficult problems more likely, and will enlarge the class of allowable performance functionals, as well as having an effect on speed.

Thirdly, and perhaps most importantly, will be the increasing exploitation of the modern computing environment. Work is already under way on the development of integrated environments which will make it easy for the casual or inexperienced user to connect software for simulation and optimization and thus formulate a particular design problem, without needing to be an expert in either simulation or optimization – and without being an expert programmer, either. A relatively unexplored area with much promise is that of displaying information to the user which allows him to monitor the progress of an optimization exercise. For instance, referring to the example discussed in Section 7.4.3, how should one inform the user of progress on the optimization of forty-two objectives simultaneously? These questions are addressed by Nye and Tits (1986) and Ng (1988).

SUMMARY

In this chapter we have presented three relatively straightforward methods of optimizing the parameters of a controller for a feedback system, assuming that its structure has already been chosen.

Edmunds' algorithm solves a specific frequency-domain optimization problem, using a rather crude algorithm. Nevertheless, when used intelligently it is capable of finding very good controllers of relatively low complexity, as the design example showed.

The method of inequalities and multi-objective optimization are both capable of solving very general design problems, and are not inherently restricted to the frequency domain. Both can be implemented quite easily, using standard optimization subroutines. Their effectiveness depends, however, on good user interfaces, and the development of these is an area of current research.

Note

1.   Gill *et al.* (1981) use the term 'polytope' in place of 'simplex'.

EXERCISES

7.1   (a)   Show that Edmunds' algorithm can be used to tune the controller parameters if a target-sensitivity function S_t is specified (where $S = (I + GK)^{-1}$).

   (b)   Show that Edmunds' algorithm can be used to obtain designs which represent a compromise between achieving a specified closed-loop transfer function and a specified behaviour of the control signals (plant inputs).

(*Hint*: Augment the plant with suitable additional outputs.)

7.2   A SISO plant has transfer function

$$G(s) = \frac{s-1}{s+3}$$

A desired, but unattainable, closed-loop transfer function for it is

$$T_d(s) = \frac{10}{s+10}$$

Find an attainable closed-loop transfer function $T_t(s)$, which has the same gain behaviour as $T_d(s)$, and the corresponding feedback controller. Compare the characteristics of the two sensitivity functions $S_d(s) = 1 - T_d(s)$ and $S_t(s) = 1 - T_t(s)$.

Note: The following exercises probably require considerable software development. A language such as Matlab should be used to keep this within reasonable bounds – for exercises 7.4 and 7.5 use software which has easy-to-use optimization facilities available. In a class, these exercises are best tackled by teams of two or three people, each team solving one exercise.

7.3   Apply Edmunds' algorithm to solve Exercise 5.3, 5.4 or 5.5.

7.4   Apply the method of inequalities to solve Exercise 5.3, 5.4 or 5.5.

7.5   Apply multi-objective optimization to solve Exercise 5.3, 5.4 or 5.5.

References

Becker R.G., Heunis A.J. and Mayne D.Q. (1979). Computer-aided design of control systems via optimization. *Proceedings of the Institution of Electrical Engineers* **126**, 573–8.

Boyd S.P., Balakrishnan V., Barratt C.H., *et al.* (1988). A new CAD method and associated architectures for linear controllers. *IEEE Transactions on Automatic Control*, **AC-23**, 268–83.

Edmunds J.M. (1979). Control system design and analysis using closed-loop Nyquist and Bode arrays. *International Journal of Control*, **30**, 773–802.

Gill P.E., Murray W. and Wright M.H. (1981). *Practical Optimization*. London: Academic Press.

Hung Y.S. and MacFarlane A.G.J. (1982). *Multivariable Feedback: A Quasi-classical Approach*, Lecture Notes in Control and Information Sciences, Vol. 40. Berlin: Springer-Verlag.

Kreisselmeier G. and Steinhauser R. (1979). Systematic control design by optimizing a vector performance index. In *Proc. IFAC Symp. on Computer-Aided Design of Control Systems*, Zurich, pp. 113–17.

Kreisselmeier G. and Steinhauser R. (1983). Application of vector performance optimization to a robust control loop design for a fighter aircraft. *International Journal of Control*, **37**, 251–84.

Lawson C.L. and Hanson R.J. (1974). *Solving Least-Squares Problems*. Englewood Cliffs NJ: Prentice-Hall.

Mayne D.Q., Polak E. and Sangiovanni-Vincentelli A. (1982). Computer-aided design via optimization: A review. *Automatica*, **18**, 147–54.

Nelder J.A. and Mead R. (1965). A simplex method for function minimization. *Computer Journal*, **7**, 308–13.

Nett C.N. (1988). Personal communication.

Ng W.-Y. (1988). A decision support system for multi-objective design of practical controllers. In *Proc. American Control Conf.*, Atlanta, GA, pp. 713–18.

Nye W.T. and Tits A.L. (1986). An application-oriented optimization-based methodology for interactive design of engineering systems. *International Journal of Control*, **43**, 1693–721.

Polak E., Mayne D.Q. and Stimler D.M. (1984). Control system design via semi-infinite optimization: A review. *Proc. IEEE*, **72**, 1777–94.

Rosenbrock H.H. (1960). An automatic method for finding the greatest or least values of a function. *Computer Journal*, **3**, 175–84.

Swann W.H. (1974). Constrained optimization by direct search. In *Numerical methods for constrained optimization* (Gill P.E. and Murray W., eds), pp. 191–218. London: Academic Press.

Zakian V. (1979). New formulation for the method of inequalities. *Proceedings of the Institution of Electrical Engineers*, **126**, 579–84.

Zakian V. (1986). A performance criterion. *International Journal of Control*, **43**, 921–31.

Zakian V. and Al-Naib U. (1973). Design of dynamical and control systems by the method of inequalities. *Proceedings of the Institution of Electrical Engineers*, **120**, 1421–7.

CHAPTER 8

Computer-aided Design

8.1 Introduction
8.2 Elements of numerical algorithms
8.3 Applications to linear systems

8.4 Software for control engineering
Summary
Exercises
References

8.1 Introduction

It is apparent that computer assistance is essential if one is to do any multivariable analysis or design, and even for SISO problems the use of the traditional graphical methods to obtain accurate results can no longer be justified (although these methods do have some didactic value). The nature of the assistance required can for convenience be divided into two categories: numerical algorithms for performing basic tasks such as evaluation of frequency responses, and higher-level software which makes the numerical algorithms available to the user (invariably in an interactive environment), allows results to be passed from one algorithm to another, presents results graphically when required, and so on. The units in which the higher-level software is organized are often referred to as 'packages'.

The basic ideas of some numerical algorithms have already been described in earlier chapters. In this chapter we shall present some of the more important algorithms which are widely used in the analysis and design

355

of control systems, and which have not been described in earlier chapters. In particular, we shall describe how to compute frequency responses and characteristic loci, how to compute interconnections and inverses of systems, how to obtain minimal state-space realizations, how to split a system into its stable and unstable parts (and, more generally, into its 'partial fractions'), how to compute the root loci of multivariable systems, how to compute transmission zeros and how to compute balanced realizations and low-order approximate systems. Finally, we shall have more to say about the solution of Riccati equations (in addition to what has already been said in Chapter 5). The reader will also find material on algorithms for diagonalizing systems in Chapter 4, for computing low-order approximations in Chapter 6, and for solving least-squares and other optimization problems in Chapter 7.

We shall then go on to examine the software actually available for control-system design, and how it is likely to develop over the next few years.

A word of warning is in order. We can only outline the basic ideas of each algorithm in this book. Turning these ideas into reliable mathematical software requires a major development effort, as the experience of several groups, both academic and commercial, has repeatedly shown.

8.2 Elements of numerical algorithms

8.2.1 Conditioning and numerical stability

Since computation with real numbers can be performed with only limited precision, and is constrained to use a particular machine's number system, numerical algorithms which operate on real numbers (as opposed to integers) almost always give wrong answers, (the only exceptions occurring in (some of) those rare instances when both the data and the results are exactly representable in the machine's number system). Numerical analysts have therefore expended much effort in devising algorithms which are 'good' in the sense that they do not give unnecessarily large errors, and most of the basic algorithms available in modern subroutine libraries, particularly those concerned with linear algebra, are good in this sense. We cannot say that a good algorithm never gives large errors, because even a good algorithm may on occasions be used to solve a problem whose solution is· so sensitive to specification errors that large errors result, simply from expressing the input data with finite precision. Problems which exhibit such sensitivity are said to be **ill-conditioned**.

While we cannot say that an algorithm never gives large errors, we may be able to say that a particular algorithm always gives the exact solution of some problem which is 'close' to the original problem. Such algorithms are called **numerically stable** or **backwards-stable**. If we use a numerically stable algorithm on a well-conditioned problem, then we can be sure that the result will be close to the true result (see Exercise 8.1).

The investigation of the numerical stability of an algorithm is usually very complicated, requiring ingenuity as well as attention to a myriad of details. But one can pinpoint the common causes of numerical instability: subtraction of one large number from another, possibly resulting in a completely false result because of limited precision, division by a number whose magnitude may be very small (or even zero), and multiplication of two small numbers. The unnecessary inversion of matrices is something which is particularly to be avoided, since such an inversion may well be ill-conditioned; if this occurs inside an algorithm, the algorithm is not likely to be numerically stable. An exception to this is the inversion of orthogonal matrices (namely those for which $T^{-1} = T^T$), since it can be accomplished simply by transposition, without introducing any error. For this reason most of the numerically stable algorithms for linear algebra make much use of orthogonal transformations.

When we analyse feedback systems we frequently require transformations of complex matrices (usually frequency-response matrices), in which case orthogonal matrices are generally replaced by unitary matrices, namely those for which $T^{-1} = T^H$.

Many of the techniques presented in text-books and papers on control, which work perfectly well on small problems which can be solved exactly by hand, lead to very poor algorithms when used as the basis of machine software. Examples are the usual rank tests for controllability and observability, and evaluations of $\exp(At)$ based on spectral decomposition of A. Fortunately much work has been done since about 1975 on developing numerically stable algorithms for control applications, and today we can perform most of the important computations reliably. The reader is referred to van Dooren (1981a, b), Emami-Naeini and van Dooren (1982), Laub (1985) and Arnold and Laub (1984), and the references therein, for detailed discussions and analyses.

Successful algorithms in the fields of systems and control depend, for the most part, on reliable, numerically stable algorithms for linear algebra. We shall now summarize the most important of these, but the reader should consult a numerical analysis text such as Golub and van Loan (1983) or Stewart (1973) for more details. There is also an excellent summary in Gill et al. (1981).

8.2.2   Solution of $Ax = b$ when A is square

The exact solution is $x = A^{-1}b$, but explicit computation of A^{-1} gives a numerically unstable algorithm. A standard algorithm, which can be proved to be numerically stable, uses Gaussian elimination with a so-called **pivoting**

strategy to obtain the factorization

$$PA = LU \tag{8.1}$$

in which L is a lower-triangular matrix, U is an upper-triangular matrix, and P is a permutation matrix – namely one whose elements are all zero, except for exactly one 1 in each column and row. (Numerical analysts call (8.1) an *LU* **factorization** of A.) Then the equation

$$Ly = Pb \tag{8.2}$$

is solved for y, which can be done easily by substitution since L is lower-triangular, first solving for y_1, then for y_2, and so on. Finally the equation

$$Ux = y \tag{8.3}$$

is solved for x, which again can be done easily by back-substitution since U is upper-triangular. Substituting (8.3) into (8.2), and noting that $P^T = P^{-1}$, gives

$$P^T LUx = b \tag{8.4}$$

which, from (8.1), is the same as

$$Ax = b \tag{8.5}$$

We shall denote the solution of (8.5) obtained using this algorithm by

$$x = A \backslash b \tag{8.6}$$

For this problem we can estimate the conditioning. In Section 4.10 the condition number of A was defined as

$$\text{cond}(A) = \frac{\bar{\sigma}(A)}{\underline{\sigma}(A)} \tag{8.7}$$

where $\underline{\sigma}(A)$ and $\bar{\sigma}(A)$ are the smallest and largest singular values of A, respectively. Note that $\text{cond}(A) \geqslant 1$. Suppose that b is perturbed to $b + \delta b$, and that the resulting (exact) solution then becomes $x + \delta x$. It can be shown that

$$\frac{\| \delta x \|}{\| x \|} \leqslant \text{cond}(A) \frac{\| \delta b \|}{\| b \|} \tag{8.8}$$

If A is perturbed to $A + \delta A$ (and b remains unperturbed), then

$$\frac{\|\delta x\|}{\|x + \delta x\|} \leqslant \text{cond}(A)\frac{\|\delta A\|}{\|A\|} \tag{8.9}$$

Thus the position with solving linear equations is very fortunate: we have a numerically stable algorithm, *and* we can tell whether a particular equation is well conditioned or not. (A useful rule of thumb is that the machine may lose the last $\log_{10}\text{cond}(A)$ decimal places of a solution because of round-off errors during the Gaussian elimination.)

A slight variation of (8.5) which arises commonly in control analysis is the linear equation

$$AX = B \tag{8.10}$$

where A is square, of dimension $n \times n$, and X and B are matrices with m columns. In this case we simply regard (8.10) as m equations of the form (8.5):

$$Ax_1 = b_1, Ax_2 = b_2, \ldots, Ax_m = b_m \tag{8.11}$$

and solve each one as before, except that the LU factorization (8.1) need be found only once. We shall use the notation

$$X = A \backslash B \tag{8.12}$$

to denote the solution of (8.10) found by this method. This solution requires about $n^3 + mn^2$ floating-point operations ('flops').

8.2.3   Householder transformations and QR factorization

A Householder transformation is a matrix of the form

$$T_k = I - \frac{2w_k w_k^H}{\|w_k\|_2^2} \tag{8.13}$$

where w_k is any non-zero vector. Such a transformation is unitary, as is easily checked. Consider a vector

$$a_k = \begin{bmatrix} a_{1k} \\ a_{2k} \\ \vdots \\ a_{nk} \end{bmatrix} \tag{8.14}$$

and

$$w_k = \begin{bmatrix} 0 \\ \vdots \\ 0 \\ a_{kk} - r_k \\ a_{k+1,k} \\ \vdots \\ a_{nk} \end{bmatrix} \left.\begin{matrix} \\ \\ \end{matrix}\right\} k-1 \tag{8.15}$$

where $|r_k| = \|a_k\|$. Then

$$T_k a_k = \begin{bmatrix} a_{1k} \\ \vdots \\ a_{k-1,k} \\ r_k \\ 0 \\ \vdots \\ 0 \end{bmatrix} \left.\begin{matrix} \\ \\ \\ \end{matrix}\right\} n-k \tag{8.16}$$

and

$$T_{k+1}(T_k a_k) = T_k a_k \tag{8.17}$$

This shows that, by applying a sequence of such Householder transformations, any $n \times n$ matrix A can be reduced to upper-triangular form:

$$T_{n-1} T_{n-2} \cdots T_1 A = R \tag{8.18}$$

where R is upper-triangular. If A is an $m \times n$ matrix, $m > n$, and rank$(A) = n$, then the same process gives

$$T_{n-1} \cdots T_1 A = \begin{bmatrix} R_1 \\ 0 \end{bmatrix} \tag{8.19}$$

where R_1 is again upper-triangular, and R_1 is non-singular.
    Writing

$$Q^H = T_{n-1} \cdots T_1 \tag{8.20}$$

and recalling that $T_k^{-1} = T_k^H$, we have

$$A = Q \begin{bmatrix} R_1 \\ 0 \end{bmatrix} \tag{8.21}$$

which is known as the **QR factorization** of A.

If $m \geqslant n$ and rank $(A) = r < n$, then an additional permutation of A may be necessary to obtain a QR factorization, so in this case we have, in general,

$$AP = QR \tag{8.22}$$

where

$$R = \begin{bmatrix} R_{11} & R_{12} \\ 0 & 0 \end{bmatrix} \tag{8.23}$$

R_{11} is an $r \times r$ upper-triangular matrix, and P is a permutation matrix; Q is unitary, as before.

We shall write

$$[Q, R] = QR(A) \tag{8.24}$$

to denote that $A = QR$, and

$$[Q, R, P] = QR(A) \tag{8.25}$$

to denote that $AP = QR$.

Most users do not have to concern themselves with the details of Householder transformations, since they are contained within subroutines. They have been described here because some knowledge of them is necessary for understanding the algorithms for partial-fraction decompositions and Riccati-equation solutions that will be described later.

The principal use of the QR factorization is in solving linear least-squares problems in which A has more rows than columns, and we wish to find the vector x that minimizes $\| Ax - b \|_2$. If A has full column rank then (since $\| . \|_2$ is invariant under unitary transformations)

$$\| Ax - b \|_2 = \| Q^H(Ax - b) \|_2$$

$$= \left\| \begin{bmatrix} R_1 \\ 0 \end{bmatrix} x - Q^H b \right\|_2 \tag{8.26}$$

If we write $Q = [Q_1, Q_2]$, with as many columns in Q_1 as there are rows in R_1, then

$$\|Ax - b\|_2 = \left\| \begin{bmatrix} R_1 x - Q_1^H b \\ -Q_2^H b \end{bmatrix} \right\|_2 \tag{8.27}$$

which is clearly minimized by

$$x = R_1 \backslash Q_1^H b \tag{8.28}$$

Since R_1 is upper-triangular, this is easily computed by back-substitution once the factors Q and R have been obtained.

If the rank of A is smaller than the number of columns, then there are either no solutions or infinitely many solutions of the equation $Ax = b$. The QR factorization can be employed to find the solution (if it exists) for which $\|x\|_2$ is minimized. Of particular interest is the 'underdetermined' problem which occurs when A has fewer rows than columns. To solve this we first find

$$[Q, R, P] = QR(A^T) \tag{8.29}$$

Then we have

$$\|x^T A^T - b^T\|_2 = \|x^T QR - b^T P\|_2 \tag{8.30}$$

since $P^T = P^{-1}$. Now we find z^T such that

$$z^T [R_{11}, R_{12}] = b^T P \tag{8.31}$$

(recall (8.23)), the solution being found by substitutions since R_{11} is upper-triangular. If R_{12} is such that no solution for z exists, then $Ax = b$ has no solutions. If a solution for z is found, then we have

$$\|x^T A^T - b^T\|_2 = \|x^T [Q_1, Q_2]R - z^T [R_{11}, R_{12}]\|_2 \tag{8.32}$$

$$= \|(x^T Q_1 - z^T)[R_{11}, R_{12}]\|_2 \tag{8.33}$$

$$= 0 \tag{}$$

if

$$x = Q_1 z \tag{8.34}$$

8.2.4   The Hessenberg form of a matrix

Any square matrix A is similar to a matrix H which is in **upper Hessenberg form,** namely such that $h_{ij} = 0$ for $i - j \geqslant 2$:

$$H = \begin{bmatrix} h_{11} & h_{12} & h_{13} & \cdots & & h_{1n} \\ h_{21} & h_{22} & h_{23} & \cdots & & h_{2n} \\ 0 & h_{32} & h_{33} & \ddots & & \vdots \\ & \ddots & \ddots & \ddots & & h_{n-1,n} \\ \mathbf{0} & & 0 & h_{n,n-1} & & h_{nn} \end{bmatrix} \qquad (8.35)$$

One can obtain H from A by using a unitary similarity transformation:

$$H = TAT^H \qquad (8.36)$$

where $T^H = T^{-1}$, and this transformation can be effected in a numerically stable way, again by obtaining T as a finite sequence of Householder transformations.

In fact, if we define w_k as

$$w_k = \begin{bmatrix} 0 \\ \vdots \\ 0 \\ a_{k+1,k} - r_k \\ a_{k+2,k} \\ \vdots \\ a_{nk} \end{bmatrix} \Bigg\} \, k \qquad (8.37)$$

with $|r_k| = \|a_k\|$ as before, and define T_k as in (8.13), then it is easily checked that

$$T_{n-2} \cdots T_2 T_1 A T_1^H T_2^H \cdots T_{n-2}^H = H \qquad (8.38)$$

We shall use the notation

$$[T, H] = \mathrm{Hess}(A)$$

to mean that H is the Hessenberg form of A, and $A = THT^H$, and $T^H T = I$.

8.2.5 The Schur form of a square matrix

Every square matrix A is unitarily similar to an upper-triangular matrix which is sometimes referred to as the **Schur form** of A. Since the diagonal elements of this matrix are the eigenvalues of A, the Schur form in general has complex elements, even if A is a real matrix.

If A is real, it is also possible to obtain a **real Schur form** in which all the elements are real, but which is not quite triangular: 2×2 blocks appear on the diagonal of the real Schur form, and each such block is associated with a complex-conjugate pair of eigenvalues. We shall call such a form **quasi-triangular**.

The Schur form cannot be computed by any finite algorithm, and must therefore be obtained approximately by an iterative algorithm. The standard means of obtaining the Schur form is the **QR algorithm**, which in essence has the following form:

$$i := 0; \qquad A_i = A;$$

repeat

    choose σ_i somehow;

    $[Q_i, R_i] := QR(A_i - \sigma_i I);$

    $A_{i+1} := R_i Q_i + \sigma_i I;$

    $i := i + 1;$

until A_i is in Schur form.

It is easy to show that each A_i is similar to the original A: since

$$A_i - \sigma_i I = Q_i R_i \tag{8.39}$$

we have

$$R_i = Q_i^H (A_i - \sigma_i I) \tag{8.40}$$

and hence

$$A_{i+1} = Q_i^H (A_i - \sigma_i I) Q_i + \sigma_i I \tag{8.41}$$

$$= Q_i^H A_i Q_i \tag{8.42}$$

From this it follows that

$$A_i = Q_{i-1}^H \cdots Q_1^H Q_0^H A Q_0 Q_1 \cdots Q_{i-1} \tag{8.43}$$

so similarity is preserved.

It is not so easy to see why this algorithm should converge to anything, let alone to the Schur form, and we refer the reader to Golub and van Loan (1983) for a detailed exposition. For our purposes we need to note the following facts. If A is first transformed to Hessenberg form, so that

$$A_0 = H = T A T^H$$

(compare equation (8.36)), then each subsequent A_i will retain the Hessenberg structure. If such an A_i has no zeros on its first sub-diagonal, then choosing the **shift** σ_i to be an eigenvalue of A (λ, say) will result in A_{i+1} having the structure

$$A_{i+1} = \begin{bmatrix} & \vdots & \vdots & \vdots \\ \ldots & x & x & x \\ \ldots & x & x & x \\ \ldots & 0 & 0 & \lambda \end{bmatrix} \tag{8.44}$$

That is, the eigenvalue will be exhibited in the (n, n) element, and the $(n, n-1)$ element will be zero. Of course, the shift cannot be chosen to be λ, since we do not know the eigenvalues of A, but it has been found that the algorithm will lead to convergence to a matrix of the form (8.44) for quite a wide choice of shifts σ_i. The best choice, which usually leads to the fastest convergence, is obtained as follows: find the eigenvalues of the 2×2 block in the bottom right-hand corner of A_i,

$$\begin{bmatrix} a^i_{n-1, n-1} & a^i_{n-1, n} \\ a^i_{n, n-1} & a^i_{n, n} \end{bmatrix} \tag{8.45}$$

and choose σ_i as the eigenvalue which is closest to $a^i_{n, n}$. When convergence has been obtained so that $a^i_{n, n-1} = 0$, the algorithm is applied to the $(n-1) \times (n-1)$ matrix formed by the first $n-1$ rows and columns of A_i, and the next eigenvalue of A is eventually exhibited in the $(n-1, n-1)$ element. This is repeated until the whole Schur form has been obtained.

If A is real, a variant of this algorithm which uses two shifts at each iteration, is commonly used to obtain the real Schur form and to avoid complex arithmetic altogether. (See Golub and van Loan (1983) for details.)

Suppose that the (complex or real) Schur form of A is partitioned so that

$$Q^H A Q = S = \begin{bmatrix} S_{11} & X \\ 0 & S_{22} \end{bmatrix} \tag{8.46}$$

where S_{11} and S_{22} are themselves in Schur form, and have no eigenvalues in common. If we partition Q conformally, so that $Q = [Q_1 \, Q_2]$, then

$$A Q_1 = Q_1 S_{11} \tag{8.47}$$

The columns of Q_1 therefore span an invariant subspace associated with the eigenvalues of S_{11}. Since Q is unitary (orthogonal in the case of the real Schur form), the columns of Q_1 form an orthonormal basis for this invariant subspace. This subspace is also spanned by the eigenvectors associated with the eigenvalues of S_{11}; but, whereas eigenvectors are difficult to compute reliably, the columns of Q are obtained from the numerically stable QR algorithm. For many applications it is enough to find this orthonormal basis of an invariant subspace, rather than the eigenvectors themselves.

We shall use the notation

$$[Q, S] = \text{Schur}(A)$$

to mean that $S = Q^{\mathrm{T}} A Q$, S is in real Schur form, and $Q^{\mathrm{T}} Q = I$.

8.2.6   Eigenvalues

The standard way of finding eigenvalues is to find the Schur form as described in the previous section, and then to read off the diagonal elements. Note that there is no particular order to these elements, although there is a tendency for the largest eigenvalues (in absolute magnitude) to be in the top left-hand corner of the Schur form, and the smallest ones to be in the bottom right-hand corner.

Although the QR algorithm is numerically stable, it is possible for eigenvalues to be ill-conditioned. Roughly speaking, two eigenvalues are ill-conditioned if they are very close to each other. (However, the remaining eigenvalues may be well-conditioned in such a case.)

8.2.7   Singular-value decomposition

For every matrix A of size $m \times n$ and rank r, one can obtain the singular-value decomposition

$$A = U \Sigma V^{\mathrm{H}} \tag{8.48}$$

in which U and V are unitary matrices of size $m \times m$ and $n \times n$, respectively, and

$$\Sigma = \begin{bmatrix} \Sigma_r & 0 \\ 0 & 0 \end{bmatrix} \tag{8.49}$$

where $\Sigma_r = \text{diag}\{\sigma_1, \sigma_2, \ldots, \sigma_r\}$, and $\sigma_1 \geqslant \sigma_2 \geqslant \ldots \geqslant \sigma_r > 0$. This finds applications in analysis and design, as we have already seen in Chapters 3 to 6, but is also useful as a constituent of several system-theoretic algorithms.

In particular, one can use the matrices U and V to obtain a **row compression** of A:

$$U^{\mathrm{H}} A = \begin{bmatrix} R \\ 0 \end{bmatrix} \tag{8.50}$$

or a **column compression** of A:

$$AV = [C \; 0] \tag{8.51}$$

as is apparent from (8.49). Here R has r rows, all of which are linearly independent (R has **full row rank**) and C has r columns, all of which are linearly independent (C has **full column rank**).

The diagonal elements of Σ_r are called **singular values**, and the columns of U and V are called the **left** and **right singular vectors**, respectively. The value of σ_r gives a measure of how close the matrix A is to a matrix of rank $r - 1$, in the very precise sense that if the matrix $A + E$ has rank $r - 1$ (or smaller), then $\|E\|_F \geqslant \sigma_r$ and an E exists for which the equality holds. Examination of the singular values therefore gives a reliable way of estimating the rank of a matrix. In practice the singular values are unlikely to be exactly zero, but one normally considers them to be zero if their magnitudes are comparable to the precision of the machine one is using (about 2×10^{-16} in implementations of IEEE Standard arithmetic).

Another reason for the usefulness of the singular-value decomposition is that a pseudo-inverse of A is easily obtained from it:

$$A^\dagger = V \begin{bmatrix} \Sigma_r^{-1} & 0 \\ 0 & 0 \end{bmatrix} U^H \tag{8.52}$$

This has the properties $AA^\dagger A = A$ and $A^\dagger A A^\dagger = A^\dagger$. If $r = m = n$ then we have $A^{-1} = V\Sigma^{-1}U^H$.

The standard way of computing the singular-value decomposition makes use of the QR algorithm, but we do not need to know the details (see Golub and van Loan, 1983). The algorithm is numerically stable, and the singular values are always well-conditioned. Note that, although the singular values are the square roots of the non-zero eigenvalues of $A^H A$, good algorithms never form this product explicitly, since this could cause an unnecessary loss of precision.

8.2.8 Generalized eigenvalues

If A and B are square, $n \times n$ matrices, and the scalar λ and the vector x are such that

$$Ax = \lambda Bx \tag{8.53}$$

then λ is a **generalized eigenvalue** and x a **generalized eigenvector** of the pair (A, B). We shall denote the set of all generalized eigenvalues of (A, B) by $\lambda(A, B)$. (Note that $\lambda(A, B) \neq \lambda(B, A)$.) If rank$(B) = n$, then $\lambda(A, B)$ has n elements, otherwise it may be empty (that is, there may be no solutions), finite or infinite.

It can be shown that one can obtain a **generalized Schur decomposition** for every pair (A, B): there exist unitary matrices Q and Z such that

$$Q^H A Z = T \quad \text{and} \quad Q^H B Z = S \qquad (8.54)$$

where T and S are both upper-triangular. Generalized eigenvalues are easily obtained from T and S. If $s_{ii} \neq 0$, then the ith generalized eigenvalue is given by

$$\lambda_i(A, B) = \frac{t_{ii}}{s_{ii}} \qquad (8.55)$$

However if, for some i, $t_{ii} = s_{ii} = 0$, then the set of generalized eigenvalues is the whole complex plane. (If $t_{ii} \neq 0$, then $\mu_i(B, A) = s_{ii}/t_{ii}$ is the ith generalized eigenvalue of the pair (B, A).) A small value of $|s_{ii}|$ in (8.55) indicates that $\lambda_i(A, B)$ is ill-conditioned.

The generalized Schur decomposition of the pair (A, B) can be obtained by the **QZ algorithm**, which is a generalization of the QR algorithm, and is numerically stable. (For details see Golub and van Loan (1983).)

8.3   Applications to linear systems

8.3.1   Frequency-response evaluation

Evaluation of the frequency response of a system requires the computation of the transfer-function matrix $G(s)$ at a number of points on the complex plane. If we have a continuous-time model these points are on the upper half of the imaginary axis, and for a discrete-time model they lie on the upper half of the origin-centred unit circle. (If w-plane rather than z-plane methods are employed for discrete-time systems, then the evaluation is again on the imaginary axis, after an appropriate transformation of the system model. See Franklin and Powell (1980) for details.)

If the system definition is already in the form of a transfer-function matrix, so that each element of the transfer function is already expressed as a function of the complex variable s, then the evaluation is, in principle, straightforward. Each element is simply evaluated at each of the required frequencies. If the system has l inputs and m outputs, and each element is a ratio of two polynomials of degree d, then this evaluation requires $4lm(d + 1)$ (complex) floating-point operations at each frequency. If it is known that all the denominator polynomials are common, the number of operations can be reduced to $lm(2d + 1) + 2d$.

A drawback of this straightforward approach is that accurate evaluation of polynomials of high degree can be surprisingly difficult (Wilkinson,

1965). It may therefore be preferable to obtain a state-space realization of the transfer-function matrix first, and then apply the algorithm described in the following paragraphs. But this is not a foolproof method either, since algorithms for obtaining realizations are not numerically stable. Fortunately, transfer-function representations are usually employed only for low-order models, which do not usually lead to numerical difficulties.

Most frequently, systems are defined by state-space models rather than transfer functions. The obvious way of computing the frequency response is to convert the model to a transfer-function representation first, then proceed as above. This is not to be recommended, however, because the only reliable way of performing the conversion to transfer-function form involves computing the frequency response first! Fortunately one can compute the frequency response directly from the state-space model, using an efficient and numerically stable algorithm proposed by Laub (1981).

We are given the matrices A, B, C and D of a state-space model, and we recall that the expression for the corresponding transfer function is

$$G(s) = C(sI - A)^{-1} B + D \tag{8.56}$$

We first obtain the Hessenberg form $\bar{A}$ of A, and the corresponding unitary transformation T:

$$[T, \bar{A}] = \text{Hess}(A) \tag{8.57}$$

so that

$$\bar{A} = T^H A T \tag{8.58}$$

We now define

$$\bar{B} = T^H B \tag{8.59}$$

and

$$\bar{C} = CT \tag{8.60}$$

and note that $(\bar{A}, \bar{B}, \bar{C}, D)$ is an alternative state-space realization of the original system.

Now we compute

$$G(s) = \bar{C}[(sI - \bar{A})\backslash\bar{B}] + D \tag{8.61}$$

at the required frequencies, the matrix $[(sI - \bar{A})\backslash\bar{B}]$ being computed (at each frequency) as described in Section 8.2.2. The point of the initial transformation to Hessenberg form is that $sI - \bar{A}$ is in Hessenberg form at each value of s, and this reduces the number of operations required to compute $[(sI - \bar{A})\backslash\bar{B}]$ from about $n^3 + ln^2$ to about ln^2, where n is the number of states. If we add the number of operations required to perform the multiplication of $\bar{C}$ and addition of D, we see that we need about $ln^2 + lm(2n + 1)$ operations at each frequency. Since a typical requirement is to compute a frequency response at 50 frequencies or more, the cost of the transformation to Hessenberg form (about $2n^3$ operations), which is performed only once, is usually very minor.

8.3.2   Characteristic loci and principal gains

The characteristic loci of a square system are just the eigenvalues of the frequency response $G(j\omega)$, and are evaluated at the same set of frequencies as the frequency response. This is done by transforming the frequency response to Schur form, as described in Section 8.2.5.

If the characteristic loci are plotted by joining corresponding eigenvalues at successive frequencies by line segments, the display may show loci crossing at points which are clearly inappropriate. This is because 'corresponding' eigenvalues sometimes appear on the diagonal of the Schur form in different positions at different frequencies. The effect is illustrated in Figure 8.1(a) for a pair of characteristic loci. If the continuous branches of the eigenvalues are labelled $\lambda_1(j\omega)$ and $\lambda_2(j\omega)$, then Figure 8.1(a) results from the eigenvalues appearing in the Schur form in the order (λ_1, λ_2) at frequency ω_1, (λ_2, λ_1) at ω_2, and (λ_1, λ_2) at ω_3. To obtain the 'correct' display, as shown in Figure 8.1(b), requires sorting the eigenvalues before displaying them. Such sorting requires a significant amount of computation.

The standard scheme of computing characteristic loci by transforming to Schur form makes no attempt to exploit the fact that the eigenvalues are continuous functions of frequency. Making use of this information should reduce the number of iterations of the QR algorithm required to obtain the Schur form, since good initial estimates of the eigenvalues are already available from the previous frequency. It should also be possible to ensure that eigenvalues are found in an order which corresponds to continuous branches, without explicit sorting. The author has experimented with various modifications of the QR algorithm (for example, using eigenvalues found at the previous frequency as shifts, or using the accumulated unitary transformations from the previous reduction to Schur form), but so far without success. The problem is that, although these modifications occasionally produce superior performance, for typical frequency responses the standard QR algorithm is very hard to beat: it normally finds the Schur form in only $2m$ or $2m + 1$ iterations, where m is the dimension of the transfer-function matrix. (This figure may not hold for large m, but it should be borne in mind that m is rarely as large as 5, and probably never as large as 10.) The published literature apparently contains no contribution to this problem.

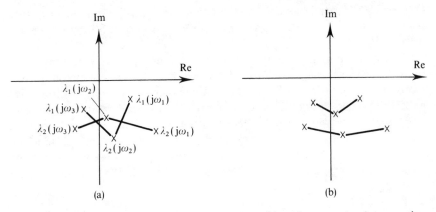

Figure 8.1 Segments of two characteristic loci, (a) without sorting into continuous loci and (b) with sorting.

Principal gains (singular values of frequency responses) are usually computed using the standard algorithm – see Section 8.2.7. This algorithm produces them already ordered by magnitude, so there is no problem in displaying them on a Bode plot.

8.3.3 Interconnections of systems

Operations which are frequently performed on systems are series, parallel and feedback connections.

If the systems are defined by their frequency responses, and the frequency response of the connection is adequate for one's needs, then all that is required is the evaluation of the frequency response at each of a set of frequencies. For example, the response of the feedback connection of $G(s)$ with $H(s)$ (see Figure 8.2) is computed as

$$T(j\omega) = [I + G(j\omega)H(j\omega)] \backslash G(j\omega) \qquad (8.62)$$

or

$$T(j\omega) = ([I + H(j\omega)G(j\omega)]^T \backslash G(j\omega)^T)^T \qquad (8.63)$$

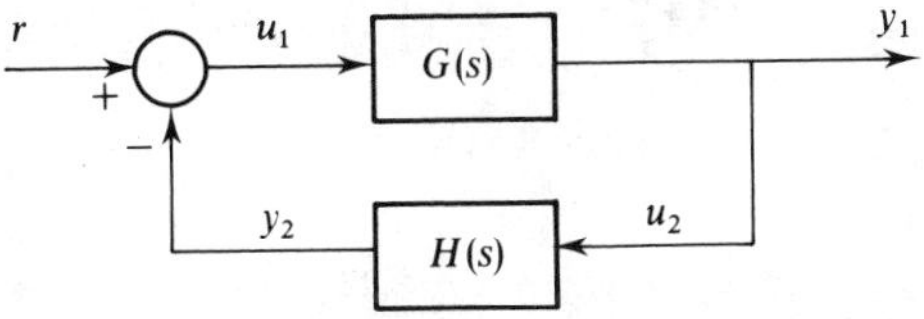

Figure 8.2 Negative-feedback connection of two systems $G(s)$ and $H(s)$.

Of these, (8.63) is to be preferred if $G(s)$ has fewer inputs than outputs, since the dimension of $I + HG$ is smaller than that of $I + GH$; otherwise (8.62) is better.

Since the manipulation of algebraic representations of transfer functions is very unreliable and therefore best avoided (except possibly in a symbolic-computation environment), frequency-response representations are inevitable when working with systems which have irrational transfer functions – most commonly, systems with time delays. But if state-space models are available (or can be obtained from rational transfer-function models), then it is much more efficient, and usually more useful, to compute the state-space representations of the connected system. We shall concentrate on doing this in the rest of this section.

SERIES CONNECTION

We suppose we are given two systems, defined by

$$\dot{x}_1 = A_1 x_1 + B_1 u_1 \tag{8.64}$$

$$y_1 = C_1 x_1 + D_1 u_1 \tag{8.65}$$

and

$$\dot{x}_2 = A_2 x_2 + B_2 u_2 \tag{8.66}$$

$$y_2 = C_2 x_2 + D_2 u_2 \tag{8.67}$$

and we connect them in series by setting

$$u_2 = y_1 \tag{8.68}$$

Substituting (8.65) into (8.66) gives

$$\dot{x}_2 = A_2 x_2 + B_2 C_1 x_1 + B_2 D_1 u_1 \tag{8.69}$$

Putting (8.64) together with (8.69) gives

$$\begin{bmatrix} \dot{x}_1 \\ \dot{x}_2 \end{bmatrix} = \begin{bmatrix} A_1 & 0 \\ B_2 C_1 & A_2 \end{bmatrix} \begin{bmatrix} x_1 \\ x_2 \end{bmatrix} + \begin{bmatrix} B_1 \\ B_2 D_1 \end{bmatrix} u_1 \tag{8.70}$$

and substituting (8.65) into (8.67) gives

$$y_2 = [D_2 C_1, C_2] \begin{bmatrix} x_1 \\ x_2 \end{bmatrix} + D_2 D_1 u_1 \tag{8.71}$$

Thus a state-space realization of the series connection is

$$\left(\begin{bmatrix} A_1 & 0 \\ B_2 C_1 & A_2 \end{bmatrix}, \begin{bmatrix} B_1 \\ B_2 D_1 \end{bmatrix}, [D_2 C_1 \quad C_2], D_2 D_1 \right)$$

There may be cancellations between the poles of one system and the transmission zeros of the other, in which case this realization will not be minimal. Often this does not matter, but if it does then a minimal realization can be obtained as described in Section 8.3.5.

PARALLEL CONNECTION

To effect a parallel connection between the two systems defined by (8.64)–(8.67), we set

$$u_1 = u_2 \tag{8.72}$$

and we form a new output

$$y = y_1 + y_2 \tag{8.73}$$

It is easy to see that a realization of the resulting system is

$$\left(\begin{bmatrix} A_1 & 0 \\ 0 & A_2 \end{bmatrix}, \begin{bmatrix} B_1 \\ B_2 \end{bmatrix}, [C_1 \quad C_2], D_1 + D_2 \right)$$

Again, this realization may not be minimal.

FEEDBACK CONNECTION

We keep the two models (8.64)–(8.67) and connect them together as shown in Figure 8.2 by setting

$$u_1 = r - y_2 \tag{8.74}$$

and

$$u_2 = y_1 \tag{8.75}$$

Since we again have a series connection of the two systems, (8.70) holds again, and the appropriate output equation is now

$$y_1 = [C_1 \quad 0] \begin{bmatrix} x_1 \\ x_2 \end{bmatrix} + D_1 u_1 \tag{8.76}$$

Now, from (8.74) and (8.71) we have

$$(I + D_2 D_1)u_1 = r - [D_2 C_1 \quad C_2] \begin{bmatrix} x_1 \\ x_2 \end{bmatrix} \qquad (8.77)$$

so (8.70) becomes

$$\begin{bmatrix} \dot{x}_1 \\ \dot{x}_2 \end{bmatrix} = \begin{bmatrix} A_1 & 0 \\ B_2 C_1 & A_2 \end{bmatrix} \begin{bmatrix} x_1 \\ x_2 \end{bmatrix} + \begin{bmatrix} B_1 \\ B_2 D_1 \end{bmatrix} (I + D_2 D_1)^{-1}$$

$$\times \left\{ r - [D_2 C_1 \quad C_2] \begin{bmatrix} x_1 \\ x_2 \end{bmatrix} \right\} \qquad (8.78)$$

$$= \begin{bmatrix} A_1 - B_1(X \backslash D_2)C_1 & -B_1(X \backslash C_2) \\ B_2[I - D_1(X \backslash D_2)]C_1 & A_2 - B_2 D_1(X \backslash C_2) \end{bmatrix} \begin{bmatrix} x_1 \\ x_2 \end{bmatrix}$$

$$+ \begin{bmatrix} B_1/X \\ B_2 D_1/X \end{bmatrix} r \qquad (8.79)$$

where

$$X = I + D_2 D_1 \qquad (8.80)$$

Here B_1/X means $(X^T \backslash B_1^T)^T$ (and D_1/X similarly), and each of the matrices $(X \backslash D_2)$ and $(X \backslash C_2)$ need be computed only once. The output equation (8.76) becomes

$$y_1 = [(I - D_1(X \backslash D_2))C_1 \quad -D_1(X \backslash C_2)] \begin{bmatrix} x_1 \\ x_2 \end{bmatrix} + (D_1/X)r \quad (8.81)$$

Again, the realization given by (8.79) and (8.81) may not be minimal. Note that when, as often happens, $D_1 = 0$ and $D_2 = 0$, the realization simplifies to

$$\left(\begin{bmatrix} A_1 & -B_1(X \backslash C_2) \\ B_2 C_1 & A_2 \end{bmatrix}, \begin{bmatrix} B_1/X \\ 0 \end{bmatrix}, [C_1 \ 0], 0 \right)$$

If the matrix $I + D_2 D_1$ is singular, the feedback connection is not well-defined, and incorrect modelling of the two systems is indicated.

CONNECTION OF GENERAL BLOCK-DIAGRAMS

At first sight, it would seem that not all possible block diagrams can be assembled by series, parallel and feedback connections. For example, the system shown in Figure 8.3(a) appears to require some other operation. In fact, it is possible to assemble all block diagrams by using one feedback and two series connections, and regarding all the subsystems as parts of a single

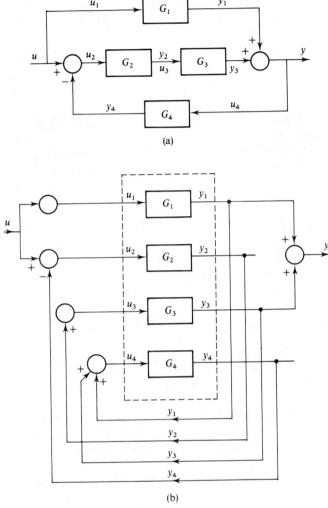

Figure 8.3   (a) A system defined by an interconnection of subsystems, and (b) a standard representation of this system.

large system, as shown by the dashed-line rectangle in Figure 8.3(b). A feedback connection is then made between this large system and a constant (non-dynamic) system in the feedback path, which, for the example shown in Figure 8.3, is

$$\begin{bmatrix} 0 & 0 & 0 & 0 \\ 0 & 0 & 0 & -I \\ 0 & I & 0 & 0 \\ I & 0 & I & 0 \end{bmatrix}$$

(and the identity matrices have appropriate dimensions). Series connections with constant systems are needed to distribute the input to the appropriate subsystems, and to form the output.

GENERAL REMARK ON INTERCONNECTIONS

When several state-space models are connected together by repeated use of series, parallel and feedback connections, the computed behaviour of the resulting system often differs significantly from its true behaviour: the conditioning of the problem has been destroyed by the successive connections. This tends to happen, for example, when a system which has only slow poles is connected to one which has only fast poles. In this case a matrix such as

$$\begin{bmatrix} A_1 & 0 \\ B_2 C_1 & A_2 \end{bmatrix}$$

has a high condition number, and so has

$$j\omega I - \begin{bmatrix} A_1 & 0 \\ B_2 C_1 & A_2 \end{bmatrix}$$

at low frequencies. Frequency responses may therefore be computed with significant errors at certain frequencies.

Sources of and cures for this problem are not yet properly understood, but empirical experience shows that the problem is largely avoided if, after every connection, a minimal realization is obtained (see Section 8.3.5) and this is then balanced (see Section 8.3.9). Connecting together only minimal, balanced models appears to be an advisable precaution. However, transformation to a minimal, balanced realization destroys any physical significance of the state variables, and this is sometimes undesirable.

8.3.4   Inverse systems

Given the system

$$\dot{x} = Ax + Bu \tag{8.82}$$

$$y = Cx + Du \tag{8.83}$$

with transfer function $G(s)$, we can obtain a state-space realization for the inverse system $G^{-1}(s)$ if the system is square, and D is non-singular. In this case we write (8.83) as

$$u = -D^{-1}Cx + D^{-1}y \tag{8.84}$$

which we substitute into (8.82):

$$\dot{x} = (A - BD^{-1}C)x + BD^{-1}y \tag{8.85}$$

Since the required inverse system has y as its input and u as its output, it is clear that the required realization can be computed as $([A - B(D\backslash C)], B/D, -D\backslash C, D\backslash I)$, where again B/D means $(D^T\backslash B^T)^T$, and the '$\backslash$' operation is defined as in Section 8.2.2.

If the system has l inputs and m outputs, and $l > m$ and rank$(D) = m$, then a **right inverse** $D^\dagger$ of D exists such that

$$DD^\dagger = I_m \tag{8.86}$$

It turns out that we can obtain a realization of the right-inverse system $G^\dagger(s)$ such that

$$G(s)G^\dagger(s) = I_m \tag{8.87}$$

simply by replacing D^{-1} by $D^\dagger$ in (8.84) and (8.85). We can prove this as follows. Using our earlier results on series connections, we see that the realization of the system formed by connecting the output of $G^\dagger(s)$ to the input of $G(s)$ is

$$\left(\begin{bmatrix} A - BD^\dagger C & 0 \\ -BD^\dagger C & A \end{bmatrix}, \begin{bmatrix} BD^\dagger \\ BD^\dagger \end{bmatrix}, [-C \quad C], I_m \right) \tag{8.88}$$

If we call the matrices in this realization $\tilde{A}$, $\tilde{B}$, $\tilde{C}$ and $\tilde{D}$, and apply the state-coordinate transformation

$$T = \begin{bmatrix} I & 0 \\ I & I \end{bmatrix}$$

we obtain the realization

$$(T^{-1}\tilde{A}T, T^{-1}\tilde{B}, \tilde{C}T, I_m) = \left(\begin{bmatrix} A - BD^\dagger C & 0 \\ 0 & A \end{bmatrix}, \right.$$

$$\left. \begin{bmatrix} BD^\dagger \\ 0 \end{bmatrix}, [0 \quad C], I_m \right) \tag{8.89}$$

for the series connection. Applying the usual formula (8.56), we see that this has transfer function I_m, so (8.87) is verified.

The right inverse $D^\dagger$ is not unique. One choice for it is the pseudo-inverse which can be obtained from the singular-value decomposition of D, as in equation (8.52). Alternatively, if we let $X = D^\dagger C$, so that $DX = C$, we see that this is an undetermined linear equation which can be solved using the QR factorization as given by equations (8.29), (8.31) and (8.34). Applying these equations in this case, we obtain

$$[Q, R, P] = QR(D^T) \tag{8.90}$$

$$Z^T[R_{11}, R_{12}] = C^T P \tag{8.91}$$

$$X = Q_1 Z \tag{8.92}$$

To compute $BD^\dagger$ and $D^\dagger$ itself, we can regard these as $B(D^\dagger I_m)$ and $(D^\dagger I_m)$, respectively, let $Y = D^\dagger I_m$, so that $DY = I_m$, and proceed similarly. If D has a high condition number, then the singular-value decomposition is to be preferred. If the system has more outputs than inputs, so that $l < m$ and rank$(D) = l$, then a **left-inverse** system $G^\dagger(s)$ can be obtained such that $G^\dagger(s)G(s) = I_l$, in the same way.

When working with continuous-time models, the rank conditions on D, which are required for these inverse realizations to exist, do not usually hold. Indeed, very often $D = 0$. (The inverses can still be defined by transfer functions, but these are now improper and so do not have state-space realizations.) However, one often requires a system which shows inverse characteristics over some finite frequency range only. In this case, it may be possible to modify D in such a way that the original system's frequency response is almost unchanged over the frequency range of interest, but with D having the required rank. This should be done in such a way that D is not too ill-conditioned, which should be easy if originally $D = 0$, but may be difficult or impossible otherwise.

Note that inverse systems are rarely of use as compensators since they produce large control signals, amplify high-frequency noise and may introduce unstable pole–zero cancellations. They can be useful inside design algorithms, however.

8.3.5   Minimal realizations

Given a state-space model (A, B, C, D), we can obtain a minimal realization having the same transfer function by removing first any uncontrollable modes, and then any unobservable modes. The theoretical basis of the algorithm which we shall describe is the Popov–Belevitch–Hautus (PBH) rank test for controllability and observability (Kailath, 1980):

PBH rank test: The pair (A, B) is controllable if and only if $[sI - A, B]$ has full row rank for all (complex) s.
The pair (C, A) is observable if and only if $[sI - A^T \quad - C^T]^T$ has full column rank for all (complex) s.

We shall now describe the **staircase algorithm** (van Dooren, 1981), which can be proved to be numerically stable.

First obtain a singular-value decomposition of B (see Section 8.2.7):

$$B = U_1 \Sigma_1 V_1^H \tag{8.93}$$

and hence obtain a row compression of B:

$$U_1^H B = \begin{bmatrix} Z_1 \\ 0 \end{bmatrix} \tag{8.94}$$

in which Z_1 has full row rank. Let

$$U_1^H A U_1 = \begin{bmatrix} Y_1 & X_1 \\ B_1 & A_1 \end{bmatrix} \tag{8.95}$$

in which X_1 and Y_1 have the same number of rows as Z_1, and A_1 and Y_1 are square. Now obtain a singular-value decomposition of B_1:

$$B_1 = U_2 \Sigma_2 V_2^H \tag{8.96}$$

and, if B_1 has neither zero nor full row rank, let

$$U_2^H B_1 = \begin{bmatrix} Z_2 \\ 0 \end{bmatrix} \tag{8.97}$$

Now observe that

$$\begin{bmatrix} I & 0 \\ 0 & U_2^H \end{bmatrix} \begin{bmatrix} Y_1 & X_1 \\ B_1 & A_1 \end{bmatrix} \begin{bmatrix} I & 0 \\ 0 & U_2 \end{bmatrix} = \begin{bmatrix} Y_1 & X_1 U_2 \\ U_2^H B_1 & U_2^H A_1 U_2 \end{bmatrix} \tag{8.98}$$

$$= \left[\begin{array}{c|c} Y_1 & X_1 U_2 \\ \hline \begin{bmatrix} Z_2 \\ 0 \end{bmatrix} & \begin{bmatrix} Y_2 & X_2 \\ B_2 & A_2 \end{bmatrix} \end{array} \right] \tag{8.99}$$

so that the state-coordinate transformation

$$U_1 \begin{bmatrix} I & 0 \\ 0 & U_2 \end{bmatrix}$$

transforms the pair (B, A) into the pair

$$\left(\begin{bmatrix} Z_1 \\ 0 \end{bmatrix}, \begin{bmatrix} Y_1 & X_1 U_2 \\ \begin{bmatrix} Z_2 \\ 0 \end{bmatrix} & \begin{bmatrix} Y_2 & X_2 \\ B_2 & A_2 \end{bmatrix} \end{bmatrix} \right) \tag{8.100}$$

This process can be repeated until eventually it terminates with $B_k = 0$, or B_k having full row rank. If $B_k = 0$ then the pair (B, A) is eventually transformed into the 'staircase' form

$$\left(\begin{bmatrix} Z_1 \\ 0 \\ 0 \\ \vdots \\ 0 \\ \hline 0 \end{bmatrix}, \begin{bmatrix} Y_1 & * & \dots & * & * & \vdots & * \\ Z_2 & * & \dots & * & * & \vdots & * \\ 0 & Z_3 & \dots & * & * & \vdots & * \\ \vdots & \vdots & & \vdots & \vdots & \vdots & \vdots \\ 0 & 0 & \dots & Z_k & Y_k & \vdots & X_k \\ \hline 0 & 0 & \dots & 0 & 0 & \vdots & A_k \end{bmatrix} \right) \tag{8.101}$$

in which each asterisk denotes a matrix whose value is not important.

If we partition this pair into

$$\left(\begin{bmatrix} B_c \\ \hline 0 \end{bmatrix}, \begin{bmatrix} A_c & \vdots & * \\ \hline 0 & \vdots & A_k \end{bmatrix} \right)$$

then we see that $[B_c, sI - A_c]$ has full row rank for all s (since the Z_i have full row rank), and $[0, sI - A_k]$ loses rank whenever s is an eigenvalue of A_k. This partitioning therefore separates out the controllable and uncontrollable 'parts' of the pair (A, B).

To be more precise, let

$$T_i = \begin{bmatrix} I & 0 \\ 0 & U_i \end{bmatrix} \tag{8.102}$$

in which the dimensions of I are the same as those of Y_{i-1}, and let

$$T = T_1 T_2 \dots T_k \tag{8.103}$$

Then

$$
\begin{bmatrix} I & 0 \\ 0 & T^{-1} \end{bmatrix} \begin{bmatrix} D & -C \\ B & sI - A \end{bmatrix} \begin{bmatrix} I & 0 \\ 0 & T \end{bmatrix}
$$

$$
= \left[\begin{array}{cc|c} D & -C_c & * \\ B_c & sI - A_c & * \\ \hline 0 & 0 & sI - A_k \end{array} \right] \tag{8.104}
$$

so that (A_c, B_c, C_c, D) is a controllable realization with the same transfer function as (A, B, C, D).

Note, in passing, that the row dimensions of the Z_i are the **controllability (Kronecker) indices** of the pair (A, B), which play an important role in multivariable-system identification. Also, the eigenvalues of A_k are the so-called **input decoupling zeros**.

The realization (A_c, B_c, C_c, D) may not be minimal, since the pair (C, A) may be unobservable. Since (C, A) is observable if and only if (A^T, C^T) is controllable, the algorithm described above can be used to remove the unobservable modes. By this means (and by employing conjugate transposes in (8.102) as appropriate) the pair $[C_c^T \ A_c^T]^T$ is transformed into the pair

$$
\begin{bmatrix} [C_{cO} & 0] \\ \begin{bmatrix} A_{cO} & 0 \\ * & A_l \end{bmatrix} \end{bmatrix} \tag{8.105}
$$

and the matrix 'pencil' $\begin{bmatrix} D & -C_c \\ B_c & sI - A_c \end{bmatrix}$ is transformed into

$$
\left[\begin{array}{cc|c} D & -C_{cO} & 0 \\ B_{cO} & sI - A_{cO} & 0 \\ \hline * & * & sI - A_l \end{array} \right] \tag{8.106}
$$

so that $(A_{cO}, B_{cO}, C_{cO}, D)$ is a minimal realization with the same transfer function as (A, B, C, D).

When we compute the minimal realization, we have to decide whether certain quantities are zero or not. The rank of each of the matrices $B, B_1, \ldots, B_k$ in equations (8.94) and (8.96) must be determined. This is done by examining the singular values in $\Sigma_1, \Sigma_2, \ldots$ and deciding how many of them are zero, either in an 'automatic' way, by comparing their magnitudes with the machine precision, or by comparison with a user-defined threshold value.

8.3.6 Partial-fraction decomposition

We shall now describe how a state-space system can be decomposed into two (or more) parallel systems, the eigenvalues (poles) of the original system being divided between the two systems, and no new eigenvalues being introduced. The most common application of this is to the decomposition of a system into its stable and unstable parts. If the original system has state dimension n, and all its eigenvalues are distinct (and sufficiently distinct to be numerically distinguishable), then it can be decomposed into n parallel first-order systems, each corresponding to one of the eigenvalues. Such a decomposition corresponds to a partial-fraction decomposition of the transfer function. (In this case some of the matrices A may be complex, and it may be more useful to combine complex-conjugate pairs of eigenvalues into real second-order systems.) However, if some of the eigenvalues are repeated then we cannot reliably split up the repeated eigenvalues. In this case it is more accurate to speak of a 'direct-sum decomposition'.

In Section 8.3.3 we saw that one realization of a parallel connection of state-space systems has a block-diagonal matrix A. So the problem of decomposing a system into a parallel connection of subsystems is equivalent to the problem of finding a state-coordinate transformation which results in the matrix A being block-diagonal, each of the diagonal blocks having the required eigenvalues. This problem can be solved in three steps: first reduce the matrix A to Schur form (usually to real Schur form), secondly reorder the Schur form so that the eigenvalues (diagonal elements) are grouped as required, and thirdly reduce the remaining off-diagonal blocks to zero without disturbing the diagonal blocks.

To describe this procedure in detail, we assume that the decomposition is to be into two (real) subsystems. Each resulting subsystem can then be decomposed further, if required. First, find a unitary similarity transformation Q_0 which reduces A to real Schur form (see Section 8.2.5):

$$Q_0^{\mathrm{H}} A Q_0 = \begin{bmatrix} A_{11} & A_{12} \\ 0 & A_{22} \end{bmatrix} \tag{8.107}$$

where A_{11} and A_{22} are upper quasi-triangular, and their dimensions are determined by the required state dimensions of the parallel subsystems – for example, if A has five eigenvalues, of which two are stable, and a stable/ unstable decomposition is required, then A_{11} has dimension 2 and A_{22} has dimension 3. After this transformation the eigenvalues are distributed randomly between A_{11} and A_{22}, so we now have to find a state-coordinate transformation which distributes them as required (to continue the example, the two stable eigenvalues in A_{11}, the remaining three in A_{22}).

We proceed by finding a sequence of state-coordinate transformations, each of which interchanges a pair of adjacent diagonal elements of $Q_0^H A Q_0$, by means of so-called **Givens rotations**. A Givens rotation $J(i, j, \theta)$ is an orthogonal matrix which is almost an identity, but with the elements (i, i), (i, j), (j, i) and (j, j) replaced by the elements of the rotation matrix

$$\begin{bmatrix} c & s \\ -s & c \end{bmatrix}$$

where $c = \cos \theta$ and $s = \sin \theta$:

$$J(i, j, \theta) = \begin{bmatrix} 1 \\ & \ddots \\ & & 1 \\ & & & c & 1 & & s \\ & & & & \ddots \\ & & & -s & & 1 & c & 1 \\ & & & & & & & \ddots \\ & & & & & & & & 1 \end{bmatrix} \begin{matrix} \\ \\ \\ \leftarrow \text{ row } i \\ \\ \leftarrow \text{ row } j \\ \\ \\ \end{matrix} \qquad (8.108)$$

$$\underset{\text{column } i \quad \text{column } j}{\uparrow \qquad \uparrow}$$

Suppose that the 2×2 block at the intersection of rows and columns k and $k + 1$ of $Q_0^H A Q_0$ has the form

$$\begin{bmatrix} t_{kk} & t_{k, k+1} \\ 0 & t_{k+1, k+1} \end{bmatrix}$$

Find θ such that

$$\begin{bmatrix} c & s \\ -s & c \end{bmatrix} \begin{bmatrix} t_{k, k+1} \\ t_{kk} - t_{k+1, k+1} \end{bmatrix} = \begin{bmatrix} * \\ 0 \end{bmatrix} \qquad (8.109)$$

(where the value of $*$ is unimportant). Then

$$J(k, k+1, \theta) Q_0^H A Q_0 J(k, k+1, \theta)^T$$

(which remains similar to A) will remain quasi-triangular, it will differ from $Q_0^H A Q_0$ only in the kth and $(k+1)$th rows and columns, and the 2×2 block at the intersection of these rows and columns will have the form

$$\begin{bmatrix} t_{k+1,k+1} & \pm t_{k,k+1} \\ 0 & t_{kk} \end{bmatrix}$$

If $Q_0^H A Q_0$ is triangular – namely, if all the eigenvalues of A are real – then the following algorithm will give the required transformation, if Λ denotes the set of eigenvalues which is to appear in the top left-hand block, and $T = Q_0^H A Q_0$:

$p := \dim(A_{11})$;

$Q := Q_0$;

while $\{t_{11}, t_{22}, \ldots, t_{pp}\} \neq \Lambda$ **do**

begin for $k := 1$ **to** $n - 1$ **do**

if $t_{kk} \notin \Lambda$ and $t_{k+1,k+1} \in \Lambda$ **then**

begin

find θ as above;

$Q := Q J(k, k+1, \theta)^T$;

$T := J(k, k+1, \theta) T J(k, k+1, \theta)^T$;

end.

end.

When this algorithm terminates, we have $T = Q^H A Q$, with T upper-triangular, the top left-hand $p \times p$ block having Λ as its set of eigenvalues.

If A has complex eigenvalues, so that T is quasi-triangular but not triangular, then the algorithm becomes more complicated and rather different in nature. In this case, if we wish to rearrange the eigenvalues of the blocks

$$\begin{bmatrix} T_{ii} & T_{i,i+1} \\ 0 & T_{i+1,i+1} \end{bmatrix}$$

where one or both of T_{ii}, $T_{i+1,i+1}$ are 2×2 matrices, one step of the QR algorithm (see Section 8.2.5) is performed, with the shift chosen to reverse the ordering of the eigenvalues – here we can choose the shift to accomplish this reversal, since we already know the eigenvalues of T_{ii}. The algorithm generally used for this, and for performing Givens rotations on 1×1 blocks, is known as EXCHNG, and is given by Stewart (1976).

In practice it is never necessary to isolate single complex eigenvalues, so it is never necessary to work with the complex Schur form. But in principle it is possible to do so; in this case the Givens rotations have to be generalized to unitary matrices of the form

$$\begin{bmatrix} c & \bar{s} \\ -s & c \end{bmatrix}$$

in which c is real, and $c^2 + s\bar{s} = 1$.

We have now obtained

$$Q^H A Q = T = \begin{bmatrix} T_{11} & T_{12} \\ 0 & T_{22} \end{bmatrix} \tag{8.110}$$

and must find a further state-coordinate transformation to reduce T_{12} to zero. If T_{11} and T_{12} have no common eigenvalues then this can always be done, using a transformation

$$Y = \begin{bmatrix} I & Y_{12} \\ 0 & I \end{bmatrix} \tag{8.111}$$

in which the dimensions of the blocks are the same as those in T (Golub and van Loan, 1983). Suppose that

$$Y^{-1} T Y = \bar{T} = \begin{bmatrix} T_{11} & 0 \\ 0 & T_{22} \end{bmatrix} \tag{8.112}$$

then

$$TY = Y\bar{T} \tag{8.113}$$

leads to the **Sylvester equation**

$$T_{11} Y_{12} - Y_{12} T_{22} = - T_{12} \tag{8.114}$$

which must be solved for Y_{12}.

Since T_{11} and T_{22} are quasi-triangular, or even triangular, this equation can be solved efficiently by a kind of back-substitution process. The best algorithm for doing this is the one proposed by Bartels and Stewart (1972), and described by Golub and van Loan (1983).

Once equation (8.114) has been solved, the state-coordinate transformation required to decompose the original system is

$$R = QY \tag{8.115}$$

so that the decomposition is obtained by simply partitioning the matrices of the realization $(R^{-1}AR, R^{-1}B, CR, D)$. The matrix D is not partitioned, of course, but may be regarded as a separate parallel subsystem. Note that, although Y is not orthogonal, Y^{-1} can be computed stably since

$$Y^{-1} = \begin{bmatrix} I & -Y_{12} \\ 0 & I \end{bmatrix} \qquad (8.116)$$

The process of rearranging eigenvalues in the Schur form can lead to significant changes in the values of any ill-conditioned eigenvalues. It is therefore not advisable to try to obtain a decomposition into subsystems which have nearly equal eigenvalues. This can create problems when one is trying to separate out the stable and unstable parts of a system, if there are eigenvalues which are close to each other but lie on opposite sides of the stability boundary.

8.3.7 Transmission zeros

We now consider the computation of transmission zeros from the state-space realization (A, B, C, D) of a linear system. If D is invertible, these zeros are just the eigenvalues of $A - BD^{-1}C$, but any attempt to use this fact directly for computation leads to a numerically unstable algorithm. It is also of limited use, since D is frequently singular. If D is singular then there are fewer than n finite transmission zeros (where n is the dimension of A), and zeros appear 'at infinity'. Naive attempts to compute the zeros usually end up transforming these infinite zeros to arbitrary locations in the complex plane, where their presence may destroy the conditioning of the true (finite) zeros.

The matrix

$$P(s) = \begin{bmatrix} sI - A & B \\ -C & D \end{bmatrix} \qquad (8.117)$$

which appeared in equation (8.104) (with its rows and columns rearranged) is known as **Rosenbrock's system matrix** (Rosenbrock, 1970). It loses rank at those points in the complex plane which are either eigenvalues of A corresponding to uncontrollable or unobservable modes (input or output decoupling zeros), or transmission zeros of $G(s) = C(sI - A)^{-1}B + D$. If (A, B, C, D) is a minimal realization, then the system matrix loses rank only at the transmission zeros, so we shall assume that we start with a minimal realization (obtained as in Section 8.3.5).

A numerically stable algorithm for finding transmission zeros is obtained by first performing row and column compressions on (parts of) the system matrix, in order to remove any zeros at infinity, and then finding the zeros by solving a generalized eigenvalue problem.

We begin by obtaining a singular-value decomposition of D:

$$D = U_1 \Sigma_1 V_1^H \tag{8.118}$$

and using it to compress the rows of D, simultaneously transforming C:

$$U_1^H [C, D] = \begin{bmatrix} \bar{C}_0 & \bar{D}_0 \\ \tilde{C}_0 & 0 \end{bmatrix} \tag{8.119}$$

If $D = 0$ we set $\tilde{C}_0 = C$.

Next we compress the columns of $\tilde{C}_0$, by first obtaining another singular-value decomposition:

$$\tilde{C}_0 = P_1 \Lambda_1 Q_1^H \tag{8.120}$$

$$\tilde{C}_0 Q_1 = [0 \quad S_1] \tag{8.121}$$

(where we have assumed that the singular values in Λ_1 have been reordered to place the zero singular values in the top left-hand corner, in order to compress the columns of $\tilde{C}_0$ to the right rather than the left).

Let σ_1 be the number of rows in $\bar{C}_0$ and $\bar{D}_0$, and p be the number of columns of D (that is, the number of system inputs). Now define

$$\begin{bmatrix} A_1 & * & B_1 \\ \hline C_1 & * & D_1 \end{bmatrix} = \begin{bmatrix} Q_1^H & 0 \\ \hline 0 & I_{\sigma_1} \end{bmatrix} \begin{bmatrix} A & B \\ \hline \bar{C}_0 & \bar{D}_0 \end{bmatrix}$$

$$\times \begin{bmatrix} Q_1 & 0 \\ \hline 0 & I_p \end{bmatrix} \tag{8.122}$$

where the blocks denoted by $*$ have the same number of columns as S_1; A_1 is square, with the same number of columns as the zero block in (8.121); and B_1 and D_1 have p columns.

This whole process is now repeated, starting with (A_k, B_k, C_k, D_k), where $k = 1$ initially, to generate the sequence $\bar{C}_k$, $\tilde{C}_k$, $\bar{D}_k$, S_{k+1}, A_{k+1}, B_{k+1}, C_{k+1}, D_{k+1} ($k = 1, 2, \ldots$), and is terminated either when D_k has full row rank, or when $\tilde{C}_k$ has either zero or full column rank. If at any stage $\tilde{C}_k$ has full column rank, then the system has no (finite) transmission zeros, and there is no more to be done.

When either D_k has full row rank, or $\tilde{C}_k = 0$, we set $A_r := A_k$, $B_r := B_k$, $C_r := \bar{C}_k$ and $D_r := \bar{D}_k$, and proceed to the dual process described below. Note that D_r has full row rank.

In the jth step of the above process we effectively perform the following transformation, if we adopt the convention that (A_0, B_0, C_0, D_0) $= (A, B, C, D)$:

$$\begin{bmatrix} Q_j^H & 0 \\ 0 & U_j^H \end{bmatrix} \begin{bmatrix} sI - A_{j-1} & B_{j-1} \\ -C_{j-1} & D_{j-1} \end{bmatrix} \begin{bmatrix} Q_j & 0 \\ 0 & I_p \end{bmatrix}$$

$$= \begin{bmatrix} sI - A_j & * & B_j \\ -C_j & * & D_j \\ 0 & -S_j & 0 \end{bmatrix} \tag{8.123}$$

so that on termination we have transformed the original system matrix to

$$\begin{bmatrix} sI - A_r & * & * & \cdots & * & B_r \\ -C_r & * & * & \cdots & * & D_r \\ 0 & -S_k & * & \cdots & * & 0 \\ & 0 & -S_{k-1} & \ddots & \vdots & \vdots \\ \mathbf{0} & & \ddots & \ddots & * & 0 \\ & & & 0 & -S_1 & 0 \end{bmatrix} \tag{8.124}$$

This matrix loses rank at the same values of s as the original system matrix, since the transformations employed at each step (8.123) are non-singular. But each S_j has full column rank, so (8.124) loses rank at the same points at which

$$\begin{bmatrix} sI - A_r & B_r \\ -C_r & D_r \end{bmatrix} \tag{8.125}$$

loses rank. In other words, the (finite) transmission zeros of (A, B, C, D) are the same as those of (A_r, B_r, C_r, D_r).

We now submit (A_r, B_r, C_r, D_r) to the dual process of that described above, in which the roles of rows and columns are systematically inter-changed. (Equivalently, we could submit $(A_r^T, C_r^T, B_r^T, D_r^T)$ to the same process as that above, and apply a corresponding transformation to the result.) This will either show that there are no transmission zeros, or termin-ate yielding $(A_{rc}, B_{rc}, C_{rc}, D_{rc})$ in which D_{rc} has full column rank, by con-struction. But the full row rank of D_r is not destroyed by the dual algorithm, so D_{rc} must be square and invertible. Hence $(A_{rc}, B_{rc}, C_{rc}, D_{rc})$ has finite transmission zeros only, and these are the same as the transmission zeros of (A, B, C, D).

It remains to find these transmission zeros. For this we need yet another singular-value decomposition, this time of $[B_{rc}^T\, D_{rc}^T]^T$:

$$\begin{bmatrix} B_{rc} \\ D_{rc} \end{bmatrix} = X \Xi Y \tag{8.126}$$

and we use this to transform the system matrix of $(A_{rc}, B_{rc}, C_{rc}, D_{rc})$:

$$X^H \begin{bmatrix} sI - A_{rc} & B_{rc} \\ -C_{rc} & D_{rc} \end{bmatrix} = \begin{bmatrix} s\tilde{B} - \tilde{A} & 0 \\ * & \tilde{D} \end{bmatrix} \tag{8.127}$$

(where we have again used a reversed ordering of the singular values in Ξ, with respect to the usual ordering). Now, this transformed system matrix clearly loses rank only when $s\tilde{B} - \tilde{A}$ loses rank – in other words, at the generalized eigenvalues of the pair $(\tilde{A}, \tilde{B})$ (see Section 8.2.8). That is, the generalized eigenvalues of the pair $(\tilde{A}, \tilde{B})$ are the transmission zeros of the original system (A, B, C, D).

Computation of the eigenvalues of $A_{rc} - B_{rc} D_{rc}^{-1} C_{rc}$ has been avoided, since D_{rc} may be ill-conditioned with respect to inversion.

The algorithm described in this section is taken from van Dooren (1981a). A faster version, which uses Householder transformations instead of singular-value decompositions, is described by Emami-Naeini and van Dooren (1982), who give an extensive discussion of the performance of the algorithm and compare it with other proposed methods.

In the course of this algorithm the rank of certain matrices must be determined at each step. As discussed in Section 8.2.7, these decisions are usually taken by comparing singular values with the machine precision. However, there may be grounds for considering larger singular values to be zero, for example if one's data (namely the matrices A, B, C and D) is known to be approximate, and information about its precision is available.

8.3.8   Root loci

It is occasionally useful to follow the variations of the closed-loop characteristic roots when negative feedback of the form kI is placed around a square, multivariable system, as shown in Figure 8.4. Here k is a scalar parameter which is usually real and takes values in the range $0 \leqslant k < \infty$, or (less commonly) $-\infty < k < \infty$. From (8.79) we find that the closed-loop 'A' matrix in this case is

$$A_c = A - B(I + kD)^{-1} kC \tag{8.128}$$

if (A, B, C, D) is a realization of the system in the forward path. If we let

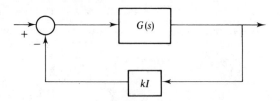

Figure 8.4 Multivariable root loci are usually investigated for this feedback configuration, as the gain *k* is varied.

$g = 1/k$, we obtain

$$A_c = A - B(gI + D)^{-1}C \tag{8.129}$$

and the closed-loop characteristic roots are simply the eigenvalues of A_c.

Note the identity of form between the two expressions

$$A_c(g) = A - B(gI + D)^{-1}C \tag{8.130}$$

and

$$G(s) = D + C(sI - A)^{-1}B \tag{8.131}$$

It implies that, if we have an algorithm for finding the characteristic loci of a system with realization (A, B, C, D), then we can find the multivariable root loci by applying the same algorithm to the realization $(-D, C, -B, A)$. (If we choose a positive-feedback convention we do not even need to change the signs of B and D.)

We have not assumed (A, B, C, D) to be minimal. Any uncontrollable or unobservable modes simply correspond to eigenvalues of $A_c(g)$ which do not vary with g. This can be used to give a simple visual test for minimality, when appropriate display facilities are available.

For a theoretical investigation of multivariable root loci see Postlethwaite and MacFarlane (1979) and the references cited therein.

8.3.9 Balanced realizations and model approximation

For a continuous-time, asymptotically stable system, a balanced realization (A_b, B_b, C_b, D_b) is one for which the Lyapunov equations

$$A_b P + P A_b^T + B_b B_b^T = 0 \tag{8.132}$$

and

$$A_b^T Q + Q A_b + C_b^T C_b = 0 \tag{8.133}$$

have a common solution

$$P = Q = \Sigma \tag{8.134}$$

and Σ is diagonal. Such a realization always exists.

Each state variable is, in a sense, equally strongly coupled to the input and the output, in a balanced realization, and this lies at the root of the usefulness of these realizations, particularly for model approximation.

We assume that we are given a minimal realization (A, B, C, D), which is not balanced. Exercise 6.12 has already given the basis of an algorithm, which could be implemented in four steps:

(1)   Solve the two Lyapunov equations, giving solutions P and Q.

(2)   Obtain the factorization $Q = R^T R$; this can be done using Cholesky decomposition (Golub and van Loan, 1983).

(3)   Find the singular-value decomposition

$$RPR^T = U\Sigma^2 U^T \tag{8.135}$$

(4)   Form the balanced realization

$$(A_b, B_b, C_b, D_b) = (TAT^{-1}, TB, CT^{-1}, D) \tag{8.136}$$

where

$$T = \Sigma^{-1/2} U^T R \tag{8.137}$$

This algorithm can be improved as follows. In step 1 use the algorithm proposed by Hammarling (1982), which never forms the Lyapunov solutions P and Q, but only their 'square roots', R and S respectively, such that

$$P = S^T S \quad \text{and} \quad Q = R^T R \tag{8.138}$$

Hammarling's algorithm is based on finding the real Schur form of A, then using a back-substitution process to solve for R or S, without ever forming the products $C^T C$ or BB^T. The resulting factor is triangular. Use of this algorithm clearly avoids the need for step 2, and one can go directly to step 3. Here again there is room for improvement, because one now needs the singular-value decomposition of the product

$$RS^T = U\Sigma V^T \tag{8.139}$$

which is best found by using the algorithm developed by Heath *et al.* (1986), without actually forming the product RS^T. Finally we set $T = \Sigma^{-1/2} U^T R$, as before, but we can now obtain T^{-1} as

$$T^{-1} = S^T V \Sigma^{-1/2} \tag{8.140}$$

This improved algorithm was put forward by Laub *et al.* (1987).

MODEL APPROXIMATION

One of the principal uses of balanced realizations is in the approximation of state-space systems by simpler ones which approximate the input/output behaviour of the original systems. The need for such approximations arises constantly when working with multivariable systems. As mentioned at the end of Section 6.6, Hankel approximation theory can be used to obtain approximations which are optimal in the sense that they minimize the Hankel norm of the approximation error, and of course balanced realizations play a crucial role in obtaining these approximations.

But in practice very good approximations can be obtained by using a rather simpler procedure (Moore, 1981) in which a balanced realization of the original system is obtained, and then simply 'truncated' by discarding those parts relating to the state variables which are most weakly coupled to the inputs and outputs. Suppose that we have a minimal balanced realization of an asymptotically stable system G, of McMillan degree N. Let the common solution of (8.132) and (8.133) be

$$\Sigma = \begin{bmatrix} \Sigma_1 & 0 \\ 0 & \Sigma_2 \end{bmatrix}$$

where

$$\Sigma_1 = \text{diag}\{\sigma_1, \sigma_2, \ldots, \sigma_k\}$$

and Σ_1, Σ_2 have no common eigenvalues. Now let the matrices of this realization be partitioned conformally with Σ:

$$G: \left(\begin{bmatrix} A_{11} & A_{12} \\ A_{21} & A_{22} \end{bmatrix}, \begin{bmatrix} B_1 \\ B_2 \end{bmatrix}, [C_1 \ C_2], D \right)$$

and let $\tilde{G}_k$ be the system with realization

$$\tilde{G}_k: (A_{11}, B_1, C_1, D)$$

Then it can be shown that $\tilde{G}_k$ is asymptotically stable, (A_{11}, B_1, C_1, D) is a minimal and balanced realization of $\tilde{G}_k$, and

$$\| G - \tilde{G}_k \|_\infty \leqslant 2 \sum_{i=1}^{N-k} \sigma_{k+i}$$

8.3.10   Algebraic Riccati equations

We have already seen, in Chapter 5, that the continuous-time algebraic
Riccati equation

$$A^\mathrm{T}P + PA - PBR^{-1}B^\mathrm{T}P + Q = 0 \tag{8.141}$$

in which $Q \geqslant 0$ and $R > 0$, can be solved by performing an
eigenvalue–eigenvector analysis of the Hamiltonian matrix

$$H = \begin{bmatrix} A & -BR^{-1}B^\mathrm{T} \\ -Q & -A^\mathrm{T} \end{bmatrix} \tag{8.142}$$

This method is not to be recommended, however, because of the difficulties
associated with computing eigenvectors. A better algorithm is obtained by
recognizing that it is sufficient to find a basis for the subspace spanned by the
eigenvectors which correspond to the stable eigenvalues of H (Laub, 1979).
Such a subspace can be determined reliably by reducing H to real Schur form,
and then reordering the eigenvalues in this form.

In the notation used in Section 8.2.5, we write

$$[X, S] = \mathrm{Schur}(H) \tag{8.143}$$

We then reorder the diagonal elements (and 2×2 blocks) of S, using the
algorithm described in Section 8.3.6, to get the partitioning

$$Z^\mathrm{T}SZ = \begin{bmatrix} S_{11} & * \\ 0 & S_{22} \end{bmatrix} \tag{8.144}$$

in which all the eigenvalues of S_{11} lie in the left half-plane, and all those of S_{22}
in the right half-plane. (Eigenvalues on the imaginary axis are precluded by
the standard assumption that $(A, B, Q^{1/2})$ is minimal.) Recall that there are
equal numbers of eigenvalues in each half-plane. Now let $T = XZ$, and
partition T into

$$T = \begin{bmatrix} T_{11} & T_{12} \\ T_{21} & T_{22} \end{bmatrix} \tag{8.145}$$

so that

$$H \begin{bmatrix} T_{11} \\ T_{21} \end{bmatrix} = \begin{bmatrix} T_{11} \\ T_{21} \end{bmatrix} S_{11} \tag{8.146}$$

Then the unique positive-semidefinite solution of (8.141) is given by

$$P = T_{21}T_{11}^{-1} \tag{8.147}$$

A weakness of this algorithm is that inversion of R is required in order to form the Hamiltonian H. If R is ill-conditioned this can introduce large errors. To overcome this, van Dooren (1981b) has shown that P can also be computed from the generalized Schur decomposition (see Section 8.2.8) of the pair $(I, \tilde{H})$, where

$$\tilde{H} = \begin{bmatrix} A & 0 & B \\ -Q & -A^\mathrm{T} & 0 \\ 0 & B^\mathrm{T} & R \end{bmatrix} \qquad (8.148)$$

This avoids the need to invert R.

8.4   Software for control engineering

8.4.1   Mathematical software

A good numerical algorithm is not the same thing as good mathematical software. A good implementation is necessary to transform one into the other. Attention must be paid to many small details to guard against unexpected conditions which may lead to overflows or underflows, to ensure that the sequence of computations is ordered so that it does not increase the sensitivity of the calculation unnecessarily, that the software will produce similar results when processed by various compilers on various machines, and to avoid unnecessarily lengthy computation. Even careful attention to all these details is of little use without good documentation, and without a good, well-defined user interface. Obviously, such documentation should specify the data required by the software and the results produced by it. But it should also define the class of problems for which the software produces reliable results, what happens if the boundary of this class is approached, and how failures and difficulties are signalled to the user.

Many of the 'building-block' algorithms which are needed for control-engineering software are already available in the form of high-quality mathematical software, and have been assembled into coherent subroutine libraries. The best-known of these are the Eispack and Linpack libraries for eigenproblems and linear systems of equations, respectively, and the NAG (Numerical Algorithms Group Ltd) and IMSL (International Mathematical and Statistical Libraries, Inc.) libraries, each of which covers the whole field of numerical analysis.

Similar libraries of routines, but aimed specifically at the needs of control engineers, have recently appeared. Examples of these are Slice, which was developed in the United Kingdom and is distributed by NAG, and Oracls, which was developed in the USA by NASA. Details of these and other libraries, and also of other software relevant to control engineering, can be found in the *Extended List of Control Software* (ELCS), which is a collection of one-page summaries of subroutine libraries and software packages, and which is revised regularly (Frederick *et al.*, 1987).[1] The Benelux Working Group on Software has also compiled an inventory of software for control (WGS, 1985).

Some tools are available to help the development of good mathematical software, such as the PFORT verifier which detects the presence of non-standard FORTRAN code in a program. Implementation is also made easier by the existence of IEEE Standard 754 for binary floating-point arithmetic (IEEE, 1981), and this is being adhered to by increasingly many processors. But this still leaves the development of good mathematical software as a very expensive activity which should not be undertaken lightly. Of course, this is not an argument for the development of low-quality software; that inevitably turns out to be even more expensive in the long run. Rather, it implies that low-level software development should be undertaken only as a last resort – the software you need is probably already available somewhere. For most projects, it should not even be necessary to write high-level software nowadays, except perhaps macros in some control-specific language.

8.4.2   Software packages

Of course, a control engineer (and any other user of software) does not want to work at the level of subroutine libraries – he needs a complete working program, or set of programs. The nature of control-system analysis and design, as of any engineering design process which contains unpredictable sequences of cut-and-try activities and relies heavily on graphical indicators of performance (time and frequency responses, mostly), dictates that such a program should be interactive, and should have graphical output facilities.

A number of software 'packages' now exist which integrate basic algorithms with data management, an interactive user interface and graphical output capability, and which are aimed specifically at control engineering. Such packages were first developed in Europe in the 1970s by university research groups. An early example of a very successful family of such packages, which still compares well with some more recent offerings, is the Intrac-based software developed at the Lund Institute of Technology in Sweden (Åström and Wieslander, 1981).

Of more direct interest to the topics covered in this book are the software packages which were developed in the UK to support research into multivariable frequency-domain techniques. One of these, the Cambridge Linear Analysis and Design Programs (CLADP), has now become a properly supported commercial product and offers a comprehensive software facility for multivariable feedback analysis and design (Edmunds, 1979; Maciejowski and MacFarlane, 1982). It provides not only basic facilities such as computation and display of Nyquist arrays, Gershgorin bands, characteristic loci and so on, but also facilities for manipulating polynomials and matrices (which makes it possible to perform LQG design, for example) and useful tools for performing tasks such as converting from transfer-function to state-space models, or from continuous time to discrete time, or converting block diagrams into equivalent single systems.

In about 1980 a new family of software packages began to appear in the USA. The Eispack and Linpack subroutine libraries had been provided with an interactive 'front end', named Matlab (Matrix Laboratory), which made it possible to enter command lines such as

$$V = \text{EIG}(A * B')$$

at a computer terminal, and obtain the eigenvalues of the matrix product AB^{T} almost immediately, stored in the vector variable V. Matlab also allowed 'macros' to be written, namely collections of Matlab statements which could be saved on file and executed at will. Since these statements included flow control statements such as

'if . . . then . . . else . . .'

this effectively constituted a high-level programming language specifically tailored to numerical linear algebra. This language was used as the basis of the control-oriented packages, CTRL-C (Little *et al.*, 1985) and Matrix$_x$ (Shah *et al.*, 1985). Basically, these packages added control-specific macros to Matlab, added good graphical output facilities, and provided interfaces to model-building and simulation software. With one of these packages available, the user was able to enter statements such as

$$K = \text{OPTREG}(A, B, Q, R)$$

to obtain the state-feedback matrix which implements a linear–quadratic optimal regulator, for example. The resulting time and frequency responses could then be displayed easily.

This approach to producing control software has proved very popular. Its major attraction is that the user is faced with a very uniform interface, he can easily mix high-level control-specific commands with low-level numerical or graphical commands, and he can easily tailor the package to his own needs by writing appropriate macros. Whereas the original Matlab was freely available in the public domain, a new commercial version is now available (for personal computers, called PC-Matlab, and for minicomputers and workstations, called Pro-Matlab). In this version, great attention has been paid to making it possible for the user to tailor the software for his own specific needs, and to extend its capabilities. For example, it is very easy to write one's own macros, with formal input and output parameters, some of which may be omitted and replaced by default values. Also, this version has simple but powerful graphics facilities (*plot* (*c*) will plot the vector *c* of complex numbers on the complex plane, for example), and the code has been optimized so that the speed of execution is impressively high, even on a personal computer. Collections of macros for use with this new version are called **toolboxes**, and at the time of writing a *Control System Toolbox* (Little and Laub, 1985), a *System Identification Toolbox*, a *Multivariable Frequency Domain Toolbox* (Boyle *et al.*, 1989) and a *Robust Control Toolbox* are available.

An inherent limitation of Matlab is that the only data type which it supports is the (complex) matrix. Although this is enough for most of state-space based linear systems theory, there are clearly objects which arise in control engineering which cannot be naturally represented as matrices. Obvious examples are multivariable time and frequency responses, which are more naturally represented as three-dimensional arrays. It is well known that the development of software is greatly eased if powerful data-structuring facilities are available, and such facilities form the central feature of modern programming languages such as Pascal or Ada. An interesting new package, which uses the Matlab style of interaction with the user, but which is written in Ada and offers the user an Ada-like interactive language, is Impact (Rimvall, 1987). This supports control-specific data types such as transfer-function matrices, linear-system descriptions and trajectories. However, this package is still under development, and for the time being must be viewed as a promising experiment in computer-aided design. Another package which aims to provide Matlab-like features with the addition of control-specific data structures is Basile (Delebecque *et al.*, 1988).

All the packages mentioned in this section, and many others, are listed in the ELCS, mentioned in Section 8.4.1. The ELCS includes information on the availability of packages, their functionality, hardware requirements and contact addresses. In this section we have discussed only those software packages which are specifically designed to support design techniques of the kind described in this book. There are, of course, many packages aimed at other control-engineering activities, the principal one being simulation. For information about these the reader is again referred to the ELCS.

Other sources of information about the state of computer-aided control-system design up to 1988 are: Herget and Laub (1982), Herget and Laub (1984), Jamshidi and Herget (1985), Hansen and Larsen (1985), IEEE (1986) and Chen (1988).

8.4.3   Current trends

Most of the activity in software development for control-system design which is currently going on, is concerned with taking full advantage of the rapid changes which are occurring in the computing environment, such as the advent of very powerful personal workstations. This activity is concentrated in three areas: integration of software packages into unified control-engineering 'environments'; increased use of high-resolution graphics, particularly for data input and problem specification; and the use of techniques from artificial intelligence. There is also some activity which is driven by theoretical advances – software packages for H_∞ design and adaptive control are beginning to appear, for example, and research is being pursued into the use of parallel computing for handling models with very large numbers of states, such as those obtained from finite-element software.

The interest in integration arises from the difficulties which exist at present with following a project through the stages of modelling and system identification, analysis and design, simulation and implementation. Software is available to support each of these stages, but there is frequently difficulty in transferring data between the various software packages in use. In a large project there is also the difficulty of keeping track of the data generated during the lifetime of a project. Several of the commercially available packages offer interfaces to other packages, two-way transfer of data between design and simulation software being quite common.

Within a few years we are likely to see control-oriented software environments which offer sophisticated data-management facilities, uniform interfaces to constituent packages, and considerable freedom to tailor such an environment to one's own needs by importing packages of one's own choice. To facilitate such a development, the International Federation of Automatic Control (IFAC) has recently begun the process of defining standards for data structures, user interfaces and numerical algorithms. Some engineering corporations have already made progress in this direction (Spang, 1984), and in the UK an integrated environment of this kind, called Ecstasy, has been developed (Munro, 1988).

Several proposals have been made to use techniques developed for expert systems to help in the design of control systems (James *et al.*, 1985; Birdwell *et al.*, 1985; Boyle *et al.*, 1989; Pang and MacFarlane, 1987). These proposals recognize that control design is a complex activity, and that mathematical design of the kind described in this book is performed relatively rarely, even by practising control engineers – most of their efforts are devoted to modelling and implementation. There is therefore much scope for using an expert system to guide a designer through the design process, for example in the following sequence:

(1)   Examine plant characteristics, stability, and so on.

(2)   Define specification.

(3)   On the basis of 1 and 2, choose a suitable design technique – start with the simplest.

(4)   Attempt a design; if unsuccessful, return to 3 and try another technique.

(5)   If iterations through 3 and 4 do not terminate, return to 2 and consider modifying the specification to remove inconsistencies.

(6)   If successful, move on to validation by means of simulation.

At each of these steps the expert system may provide detailed expertise on how to use a particular software package, how to choose parameters for various algorithms, and so on.

It is sometimes claimed that expert systems have the potential to solve difficult control problems, such as the problem of loop assignment in process control systems. The reader is advised to treat such claims with scepticism. Current expert systems have very little ability to learn, in the sense of being able to infer rules from observed behaviour; they can do little other than encapsulate knowledge which is already available. And at present we do not know how to solve 'difficult' control problems – that is why they are difficult. The solution of such problems must await suitable theoretical advances.

The field of computer-aided control-system design is developing at such a rate that speculation about the future is of little value. To keep up with developments in this field there is no alternative to looking regularly at the relevant journals, particularly the IEEE's *Control Systems Magazine* (including the advertisements), occasional publications such as the *Benelux WGS Newsletter*, and the proceedings of relevant conferences. The two regular international events in this field are the IFAC and IEEE Symposia on Computer-Aided Control System Design (CACSD), which are held at intervals of 3 years and about 18 months, respectively.

SUMMARY

Much of this chapter has been devoted to numerical analysis. The notions of the conditioning of a problem and the numerical stability of a solution algorithm have been introduced, and some fundamental algorithms from numerical linear algebra have been presented. These have then been applied to solving some problems which occur when analysing and designing feedback systems. The algorithms which have been presented range from very straightforward ones, for performing tasks such as representing interconnections of systems, to very sophisticated ones, such as the algorithm for finding transmission zeros.

These algorithms are nowadays available in software products, usually suitable for interactive use, so that the control engineer no longer needs to work with them as subroutines. But experience has shown that some knowledge of how these algorithms work is needed, since it is often necessary to be aware of their limitations. Some of the available software products have been described, and current trends in computer-aided control system design have been reviewed.

Note

1.   For more information on ELCS, contact Professor D. K. Frederick, Rensselaer Polytechnic Institute, Troy, NY 12180, USA; Dr C. J. Herget, Lawrence Livermore National Laboratory, Livermore, CA 94550, USA; or R. Kool, Eindhoven University of Technology, NL-6600 MB Eindhoven, The Netherlands.

EXERCISES

8.1   Let $p(x)$ denote the exact solution of a problem p when given the data x, and let $m(x)$ denote the solution when computed by a particular algorithm.

Devise formal definitions of **conditioning** of the problem and **numerical stability** of the algorithm, and show from your definitions that, if p is well-conditioned and m is numerically stable, then the computed solution is close to the true solution. Assume that a suitable norm can be defined on the set of possible solutions.

8.2   A state-space model has two inputs, three outputs, and n states. Its frequency response is to be evaluated at 50 frequencies, using Laub's algorithm. How large does n have to be before the overhead of transforming to Hessenberg form becomes significant? If your machine can perform 10^5 complex floating-point operations per second, how long does the evaluation take with this number of states? How long would it take if the transformation to Hessenberg form were omitted?

8.3   Evaluate and plot the frequency response and characteristic loci of the AIRC model, which is defined in the Appendix, at 50 frequencies between 0.01 and 100 rad s^{-1}. Try this with both linear and logarithmic spacing of frequencies.

8.4   For the three-input, two-output system defined by

$$\dot{x} = -x + [1\ 1\ 1]u$$

$$y = \begin{bmatrix} 1 \\ 1 \end{bmatrix} x + \begin{bmatrix} 1 & 0.1 & 0 \\ 0 & 0 & 0 \end{bmatrix} u$$

find an approximate right-inverse system which gives a good approximation (in frequency response) up to

(a) $\omega = 10$

(b) $\omega = 10^4$

What is the condition number of the modified matrix D in each case?

Check the frequency and step responses of the series connection of the inverse system with the original system in each case.

8.5 Describe in detail how van Dooren's 'staircase' algorithm is used to remove the unobservable modes of a state-space model.

8.6 Implement an algorithm for computing transmission zeros, given a state-space realization. (To do this without an inordinate amount of effort, you should use a matrix-oriented language such as Matlab.)

Use your algorithm to find the zeros of the system (Kouvaritakis and MacFarlane, 1976)

$$
A = \begin{bmatrix}
-2 & -6 & 3 & -7 & 6 \\
0 & -5 & 4 & -4 & 8 \\
0 & 2 & 0 & 2 & -2 \\
0 & 6 & -3 & 5 & -6 \\
0 & -2 & 2 & -2 & 5
\end{bmatrix}, \quad
B = \begin{bmatrix}
-2 & 7 \\
-8 & -5 \\
-3 & 0 \\
1 & 5 \\
-8 & 0
\end{bmatrix}
$$

$$
C = \begin{bmatrix}
0 & -1 & 2 & -1 & -1 \\
1 & 1 & 1 & 0 & -1 \\
0 & 3 & -2 & 3 & -1
\end{bmatrix}, \quad
D = \begin{bmatrix}
0 & 0 \\
0 & 0 \\
0 & 0
\end{bmatrix}
$$

Note that this system is not square, so arbitrary perturbations of its elements will destroy any finite zeros. However, correct implementation of the required numerical rank decisions will result in (almost) the correct solution being found.

(*Solution*: $-3, +4$)

8.7 Verify that, if

$$
\begin{bmatrix}
A & -BR^{-1}B^{\mathrm{T}} \\
-Q & -A^{\mathrm{T}}
\end{bmatrix}
\begin{bmatrix}
T_{11} & T_{12} \\
T_{21} & T_{22}
\end{bmatrix}
=
\begin{bmatrix}
T_{11} & T_{12} \\
T_{21} & T_{22}
\end{bmatrix}
\begin{bmatrix}
S_{11} & S_{12} \\
0 & S_{22}
\end{bmatrix}
$$

and T_{11} is invertible, then $P = T_{21}T_{11}^{-1}$ is a solution of the algebraic Riccati equation

$$
A^{\mathrm{T}}P + PA - PBR^{-1}B^{\mathrm{T}}P + Q = 0
$$

If $K = R^{-1}B^{\mathrm{T}}P$, verify that the eigenvalues of $A - BK$ are the same as those of S_{11}. (See Section 5.2 for the significance of this.)

8.8   Use the model approximation method given in Section 8.3.9 to find a controller of lower order than the 17-state H_∞ controller found in Section 6.8.

(*Note*: This controller has a mode at zero which is nearly unobservable; on some machines the realization will be effectively non-minimal, and the algorithm given in Section 8.3.5 may have to be used to remove this mode before a balanced realization can be obtained.) You should find that a 12-state approximation gives virtually the same performance as the original one, but that the performance begins to deteriorate as the number of states is reduced below 12.

References

Arnold W.F. and Laub A.J. (1984). Generalized eigenproblem algorithms and software for algebraic Riccati equations. *Proceedings of the Institute of Electrical and Electronics Engineers*, **72**, 1746–54.

Åström K.J. and Wieslander J. (1981). *Computer-Aided Design of Control Systems*. Final Report, STU Projects 73-3553, 75-2776 and 77-3548, Department of Automatic Control, Lund Institute of Technology. Report CODEN: LUTFD2/(TFRT-3160)/1-23/1981.

Bartels R.H. and Stewart G.W. (1972). Solution of the equation $AX + XB = C$. *Communications of the Association for Computing Machinery*, **15**, 820–6.

Birdwell J.D., Cockett J.R.B., Heller R. *et al.* (1985). Expert systems techniques in a computer-based control system analysis and design environment. In *Proc. 3rd IFAC/IFIP International Symp. on Computer Aided Design in Control and Engineering Systems* (CADCE '85), Lyngby, Denmark (Hansen N.E. and Larsen P.M., eds), pp. 1–8.

Boyle J.M., Ford M.P. and Maciejowski J.M. (1989). A multivariable toolbox for use with Matlab. *IEEE Control Systems Magazine*, **9**(1), 59–65.

Boyle J.M., Pang G.K.H. and MacFarlane A.G.J. (1989). The development and implementation of MAID: a knowledge based support system for use in control system design. *Transactions Inst. Measurement and Control*, **11**(1).

Chen, Zhen-Yu, ed. (1988). *Preprints, 4th IFAC Symp. on Computer-Aided Design of Control Systems* (CADCS '88), Beijing, Aug. 1988.

Delebecque F., Klimann C. and Steer S. (1988). *BASILE: Guide de l'Utilisateur*. Le Chesnay, France: INRIA.

Edmunds J.M. (1979). Cambridge Linear Analysis and Design Programs. In *Proc. 1st IFAC Symposium on Computer-Aided Control System Design*, Zurich, pp. 253–8.

Emami-Naeini A. and van Dooren P.M. (1982). Computation of zeros of linear multivariable systems. *Automatica*, **18**, 415–30.

Franklin G.F. and Powell J.D. (1980). *Digital Control of Dynamic Systems*. Reading MA: Addison-Wesley.

Frederick D.K., Herget C.J., Kool R. and Rimvall M. (1987). *ELCS: The Extended List of Control Software*, No. 3, Feb. 1987.

Gill P.E., Murray W. and Wright M.H. (1981). *Practical Optimization*. London: Academic Press.

Golub G.H. and van Loan C.F. (1983). *Matrix Computations*. Baltimore MD: Johns Hopkins University Press.

Hammarling, S.J. (1982). Numerical solution of the stable, non-negative definite Lyapunov equation, *IMA Journal of Numerical Analysis*, **2**, 303–23.

Hansen N.E. and Larsen P.M., eds (1985). *Proc. 3rd IFAC/IFIP International Symp. on Computer Aided Design in Control and Engineering Systems* (CADCE '85), Lyngby, Denmark.

Heath M.T., Laub A.J., Paige C.C. and Ward R.C. (1986). Computing the singular value decomposition of a product of two matrices. *SIAM Journal on Scientific and Statistical Computing*, **7**, 1147–59.

Herget C.J. and Laub A.J., eds (1982). Special issue on computer-aided control system design. *IEEE Control Systems Magazine*, **2**(4) (Dec.), 2–37.

Herget C.J. and Laub A.J., eds. (1984). Special issue on computer-aided control system design. *Proceedings of the Institute of Electrical and Electronics Engineers*, **72**, 1714–1805.

IEEE (1981). Four articles in *Computer*, **14**(3), 51–87.

IEEE (1986). *Proc. 3rd IEEE Symp. on Computer-Aided Control Systems Design*, Arlington, VA, Sept. 1986.

James J.R., Taylor J.H. and Frederick D.K. (1985). An expert system architecture for coping with complexity in computer-aided control engineering. In *Proc. 3rd IFAC/IFIP International Symp. on Computer Aided Design in Control and Engineering Systems* (CADCE '85), Lyngby, Denmark (Hansen N.E. and Larsen P.M., eds), pp. 47–52.

Jamshidi M. and Herget C.J., eds (1985). *Advances in Computer-Aided Control Systems Engineering*. Amsterdam: North-Holland.

Kailath T. (1980). *Linear Systems*. Englewood Cliffs NJ: Prentice-Hall.

Kouvaritakis B. and MacFarlane A.G.J. (1976). Geometric approach to the analysis and synthesis of system zeros. *International Journal of Control*, **23**, 167–81.

Laub A.J. (1979). A Schur method for solving algebraic Riccati equations. *IEEE Transactions on Automatic Control*, **AC-24**, 913–21.

Laub A.J. (1981). Efficient multivariable frequency response calculations. *IEEE Transactions on Automatic Control*, **AC-26**, 407–8.

Laub A.J. (1985). Numerical linear algebra aspects of control design computations. *IEEE Transactions on Automatic Control*, **AC-30**, 97–108.

Laub A.J., Heath M.T., Paige C.C and Ward R.C. (1987). Computation of system balancing transformations and other applications of simultaneous diagonalization algorithms. *IEEE Transactions on Automatic Control*, **AC-32**, 115–22.

Little J.N., Emami-Naeini A. and Bangert S.N. (1985). Ctrl-C and matrix environments for the computer-aided design of control systems. In *Advances in Computer-Aided Control Systems Engineering* (Jamshidi M. and Herget C.J., eds), pp. 111–24. Amsterdam: North-Holland.

Little J.N. and Laub A.J. (1985). *Control System Toolbox: User's Guide*. Sherborne MA: The Mathworks, Inc.

Maciejowski J.M. and MacFarlane A.G.J. (1982). CLADP: The Cambridge Linear Analysis and Design Programs. *IEEE Control Systems Magazine*, **2**(4), 3–8. (Reprinted in Jamshidi and Herget, 1985.)

Moore B.C. (1981). Principal component analysis in linear systems: controllability, observability, and model reduction. *IEEE Transactions on Automatic Control*, **AC-26**, 17–27.

Munro N. (1988). Ecstasy–A control system CAD environment. In *Proc. International Conf. 'Control 88'*, IEE Conf. Publ. no. 285, pp. 76–80. London: Institution of Electrical Engineers.

Pang G.K.H. and MacFarlane A.G.J. (1987). *An Expert Systems Approach to Computer-Aided Design of Multivariable Systems*. Lecture Notes in Control and Information Sciences, Vol. 89. Berlin: Springer-Verlag.

Postlethwaite I. and MacFarlane A.G.J. (1979). *A Complex Variable Approach to the Analysis of Linear Multivariable Feedback Systems*. Berlin: Springer-Verlag.

Rimvall M. (1987). Man–machine interfaces and implementation issues in computer-aided control system design. *PhD Dissertation*, No. 8200, ETH Zurich.

Rosenbrock H.H. (1970). *State-Space and Multivariable Theory*. New York: Wiley.

Shah S.C., Floyd M.A. and Lehman L.L. (1985). Matrix-X: Control and model-building capability. In *Advances in Computer-Aided Control Systems Engineering* (Jamshidi M. and Herget C.J., eds), pp. 181–207. Amsterdam: North-Holland.

Spang H.A. (1984). The federated computer-aided control design system. *Proceedings of the Institute of Electrical and Electronics Engineers*, **72**, 1724–31.

Stewart G.W. (1976). Algorithm 406: HQR3 and EXCHNG: Fortran subroutines for calculating and ordering the eigenvalues of a real upper Hessenberg matrix. *ACM Transactions on Mathematical Software*, **2**, 275–80.

Stewart G.W. (1973). *Introduction to Matrix Computations*. New York: Academic Press.

van Dooren P.M. (1981a). The generalized eigenstructure problem in linear system theory. *IEEE Transactions on Automatic Control*, **AC-26**, 111–29.

van Dooren P.M. (1981b). A generalized eigenvalue approach for solving Riccati equations. *SIAM Journal on Scientific and Statistical Computing*, **2**, 121–35.

WGS: Benelux Working Group on Software (1985). *An Inventory of Basic Software for Computer Aided Control System Design (CACSD)*. WGS Report 85-1, Eindhoven University of Technology, Department of Mathematics and Computer Science.

Wilkinson J.H. (1965). *The Algebraic Eigenvalue Problem*. Oxford: The Clarendon Press.

APPENDIX

Models used in Examples and Exercises

A.1 Aircraft model AIRC

This model has been taken from Appendix F of Hung and MacFarlane (1982). It represents a linearized model of the vertical-plane dynamics of an aircraft, and has three inputs, three outputs, and five states.

The inputs are:

$u1$: spoiler angle (measured in tenths of a degree)

$u2$: forward acceleration (m s^{-2})

$u3$: elevator angle (degrees)

The states are:

$x1$: altitude relative to some datum (m)

$x2$: forward speed (m s^{-1})

$x3$: pitch angle (degrees)

$x4$: pitch rate (deg s^{-1})

$x5$: vertical speed (m s^{-1})

The three outputs are just the first three states, which are to be controlled. The matrices A, B, C and D of the state-space model are:

$$A = \begin{bmatrix} 0 & 0 & 1.1320 & 0 & -1.000 \\ 0 & -0.0538 & -0.1712 & 0 & 0.0705 \\ 0 & 0 & 0 & 1.0000 & 0 \\ 0 & 0.0485 & 0 & -0.8556 & -1.013 \\ 0 & -0.2909 & 0 & 1.0532 & -0.6859 \end{bmatrix}$$

$$B = \begin{bmatrix} 0 & 0 & 0 \\ -0.120 & 1.0000 & 0 \\ 0 & 0 & 0 \\ 4.4190 & 0 & -1.665 \\ 1.5750 & 0 & -0.0732 \end{bmatrix}$$

$$C = \begin{bmatrix} 1 & 0 & 0 & 0 & 0 \\ 0 & 1 & 0 & 0 & 0 \\ 0 & 0 & 1 & 0 & 0 \end{bmatrix}$$

$$D = \begin{bmatrix} 0 & 0 & 0 \\ 0 & 0 & 0 \\ 0 & 0 & 0 \end{bmatrix}$$

A.2   Turbo-generator model TGEN

This is a two-input, two-output, six-state model of a large turbo-alternator. The inputs are:

$u1$: throttle-valve position

$u2$: excitation control

The outputs are:

$y1$: generator terminal voltage

$y2$: generator load angle

The original model is described by Limebeer *et al.* (1979). A linearized, ten-state model is given by Hung and MacFarlane (1982). A multivariable controller for this model designed by the reversed-frame normalization method is given by Hung and MacFarlane (1982), one designed by the characteristic-locus method is given by Limebeer and Maciejowski (1985), and one designed by the Nyquist-array method is given by Boyle *et al.* (1989). The model is also used in the on-line tutorial examples provided with the *Multivariable Frequency-Domain Toolbox* software (Ford *et al.*, 1988).

The model we present here is a six-state approximation to that ten-state model. It is open-loop stable, but has a resonance at 6.35 rad s^{-1}, with a damping factor of only 0.05.

The matrices of the state-space model are:

$$A = \begin{bmatrix} -18.4456 & 4.2263 & -2.2830 & 0.2260 & 0.4220 & -0.0951 \\ -4.0977 & -6.0706 & 5.6825 & -0.6966 & -1.2246 & 0.2873 \\ 1.4449 & 1.4336 & -2.6477 & 0.6092 & 0.8979 & -0.2300 \\ -0.0093 & 0.2302 & -0.5002 & -0.1764 & -6.3152 & 0.1350 \\ -0.0464 & -0.3489 & 0.7238 & 6.3117 & -0.6886 & 0.3645 \\ -0.0602 & -0.2361 & 0.2300 & 0.0915 & -0.3214 & -0.2087 \end{bmatrix}$$

$$B = \begin{bmatrix} -0.2748 & 3.1463 \\ -0.0501 & -9.3737 \\ -0.1550 & 7.4296 \\ 0.0716 & -4.9176 \\ -0.0814 & -10.2648 \\ 0.0244 & 13.7943 \end{bmatrix}$$

$$C = \begin{bmatrix} 0.5971 & -0.7697 & 4.8850 & 4.8608 & -9.8177 & -8.8610 \\ 3.1013 & 9.3422 & -5.6000 & -0.7490 & 2.9974 & 10.5719 \end{bmatrix}$$

$$D = \begin{bmatrix} 0 & 0 \\ 0 & 0 \end{bmatrix}$$

A.3   Remotely piloted vehicle model RPV

The following model of a remotely piloted vehicle is given by Safonov *et al.* (1981), who use it to illustrate LQG design; Safonov and Chiang (1988) use it to illustrate H_∞ design. It has two inputs, two outputs and six states. It is *unstable*, its poles being located at -30, -30, -5.67, -0.26, $+0.69 \pm 0.25$j. The matrices of the state-space model are:

$$
A = \begin{bmatrix}
-0.02567 & -36.617 & -18.897 & -32.090 & 3.2509 & -0.76257 \\
9.257 \times 10^{-5} & -1.8997 & 0.98312 & -7.256 \times 10^{-4} & -0.1708 & -4.965 \times 10^{-3} \\
0.012338 & 11.720 & -2.6316 & 8.758 \times 10^{-4} & -31.604 & 22.396 \\
0 & 0 & 1 & 0 & 0 & 0 \\
0 & 0 & 0 & 0 & -30 & 0 \\
0 & 0 & 0 & 0 & 0 & -30
\end{bmatrix}
$$

$$
B = \begin{bmatrix}
0 & 0 \\
0 & 0 \\
0 & 0 \\
0 & 0 \\
30 & 0 \\
0 & 30
\end{bmatrix}
$$

$$
C = \begin{bmatrix}
0 & 1 & 0 & 0 & 0 & 0 \\
0 & 0 & 0 & 1 & 0 & 0
\end{bmatrix}
$$

$$
D = \begin{bmatrix}
0 & 0 \\
0 & 0
\end{bmatrix}
$$

References

Boyle J.M., Ford M.P. and Maciejowski J.M. (1989). A new multivariable toolbox for use with Matlab. *IEEE Control Systems Magazine*, **9**(1), 59–65.

Ford M.P., Maciejowski J.M. and Boyle J.M. (1988). *Multivariable Frequency Domain Toolbox: User's Guide*. Cambridge: Cambridge Control Ltd.

Hung Y.S. and MacFarlane A.G.J. (1982). *Multivariable Feedback: A Quasi-classical Approach*. Lecture Notes in Control and Information Sciences, Vol. 40. Berlin: Springer-Verlag.

Limebeer D.J.N., Harley R.G. and Schuck S.M. (1979). Subsynchronous resonance of the Koeberg turbo-generators and of a laboratory system. *Transactions of the South African Institute of Electrical Engineers*, **70**, 278–97.

Limebeer D.J.N. and Maciejowski J.M. (1985). Two tutorial examples of multivariable feedback design. *Transactions of the Institute Measurement and Control*, **7**(2), 97–107.

Safonov M.G., Laub A.J. and Hartmann G.L. (1981). Feedback properties of multivariable systems: The role and use of the return difference matrix. *IEEE Transactions on Automatic Control*, **AC-26**, 47–65.

Safonov M.G. and Chiang R.Y. (1988). CACSD using the state-space L_∞ theory – A design example. *IEEE Transactions on Automatic Control*, **AC-33**, 477–9.

Index

adjoint matrix 40, 50
admissible parameter 343, 344, 345
AIRC model 132, 155, 189, 218, 220, 244, 262, 306, 336, 400, 405
aircraft control (example) 85
algebraic function 71
algorithm for $Ax = b$ 357, 361
algorithms for linear systems 38, 357, 368–94
ALIGN algorithm 145–8, 152, 156–8, 165, 179, 188, 203, 214, 217, 238, 258
   for dynamic compensator 148
   least-squares version of 147
   multi-frequency 147, 148, 187, 188
aligned return-ratio 166
alignment of principal directions 165
all-pass transfer function 27, 30, 260, 270, 276, 294, 295, 299, 334, 335
   realization 320, 321
Al-Naib, U. 354
analytic function 95, 158
Anderson, B.D.O. 222, 263
Andres, R.P. 186, 221
approximate commutative compensator 142, 145, 147, 149, 152, 154, 158, 167, 203, 325
Arkun, Y. 218, 221

Arnold, W.F. 357, 402
artificial intelligence 398
assignment of inputs to outputs 179
Åström, K.J. 3, 24, 36, 130, 135, 222, 263, 395, 402
Athans, M. 229, 230, 243, 260, 263, 264, 402
augmentation of plant model 237, 238, 239, 240, 242, 244, 245, 261, 289
   with dummy inputs or outputs 244
   for Hankel approximation 300
autocovariance function 97
automatic controller design, dangers of 217

back-substitution 358, 362, 385, 391
backwards stability of algorithm 356
Balakrishnan, V. 100, 135, 354
balanced realization 298, 300, 322, 376, 392
   computation 390–2
   truncation 392
Balas, M.J. 315, 323
bandwidth 19, 20, 21, 24, 27, 29, 30, 31, 40, 96, 152, 154, 156, 160, 161, 163, 198, 218, 240, 248, 250, 260, 337

bandwidth (*contd.*)
  achievable   94
  at input   82
  from inverse locus   175
  of Kalman filter   236
  lower bound on   31, 96
  open-loop   27
  at output   82
  too ambitious   202
  transmission   20, 21, 30, 83, 84
  unequal   141
  upper bound on   27–30, 96
Bangert, S.N.   396, 403
Banks, S.P.   222, 263
Barratt, C.H.   354
Bartels, R.H.   385, 402
Becker, R.G.   345, 354
Benelux Working Group on Soft-
  ware   395, 404
Bezout's theorem   275
Bhattacharyya, S.P.   124, 135
Binet–Cauchy theorem   43
Birdwell, J.D.   398, 402
block relative gain array   218
Bode, H.W.   25, 32, 36
Bode plot   5, 24, 25, 129, 183, 184
Bode's integral theorem   31
Bongiorno, J.J.   315, 324
Boyd, S.P.   100, 135, 336, 354
Boyle, J.M.   397, 398, 402, 404, 407,
  408
branch points   96
Bristol, E.H.   210, 220
British school   v, 2
Butterworth filter   109

Callier, F.M.   48, 74, 315, 323
cancellation of resonant poles   198
cancellations, pole/zero in right half-
  plane   4, 13, 22, 23
canonical form   40, 44, 167
certainty equivalence   223
Chan, W.S.   71, 74
Chang, B.C.   316, 323
characteristic direction   143, 144, 150,
  152
characteristic equation   4
characteristic functions   143, 144
characteristic gain   93

  relation to principal gains   93, 94
characteristic locus   60, 61, 63, 64, 66,
  67, 68, 71, 91, 95, 113, 114, 115,
  121, 143, 148, 149, 151, 152, 155,
  161, 197, 252, 390, 395
  compensation   151, 159
  computation   370
  design method   138, 142, 154, 164,
    188, 202, 217, 258, 407
  encirclements by   153
  and principal gains   151
  and quasi-Nyquist locus   165, 166
characteristic polynomial   3
characteristic roots   3
  closed-loop   389
Chen, C.T.   3, 36, 52, 74
Chen, Z-Y.   397, 402
Chiang, R.Y.   407, 408
Cholesky decomposition   250, 305, 391
Chu, C.C.   316, 320, 323
closed-loop approach to design   22,
  314
closed-loop specifications   87, 89
closed-loop transfer function   14, 29,
  76, 81, 100, 102, 150, 171, 198, 205,
  207, 243, 272, 306, 326, 334
  parametrization   289, 317
  principal gains   248, 251, 253
  state-space realizations   292
Cockett, J.R.B.   402
column dominance   65, 69, 170, 178,
  179, 183, 186, 190
command signal *see* reference input
common denominator polynomial   327
common factors   46, 49
commutative compensator   144, 145,
  161
comparison of design methods   202,
  217, 258, 313
compensator   5, 8, 10
complementary sensitivity   14, 100,
  101, 207
  see also closed-loop transfer function
complex matrix, approximation by real
  matrix   145, 147
complexity of compensator   207
compression of rows or columns   302,
  366, 367, 379, 386, 387
computer-aided design   138, 355–404

conceptual signals 106
condition number 80, 93, 131, 215, 358, 376, 378
    optimal 215
conditional stability 161
    with LQ controller 229
conditioning of $Ax = b$ 358
conditioning of problem 356
conformal mapping 158
constraint
    envelope 343
    frequency-domain 343
    infinite-dimensional 343, 344
control signal magnitude 15, 84, 87, 91, 161, 178, 202, 313, 342
control structure design 210–18
controllability gramian 297, 392
controllability indices 381
controllability
    PBH test for 378
    rank test for 357
controllable realization 50, 52, 381
controller
    cross-coupled 139, 210, 325
    observer-based 277
    realizable 23
    state-space realization 285
    transfer function of 284
    unrealizable 22
controller parameter 346
controller stability 33, 326
controller structure 212, 213, 216, 217, 225, 277, 283, 315, 317, 326, 336, 337
    compatibility with specifications 333
    constrained 325, 326, 332
    diagonal 139
    number of 216
convex optimization problem 273
convolution 297
coprime factorization 316
    realization of 335
coprime polynomial matrices 49, 50
coprime polynomials 46
coprime transfer functions 274, 275, 277
corner (break) frequency 5
correlated perturbations 116

cost function
    in LQ problem 224, 230
    in LQG problem 223, 241
covariance 97, 223, 232, 233, 235, 239
cross-over frequency 8, 19, 20, 27, 28, 29, 94, 154, 158, 235, 239, 248
    lower bound on 30
    relation to bandwidth 19
    upper bound on 27–9

Daly, K.C. 187, 188, 190, 218, 220
damping of closed-loop poles 9
Daniel, R.W. 121, 135
data types for control engineering 397
Davis, M.H.A. 222, 263
decentralized control 210, 212, 213, 216, 220
decoupling 160, 169, 179, 217
    compensator 152, 216
    at high-frequency 152
degree of dominance 65, 183
    optimal 183
Delebecque, F. 397, 402
denominator matrix (of MFD) 49
descent direction 119
design example 155, 189, 244, 306, 336
design exercises 218, 262, 322, 353
design procedure, two-stage 76, 284
design process, outline 398
Desoer, C.A. 47, 48, 71, 74, 315, 317, 323
destabilizing perturbation 114
detuning of controller 106
detuning of systems 130
detectable system 226
determinantal divisors 41, 43
diagonal compensator 182, 184, 186, 210, 213, 214, 216
    and diagonal dominance 183
    optimal 182
    in QFT method 208
diagonal dominance 65, 66, 67, 169, 170, 173, 174, 175, 176, 177, 178, 179, 180, 186, 187, 202
    achievement 177
    cut-and-try methods 178
    measure 187, 190, 192
    optimal 219

diagonal dominance (*contd.*)
  optimization   186, 188
  and Perron–Frobenius theory   180
diagonal plant   212
diagonalization   179
difficulty of control   80, 139, 216
direct Nyquist array   170, 186, 218
  design method   176, 177, 183, 202
direct sum decomposition   382
directions of signals   85, 86
discrete-time   70, 71
disturbance   10, 12, 76, 106, 208, 225
  amplification   30
  attenuation   14
  cancellation by feed-forward   12
  confined to a cone   85
  confined to subspace   85
  directions, easy and hard   85
  effective   13
  at input   200
  at output   16, 17, 240, 245
  rejection   81, 272
  ramp   23
  signal   150
  spectrum   16
  step   19, 23
  stochastic   86
DNA *see* direct Nyquist array
Doyle, J.C.   117, 118, 119, 126, 127,
    130, 131, 134, 135, 136, 231, 244,
    263, 273, 296, 301, 304, 306, 315,
    316, 323

Edmunds' algorithm   326–41, 353
  choice of denominator poly-
    nomials   334, 335, 337
  compatibility of specifications   334
  flexibility   332
  initial design   333
  iterative   331
  line search in   332
  numerically stable solution   333
  possibility of unstable solution   332
  refinements   334
  target controller   326, 335, 337
  target transfer function   326, 333,
    335, 337
  trade-offs   335

Edmunds, J.M.   98, 116, 136, 147, 220,
    326, 331, 354, 395, 402
eigenframe   138, 148
eigenvalue   60, 63, 64, 91, 97, 363, 366
  function   143, 151
    see also characteristic locus
  conditioning   366
  closed-loop   227
  of Hamiltonian matrix   393
  of Kalman filter   227, 231, 234
  of noise model   239, 240, 241
eigenvector   143, 147, 365
  shared by plant and controller   143
elementary matrix   41, 169, 179
elementary operation   40, 41, 43, 71,
    139
Emami-Naeini, A.   36, 357, 389, 396,
    402, 403
equivalent matrices   41, 44
EXCHNG algorithm   384, 393
expectation operator   97
expert system   398
exponential stability   55, 56
extended list of control software
    (ELCS)   397, 400
external signals   106

factorization into all-pass and mini-
    mum-phase factors   335
Fan, M.K.H.   119, 135
fast loops   12, 154
fast–slow decomposition   216
feasible parameter   342, 343
feedback
  negative   4
  purpose of   22, 75, 203–4
feedback connection   38
  computation   373
feedback design problem
  elementary   3
  one degree of freedom   14
  'shape' of solution   16, 21, 25
  standard form   10
  trade-offs   14, 15, 82, 90, 236
  two degrees of freedom   14, 30
feedback systems, fundamental relations
    in   13
feedforward   12, 13, 14

Fiedler, M. 186, 220
filter, low-pass 25
final value theorem 17
finite precision of computation 356
first-order system, multivariable 74
fixed mode 220
flops (floating-point operations) 359,
    368, 370
flow measurement, error in 104
Floyd, M.A. 396, 404
Ford, M.P. 187, 188, 190, 218, 220,
    397, 402, 407, 408
FORTRAN 395
fractional representation 274, 276,
    280, 315
  coprime 276, 280
  realization 275, 277, 315, 320
  of transfer functions 105
Francis, B.A. 100, 101, 135, 301, 306,
    315, 316, 323, 324
Franklin, G.F. 3, 23, 36, 134, 135, 222,
    263, 315, 323, 368, 402
Frederick, D.K. 395, 397, 398, 402, 403
frequency-response, computation
    of 368
frequency-domain bounds from time-
    domain bounds 204
Freudenberg, J.S. 32, 33, 36, 102, 135
Frobenius norm 93
functional 342
functional analysis 209

gain 99
  of matrix 76
  of multivariable system 79, 80, 129,
    164
  open-loop 17
gain characteristic, ideal 94
gain margin 5, 23, 130
  excess 113
  from inverse locus 175
  of (LQ) optimal state feedback 229,
    259
gain-bandwidth product 27, 94, 96
  of multivariable system 96
gain-frequency behaviour 24
gain-phase relations 24, 25, 27, 32, 94,
    95, 207

for characteristic loci 96
for multivariable systems 94, 129
Gantmacher, F.R. 43, 74
gap between vectors 261
Garcia, E.C. 318, 323
Gaussian elimination 357, 359
general interconnection 374
generalized eigenvalue 146, 147, 148,
    386, 389
  conditioning 367, 368
generalized eigenvector 147, 148, 367
generalized inverse Nyquist stability
    theorem 62, 63, 64, 65, 71, 174
generalized Nyquist stability
    theorem 59, 61, 102, 113, 115,
    149, 152, 169, 333
generalized stability 69, 70, 218
Gershgorin band 64, 65, 66, 67, 68, 71,
    122, 169, 170, 174, 176, 177, 193,
    395
  generalized 122
  of inverse system 73
Gershgorin circle 64, 174, 186
Gershgorin's theorem 64
Gill, P.E. 345, 348, 354, 357, 402
Givens rotation 383, 384, 385
Glover's algorithm for Hankel approx-
    imation 298–300
Glover, K. 103, 298, 300, 301, 304,
    306, 315, 320, 323
Glover–Doyle algorithm for H_∞
    problems 301–6
Golub, G.H. 78, 135, 250, 261, 263,
    305, 357, 364, 365, 367, 368, 385,
    391, 403
Gould, L.A. 22, 36
gramian 297, 300
graph theory 71
graphical methods 355
Grosdidier, P. 153, 212, 218, 220

H_∞ and LQG 306
H_∞ controller 301
  central 305, 306, 311
  existence 304
  maximum entropy 305, 311
  parametrization 304
  and pole-zero cancellation 314

H_∞ controller (*contd.*)
  state dimension   305, 306
  state-space realization   304, 305
H_∞ design method   265–324, 326, 407
H_∞ problem   267, 269, 273, 293, 299, 315
  1-block   295, 317
  2-block   295, 308
  4-block   295, 311
  definition   267
  formulation   270, 102
  gamma iteration   293, 296, 306, 310, 317
  Glover–Doyle algorithm   301–6, 308, 317, 321
  mixed objective   272
H_∞ set of transfer functions   267
H_∞ theory   204, 316
  and complex analysis   315
  for irrational transfer functions   316
H_2 problem   267
Hadamard product   211
Hamiltonian matrix   226, 227, 393, 394
Hammarling, S.J.   391, 403
Hankel approximation   265, 293, 294, 296, 298, 317, 392
  for non-square plant   300
Hankel norm   296, 297, 321, 392
Hankel operator   297
Hansen, N.E.   397, 403
Hanson, R.J.   330, 354
Hardy space   101
Harley, R.G.   406, 408
Hartmann, G.L.   242, 244, 264, 407, 408
Hatfield, L.   402
Hawkins, D.J.   186, 220
Heath, M.T.   391, 392, 403
Heller, R.   402
Helton, J.W.   315, 316, 323
Herget, C.J.   395, 397, 402, 403
Hermite form   167
Hermitian matrix   146
Hessenberg form   362, 363, 364, 369, 370
Heunis, A.J.   345, 354
hidden dynamics   4
high gain   141, 152, 172, 177
Holt, B.R.   153, 212, 218, 220

Horowitz, I.   v, 22, 29, 36, 111, 130, 135, 203, 204, 220
Householder transformation   359, 360, 361, 363, 389
Hung, Y.S.   93, 94, 135, 165, 166, 167, 221, 294, 323, 334, 354, 405, 406, 407, 408

IEEE   222, 263, 395, 397, 403
IEEE standard arithmetic   367, 395
ill-conditioned problems   214, 356
improper systems   53, 55
INA *see* inverse Nyquist array
indentation of Nyquist contour   59
independent design of each loop   213
inequality constraints   342
inner loops   12
inner transfer function   294, 334
innovations process   261, 288
input–output pairing   139, 209, 210, 212, 213, 216, 217
integral action   154, 160, 161, 167, 213, 214, 245, 310, 337
  in LQG controller   239, 245
integral controllability   214, 218
integrity   121, 213, 214
interaction   139, 151, 152, 155, 156, 160, 161, 169, 176, 177, 181, 198, 203, 209, 212, 214, 216, 217, 258, 259, 337, 341
  reduction   139
  and scaling   181
interactive computing   346, 355, 395
interconnected system
  computation   371
  conditioning   376
internal model control   vi, 318
internal model principle   24
internal stability   55, 56, 58, 59, 71, 73, 218, 226, 227, 261, 267, 269, 276, 316
  with LQG controller   243
interpolation   316
invariant factors   41, 43
invariant subspace   365
inverse characteristic loci   64
inverse Nyquist array   64, 69, 170, 171, 184, 186

design method 170, 174, 176, 178, 183
inverse system 377, 378
  as compensator 378
  computation 376
  rank conditions 378
  state-space realization 279, 287
inversion of matrix 80, 357
inverting compensator 214
irrational transfer function 71, 144, 217, 372
irreducible MFD 50
iterative design 342

Jabr, H.A. 315, 324
Jacobson, C.A. 315, 323
James, J.R. 398, 403
Jamshidi, M. 397, 403
Jonckheere, E.A. 300, 324

Kabamba, P. 100, 135
Kailath, T. 3, 36, 50, 52, 55, 74, 130, 378, 403
Kaiser, J.F. 22, 36
Kalman filter 224, 225, 227, 231, 233, 235, 236, 237, 240, 252, 288
  example design 244
  as feedback system 236, 231
  gain matrix 226
  performance and robustness 231
  principal gains of return difference 250
  principal gains of return ratio 252
  return difference 231
  return ratio of 231, 243, 259
  uncontrollable error dynamics 232
Kalman, R.E. 229, 263
Kantor, J.C. 186, 221
Karcanias, N. 55, 74
Khargonekar, P. 301, 306, 323
Klimann, C. 397, 402
Kool, R. 395, 397, 402
Kouvaritakis, B. 120, 121, 134, 135, 143, 145, 146, 147, 220, 221, 401, 403
Kreisselmeier, G. 346, 350, 354
Kreyszig, E. 64, 74
Kronecker indices 381
Kronecker product (of matrices) 328

Kucera, V. 315, 323
Kwakernaak, H. 222, 231, 233, 260, 263

Lagrange multiplier 187
Larsen, P.M. 397, 403
Latchman, H. 120, 121, 134, 135
Laub, A.J. 242, 244, 264, 357, 369, 391, 392, 393, 397, 402, 403, 407, 408
Laurent series 70
Lawson, C.L. 330, 354
least-squares 167, 326, 328, 330, 331, 333, 361
Lebesgue space 101
Lehman, L.L. 396, 404
Li, X. 354
Limebeer, D.J.N. 294, 300, 323, 324, 406, 407, 408
linear fractional transformation 267, 317
linear quadratic Gaussian *see* LQG
linear quadratic, deterministic problem 223
Little, J.N. 396, 397, 403
Liu, R.W. 315, 323
loop assignment 139, 209, 210, 212, 213, 216, 217
  see also control structure design
loop failure 121, 122, 176, 213, 214
loop shaping 183
  using LQG 243
loop transfer recovery 231, 232, 235, 240, 244, 258–62
  controller, state-space realization of 253
  convergence 243
  design method 235, 243
  at input 240
  motivation 260
  and non-minimum-phase systems 259
  and non-square plants 244
  at output 243, 252
  pole-zero cancellation 260
  recovery step 239
  trade-offs 243, 259, 261, 262
  without LQG 260
Looze, D.P. 32, 33, 36, 102, 135

low-frequency compensation   160, 168
LQ *see* linear quadratic
LQG (linear quadratic Gaussian)   222,
   223, 225, 227, 231, 239, 240, 241,
   261
   controller
      frequency-domain interpreta-
         tion   242
      performance and robustness   231
      state dimensions   259
      state equations   227
      transfer function   232
   design method   222, 242, 267, 313,
      326, 396, 407
   trade-offs   261
   problem, solution of   225
LTR *see* loop transfer recovery
LU-factorization   358, 359
Lyapunov equation   296, 298, 300, 321,
   390, 391

M-matrix   186
M-circle   8, 9, 23, 149, 151, 152, 158,
   159, 197, 205, 206, 230
MacFarlane, A.G.J.   v, 40, 55, 71, 74,
   93, 94, 98, 116, 130, 135, 136, 142,
   146, 165, 166, 167, 221, 226, 229,
   263, 334, 354, 390, 395, 398, 401,
   403, 404, 405, 406, 407, 408
machine precision   367, 382, 389
Maciejowski, J.M.   395, 397, 402, 403,
   407, 408
manipulable variables   106
Manousiouthakis, V.   212, 218, 221
Markov parameter   55, 70
mathematical software   356
matrix fraction description   48–50, 57,
   71, 167, 274
   model of uncertainty   105
matrix inversion lemma   89, 130, 232,
   283
matrix pencil   381
Mayne, D.Q.   139, 140, 142, 221, 345,
   352, 354
McAvoy, T.J.   212, 221
McFarlane, D.   315, 323
McMillan degree   47, 50, 51, 72, 155,
   167, 392
   of controller   325

Mead, R.   345, 354
measure of skewness   93
measured signals   106
measurement error   10, 12, 83, 233,
   236
   insensitivity to   14
   sources   15
   spectrum   16, 81, 106, 240
   systematic   12
Mees, A.I.   66, 67, 74, 122, 182, 186,
   221
method of inequalities   341, 346, 348
   minimax problem   344, 345
   moving boundaries algorithm   344
   and optimization   342
   with smooth objective function   345
Meyer, D.G.   354
MFD *see* matrix fraction description
minimal realization   50, 51, 52, 54, 226,
   373, 376
   computation   378
minimality, visual test for   390
minimum-phase transfer function   24,
   25, 95, 206, 231, 274, 334
mixed performance–robustness
   problem   294
model approximation   2, 259, 265, 301,
   391, 392, 402
model, validity of   218
model-matching problem   293, 294, 316
models, definitions of   405
modes, uncontrollable unobservable   4
modified matrices formula   130
modified sensitivity   85
monic polynomial   41
Moore, B.C.   392
Moore, J.B.   222, 263
Morari, M.   vi, 104, 136, 153, 212, 218,
   220, 221, 318, 323
mu (μ) *see* structured singular value
multi-loop problem   11
multi-objective function   346
multi-objective optimization
   example   350
   feasible solutions   349
   software   348
   systematic procedure   348
   termination   348
   trade-offs   349

multi-objective problem
  as minimax problem 347
  replacement by conventional
    problem 346
  as unconstrained problem 348
multiplicity (of poles and zeros) 46, 47
Munro, N. 183, 184, 221, 398, 403
Murray, J. 315, 323
Murray, W. 345, 348, 354, 357, 402

negative feedback 58
Nehari extension problem 294
Nelder, J.A. 345, 354
Nett, C.N. 212, 221, 315, 323, 334, 354
Nevanlinna–Pick problem 316
Newton, G.C. 22, 36
Ng, W.Y. 345, 352, 354
Nichols chart 9, 24, 111, 158, 159
  with Gershgorin band 195
  with templates 205
noise process 238
noise propagation 87, 89, 197, 235,
    268
  minimization 204
non-linear plant 105
non-linear problems 210
non-linearity 225
non-minimum phase plant 94–6, 259
non-square plant 188
norm 1–215
  Euclidean vector 77
  Hankel 296, 297, 321, 392
  Hilbert 77
  induced 77
  of matrix 77, 82, 87
  of operator 99, 100, 101, 105, 124,
    130
  and performance specification 101
  and principal gains 100
  of signal 100
  spectral 77
  from state-space realization 100
  subordinate 77
  of transfer function 77, 99
  of vector 147
normal matrix 91, 165
normal rank 41, 47
normal return-ratio 93, 114, 131, 166
normality 94

normalized comparison matrix 181,
    183
normalization of objectives 347
Norman, S.A. 354
notch 240
numerical algorithms 355, 356–404
  for linear systems 38, 357, 368–94
numerical stability 356, 357, 369, 386
Nye, W.T. 352, 354
Nyquist array 64, 65, 111, 122, 138,
    177, 178, 184, 216, 395
  design method 138, 168, 176, 203,
    217, 258, 325, 407
  see also direct Nyquist array, inverse
    Nyquist array
Nyquist contour 59, 60, 65, 68, 69, 73,
    178
  indentation 157
Nyquist locus 5, 24, 59, 64, 138, 143,
    149, 159, 193, 229
  inverse 172, 174, 176
Nyquist stability theorem 4, 23, 24,
    58, 60, 207
  generalized 59, 61, 102, 113, 115,
    149, 152, 169, 333
  generalized inverse 62–5, 71, 174
Nyquist-like techniques 137

observability, PBH test for 378
observability, rank test for 357
observability gramian 297, 392
observable realization 50
observer 224, 227, 231, 275, 291, 292
  dynamics 231
observer-based controllers 285
on-line tuning 74, 214
open-loop approach to design 23, 314
open-loop compensation 22, 76
open-loop gain 87
open-loop specifications 89
Oppenheim, A.V. 15, 36
optimal loop-gain function (in QFT
    method) 206
optimal state feedback 227
optimization 100, 166, 167, 190, 191,
    203
  choice of algorithm 342
  direct search 345
  frequency-domain 326

optimization (*contd.*)
  multi-objective   346
  problem, non-smooth   347
  semi-infinite   345
  simplex (polytope) method   345
  unconstrained   348
orthogonal matrix   357, 383
orthonormal basis   365
  in place of eigenvectors   366
Osborne, E.E.   120
Ostrowski band   173, 174, 175, 176
Ostrowski circle   174
Ostrowski's theorem   173
outer loops   12
Owens, D.H.   73, 74

Paige, C.C.   391, 392, 403
Pang, G.K.H.   94, 136, 398, 404
Papoulis, A.   97, 136
parallel computing   398
parallel connection   38, 382
  computation   373
parameter optimization   2, 59, 325
  algorithms   352
  and computing environment   352
  trade-offs   342, 346
Pareto optimality   349
Parseval's theorem   15
partial fraction decomposition   361, 382
peak M-value   101
  from inverse locus   175
Pearson, J.B.   316, 323
performance   75, 76, 81, 84, 91, 166, 177, 261
  assessment, in Nyquist array methods   170
  at input   231
  limitations on   24–33, 94–7
  of optimal state feedback   227
  and relative gain array   214
  and robustness, mixed problem   272, 307
  robustness   124, 125, 126, 273
    as inequality constraint   343
  specifications   87, 102
    equivalence to robustness specifications   130

equivalent to additional perturbation block   125
  and operator norm   101, 124
permutation of rows or columns   212, 213, 214
permutation matrix   169, 358, 361
Perron–Frobenius eigenvalue   66, 67, 119, 181, 182, 183, 184
  eigenvector   66, 67, 119, 120, 181, 182, 183, 184
  theory   180, 181, 186
perturbation
  additive   271
  destabilizing   272
  feedback   202
  impossible   106
  at input   216
  multiplicative   200, 272
  permissible   113, 116
  real-valued   111
  stable   114, 116
  uncorrelated   216
  unstructured   229
phase compensation   154, 258
phase of multivariable system   164, 165
phase lag   5, 7
phase lead (phase advance)   5, 7, 154, 158, 159, 175
phase margin   5, 23, 27, 29, 102
  from inverse locus   175
  with LQ controller   229
phase-frequency behaviour   24
pivoting strategy   357–8
plant   10
  high-frequency behaviour   16
  model, validity of   342
  non-square   176
  uncertainty   129
Polak, E.   345, 352, 354
pole polynomial   46, 49, 50, 51, 63, 234
pole-placement   vi, 10, 260
pole-zero cancellation   56, 58, 59, 203, 217, 231, 243, 259, 310, 314, 341, 373
  in LQG/LTR   243
poles   4, 40, 45, 46, 50, 51, 61, 69, 155
  algorithm   47
  closed-loop   258

of compensator   176
at infinity   53
of inverse system   72
from MFD   49, 50
migration across stability
    boundary   202
in right half-plane   30, 32, 96,
    129, 271
and stability   48
unstable   30
polynomial matrix   40
Popov–Belevitch–Hautus test   378
positive feedback   40, 55
convention   271, 275, 276, 290, 283
positive matrix   66
post-compensator   180
Postlethwaite, I.   71, 74, 98, 116, 136,
    390, 404
Potter, J.E.   226, 263
Powell, J.D.   3, 23, 36, 135, 222, 263,
    368, 402
power gain   99
power limitations   218
power spectral density   3, 12, 97, 98,
    236, 238, 239, 240, 241
power of vector-valued stochastic
    process   97
pre-compensation   181
pre-filter   10, 14, 30, 76, 205, 207, 209
Priestley, M.B.   97, 136
primitive matrix   66, 67, 186
principal direction   79, 80, 165, 237
    input   80, 86
    output   80
principal gain   76–9, 81–6, 91, 98, 113,
    129, 130, 151, 161, 164, 166, 167,
    198, 200, 216, 225, 235, 237, 238,
    253, 310, 311, 313, 335
    balancing (in LQG/LTR)   237
    closed-loop   87, 243
    computation   371
    conservativeness   85
    open-loop   87, 88
    relation to characteristic loci   91
    relation to characteristic gains   93,
        94, 151
    shaping (in LQG/LTR)   235, 237
principal phase   165

principle of the argument   59
process control   14, 15, 74, 139, 325
process industry   211, 212
process noise   232, 233
process plant (example)   74, 104, 108
proper transfer function   101
pseudo-diagonalization   179, 186–8,
    190, 203, 214
    use with various design methods
        188
pseudo-inverse   79, 367, 378
Ptak, V.   186, 220

Q-parametrization   274, 317 *see also*
    Youla parametrization
QFT *see* quantitative feedback theory
QR-algorithm   366, 367, 368, 371, 384
    convergence   365
    for Schur form   364
    shift in   365
QR-factorization   130, 359, 361, 362,
    378
quality of control   21
quantitative feedback theory   130, 138,
    203–10
    design method   210, 216, 217, 325
    extension to multivariable systems
        207, 209
quasi-classical design method *see*
    reversed frame normalization
quasi-Newton algorithm   348
quasi-Nyquist decomposition   165
quasi-Nyquist locus   166, 167
quasi-triangular matrix   364, 382, 383
QZ-algorithm   368

Raggazini, J.R.   315, 323
random variable, Gaussian   86, 87
rank
    computation   367, 381, 389
    of matrix   78
rank defect   155
rational approximation   145
rational transfer function   38, 40, 44,
    261
real perturbations   124
realizable compensator   179, 183, 188,
    197, 207, 269

realization *see* state-space realization
reducible MFD   50
reference input   10, 19, 76, 83, 84, 106,
    208, 236
    spectrum   16
    tracking   87, 89, 236, 318
reference, step   19
regulator   11, 16, 76
relative gain array   210–13, 215, 216,
    217, 219, 220
    block   218
    diagonal elements   212
    evaluated over range of frequencies
        216
    large elements in   215, 216
    negative diagonal elements   213, 214
resonance   24, 161, 200
    closed-loop   9
    in noise models   240
    peak, in principal gain   151
    in plant   314
    pole, cancellation of   243
response time   19
return difference   13, 40
    at input   15
    with LQ state feedback   227
return ratio   23, 40, 61, 68, 143, 144,
    234
    with LQ state feedback   227
    of LQG control scheme   231
reversed-frame normalization   138,
    164, 166, 334, 406
    approximate   166, 167
    at low frequencies   167
RFN *see* reversed-frame normalization
RGA *see* relative gain array
Riccati equation   225, 226, 240, 244,
    253, 306, 322, 336, 361, 393
    for H_∞ problem   303, 306, 312, 317
Riemann surface   71
right half-plane zeros   69, 76
Rimvall, M.   395, 397, 402, 404
ring   274
robustness   75, 76, 91, 129, 166, 202,
    235, 243, 261, 271
    assessment   200
    at input   231
    of LQG/LTR controller   235
    of optimal state feedback   227

and relative gain array   214
    see also integrity, performance
        robustness, stability robustness
Rochelle, R.W.   402
roll-off rate   25, 32, 230, 235, 244, 253,
    311, 313, 333, 334
root-locus   10, 53
    computation   389
    optimal   260
Rosenbrock, H.H.   v, 48, 64, 68, 70, 71,
    74, 168, 169, 171, 173, 221, 345,
    354, 386, 404
round-off errors   359
row dominance   65, 69, 170, 173, 178,
    182, 183, 186
RPV model   262, 407
Rudin, W.   315, 323

Saeks, R.   315, 323
Safonov, M.G.   119, 134, 136, 229, 230,
    242, 244, 263, 264, 300, 316, 324,
    407, 408
Sangiovanni-Vincentelli, A.   345, 354
Savage, R.   218, 221
scaling   110, 212, 215, 230
    input   67
    output   67, 180
    as part of problem specification   181
Schauder's fixed-point theorem   209
Schuck, S.M.   406, 408
Schur decomposition (of matrix)   93
    generalized   368, 394
Schur form   363, 364, 365, 366, 370,
    371, 382, 385, 363
    real   364, 365, 382, 391, 393
Schur product   105, 211
Schur's formula (for determinant)   73,
    126, 317
Scott-Jones, D.F.A.   130, 136
Seneta, E.   181, 221
sensitivity   13, 29, 81, 85, 87, 89, 100,
    101, 149, 151, 204, 214, 216, 229,
    235, 239, 243, 268, 270, 272, 306,
    316, 353
    area under graph of   33
    of characteristic loci   94
    of eigenvalues   93
    increased by feedback   31
    input   15

minimization 100, 267, 270, 283, 293
to noise 8
output 15
peak 33
principal gain 237, 248, 250, 251, 253
sensor dynamics 10
sensor noise *see* measurement error
separation principle 223, 261
sequential design 12, 139
sequential loop closing 137, 138, 139, 154, 177, 209, 210, 217
sequential return difference design method 139, 141
series connection 38
computation 373
Shah, S.C. 396
shaping of principal gains 261
Shinskey, F.G. 212, 221
Shulman, J.D. 47, 74
Sidi, M. 204, 220
signal direction *see* directions of signals
Siljak, D.D. 186, 221
similarity transformation 66, 122, 363, 364, 382
simple (pole or zero) 46
simulation, software for 396, 397
simultaneous gain variations 123
singular value 76, 77, 92, 129, 130, 294, 358, 367
see also principal gain
conditioning 367
decomposition 77, 78, 92, 114, 164, 168, 237, 247, 250, 302, 366, 378, 379, 387, 389, 391
of product 391
and rank determination 389
singular vector 238, 243, 367
SISO design (single-loop techniques) 3–33, 143
Sivan, R. 222, 233, 260, 263
skew return-ratio 93, 131
skewness 94, 151
Skogestad, S. 104, 136, 212, 221
slow loops 12
small gain theorem 116, 229, 271
small phase theorem 116
Smith form 40, 41, 44, 45

Smith, M.C. 71, 74, 96, 136
Smith, S.C. 404
Smith–McMillan form 40, 44, 45, 46, 48, 51, 70, 71, 169
software vii, 2, 210
Basile 397
CLADP 395
for control engineering 394
Control System Toolbox vii, 262, 317, 397
CTRL-C 396
current developments 397
Ecstasy 398
Impact 397
Intrac 395
mathematical 394, 395
Matlab vii, 262, 317, 396
Matrix$_x$ 396
Multivariable Frequency Domain Toolbox vii, 397, 407
packages 355, 395
Robust Control Toolbox 397
Spang, H.A. 398, 404
specification 101, 155, 268, 346
closed-loop 22, 23, 97
consistency 102, 344
of feedback design problem 203, 204
open-loop 23, 94
of performance 102
spectra of external signals 21
spectral decomposition 143, 149
spectral radius 113, 117, 304
speed of response 20
square root of matrix 250
stabilising controllers, parametrization of *see* Youla parametrization
stabilising solution of Riccati equation 303
stability 3, 4, 69
closed-loop 25, 59, 65, 68, 158, 166, 170
conditional 26
internal 55–9, 61
of irrational systems 48
margin 5, 8, 9, 25, 197, 202
of LQG controllers 231
maximization 272
of multivariable system 102, 252, 253, 260

stability (*contd.*)
  robustness   111, 117, 118, 121, 124,
      216, 272, 322
    necessary and sufficient conditions
        116
    for structured uncertainty   116
    sufficient condition   113, 116, 120
    for unstructured uncertainty   111
  sufficient conditions   65
  theorem   47, 59–69
stabilizable system   226, 240
stabilizing controller   267, 301
  fractional representation   281
  observer-based   285
  parametrization   274–85
  state-space realization   289
  transfer function   282
stable–unstable decomposition   382,
    386
staircase algorithm   379, 380
standard basis vector   147, 169
standards for control engineering
    software   398
state estimate   227, 261
state feedback,   vi, 224, 225, 227, 231,
    240, 253, 256, 275, 292
  optimal (LQ)   225, 227, 230, 231,
      235
    return-ratio   259
    robustness   229
state vector   37
state-space methods   vi
state-space model   37, 223
state-space realization   38
  see also minimal realization
  of compensator   163, 197
  of inverse compensator   176
  of transfer function matrix   50, 51,
      71, 72, 142, 188, 226, 239, 260
steady-state error   23, 154, 160, 214,
    246
  from inverse locus   175
steady-state gain   212
steady-state model   211
Steer, S.   397, 402
Stein, G.   126, 127, 130, 135, 231, 243,
    244, 260, 263, 264
Steinhauser, R.   346, 350, 354

step response   19
Stewart, G.W.   78, 136, 357, 384, 385,
    402, 404
Stimler, D.M.   345, 354
stochastic process   3, 12, 21, 97
  Gaussian   223
  vector-valued   97
  white   99
stochastic signals   97
  transmission   97
strictly proper transfer function   48
structure of compensator   143, 153,
    166, 167, 168, 176, 186
structured singular value   117, 133, 273
  computation   118
  interaction measure   218
subroutine libraries for control
    engineering   394
Swann, W.H.   345, 354
Sylvester equation   385
symbolic computation   372
system
  identification   381
  linear time-invariant   3
  multivariable   37
  unstable   32, 33
system matrix, Rosenbrock's   386, 388,
    389

Taylor, J.H.   398, 403
tensor product (of matrices)   328
TGEN model   218, 220, 262, 406
time delay   38, 217, 372
time response   155, 161, 198, 200, 204,
    258, 313, 341
time-domain   vi
time-varying plant   105
Tits, A.L.   119, 135, 352, 354
tracking *see* reference input, tracking
transfer function   3
  all-pass   27
  matrix   2, 11, 37
  minimum-phase   24, 25, 50–2, 54,
      226, 373, 376
  proper   37
  rational   3, 37
  from state-space model   38, 369
transient response   8

transmission zero *see* zeros
triangle inequality   82, 87
triangular matrix   358, 360, 361, 362, 363, 368
triangular plant   212
true model, existence of   105
tuning of parameters   326
tuning parameter in LQG problem   225, 243
two degree of freedom feedback system   84, 204, 216

uncertainty   75, 76, 106, 225
   about plant   13, 21
   additive   103, 104, 113
   element-by-element   120
   frequency-dependent   103
   at input   104, 230
   input multiplicative   103
   inverse multiplicative   105
   magnitude   103
   MFD model   105
   model which allows unstable poles to vary   114
   multiplicative   103, 105, 108, 114
   and operator norms   105
   at output   104
   output multiplicative   103, 114
   quantitative description   204
   region   114
   representation   102
   standard representation   106, 107, 109, 110, 116, 122, 266
      repeated blocks   116
   structured   105, 110, 111, 120, 202
   template   111, 204, 205, 206
      boundaries   111
   unstructured   102, 103, 108
      and conservative design   106
   of valve (example)   105
uncontrollable mode   198, 226, 239, .241, 390
uncontrollable realization   50
underdetermined linear equation   362, 378
uniform plant   154
unimodular   40, 41, 48, 49, 50, 139, 274
unit circle   70

unit of a ring   274
unitary matrix   78, 117, 164, 301, 357, 359, 361, 363, 365, 366, 368, 382, 385
unmodelled phenomena   106
unobservable mode   52, 198, 226, 390
unobservable realization   50
unrealizable transfer function   53
unstabilizable modes   13, 22
unstable controller   22
unstable modes   13, 58, 59, 61, 63, 68, 217
   of plant   203
unstable perturbation   114
unstable plant   96, 259, 263, 283, 285, 407
   two-step design procedure   284

Van Dooren, P.M.   357, 379, 389, 394, 402, 404
Van Loan, C.F.   78, 135, 250, 261, 263, 305, 357, 364, 365, 367, 368, 385, 391, 403
variance
   of error   21
   of multivariable stochastic process   97, 98
Verma, M.S.   300, 316, 324
Vidyasagar, M.   vi, 100, 101, 136, 324
Vinter, R.B.   222, 263

Wall, J.E.   126, 127, 130, 135
Wang, Y.T.   71, 74
Ward, R.C.   391, 392, 403
weighted least-squares   332
weighting
   coefficient   187
   function   101, 103, 104, 125, 130, 148, 188, 190, 191
   matrix   98
weights, frequency-dependent, in LQG design 240, 242
   in Edmunds' algorithm   332, 333, 337
   in H_∞ problem   270, 272, 307, 308, 310, 314
   in LQG problem   233, 242
   in LQG/LTR design method   235

weight (*contd.*)
   implicit in LQG/LTR   243
   in representation of uncertainty   266
Wette, M.   261
white noise   223, 238, 241
Whiteley locus   174
Wiener filter   22
Wieslander, J.   395, 402
Wilkinson, J.H.   93, 136, 368, 404
Willsky, A.S.   15, 36
Wittenmark, B.   3, 24, 36, 130, 135,
   222, 263
Wonham, W.M.   24, 36
workstation   398
worst-case assessment   84
Wright, M.H.   345, 348, 354, 357, 402

Youla parameter   23, 265, 269, 270,
   274–93, 315, 318, 334, 336
   used with Edmunds' algorithm   334
   state-space realization   288
   see also stabilizing controller

Youla, D.C.   315, 324
Young, I.T.   15, 36

Zakian, V.   342, 344, 345, 354
Zames, G.   315, 316, 323, 324
zero polynomial   46, 63, 234
zeros   4, 40, 45, 46, 155
   computation   386
   at infinity   53, 55, 386
   input-decoupling   381, 386
   invariant under feedback   72
   of inverse system   72
   from MFD   49, 50
   of non-square system   52
   number of   52
   output-decoupling   386
   in right half-plane   24, 25, 27, 29, 30,
      73, 96, 129, 174, 176, 214, 231,
      233, 259, 260, 261, 316, 333, 334,
      335
   and bandwidth 27–30, 96
   transmission   40, 47, 63, 198, 233,
      277, 310, 314

Notes

Notes

Notes

Notes